Horst Bunke, Abraham Kandel, Mark Last (Eds.)

Applied Pattern Recognition

Horst Bunke
Abraham Kandel
Mark Last
(Eds.)

Applied Pattern Recognition

With 110 Figures and 20 Tables

Springer

Prof. Dr. Horst Bunke
Institute of Computer Science
and Applied Mathematics (IAM)
Neubr¨uckstrasse 10
CH-3012 Bern
Switzerland
bunke@iam.unibe.ch

Prof. Abraham Kandel
National Institute for Applied
Computational Intelligence
Computer Science &
Engineering Department
University of South Florida
4202 E. Fowler Ave., ENB 118
Tampa, FL 33620
USA
kandel@csee.usf.edu

Dr. Mark Last
Department of Information Systems Engineering
Ben-Gurion University of the Negev
Beer-Sheva 84105
Israel
mlast@bgu.ac.il

ISBN 978-3-540-76830-2 e-ISBN 978-3-540-76831-9

Studies in Computational Intelligence ISSN 1860-949X

Library of Congress Control Number: 2008921394

Cover design: Deblik, Berlin, Germany
springer.com

Bunke et. al.: Applied Pattern Recognition

First Indian Reprint 2011

ISBN 978-81-8489-872-9

This edition is published by Springer (India) Private Limited, (a part of Springer Science+Business Media), Registered Office: 3rd Floor, Gandharva Mahavidyalaya, 212, Deen Dayal Upadhyaya Marg, New Delhi – 110 002, India

Printed in India by Artxel, New Delhi.

Preface

Pattern recognition has persisted as an active research area for more than four decades. It is a combination of two basic ideas: real-world data often occurs in terms of some recurring *patterns* and computers can be taught to recognize these patterns *automatically*. Initially, pattern recognition methods were built mainly upon data analysis techniques of mathematical statistics, but over the years those were extended by such related disciplines as artificial intelligence, machine learning, optimization, data mining, and others. The diverse application areas of pattern recognition range from image analysis to character recognition and speech processing.

A sharp increase in the computing power of modern computers, accompanied by a decrease in the data storage costs, has triggered the development of extremely powerful algorithms that can analyze complex patterns in large amounts of data within a very short period of time. Consequently, it has become possible to apply pattern recognition techniques to new tasks characterized by tight real-time requirements (e.g., person identification) and/or high complexity of raw data (e.g., clustering trajectories of mobile objects). The main goal of this book is to cover some of the latest application domains of pattern recognition while presenting novel techniques that have been developed or customized in those domains. The book is divided into four parts, which are briefly described below.

Part I presents some of the latest face recognition techniques. In Chapter 1, Bourbakis and Kakumanu describe a Local-Global Graph (LGG) based method for detecting faces and recognizing facial expressions in real-world outdoor images with varying illumination. Chapter 2 by Jiang and Chen presents an overview of common facial image processing techniques such as glasses removal, facial expression synthesis, restoration of damaged or incomplete images, and caricature generation. A general statistical framework for modeling and processing head pose information in 2D images is presented by Okada and von der Malsburg in Chapter 3.

Part II deals with pattern recognition in spatio-temporal data. In Chapter 4, Abufadel et al. present a 4D spatiotemporal segmentation algorithm for fully automatic segmentation of cardiac magnetic resonance (MR) sequences. Elnekave et al. introduce in Chapter 5 a new similarity measure between mobile trajectories and then evaluate it by clustering spatio-temporal data.

Several graph-based methods of pattern recognition are covered by Part III. Thus, in Chapter 6, Bunke et al. introduce hypergraphs as a generalization of graphs for object representation in structural pattern recognition. A novel algorithm for feature-driven emergence of model graphs is presented by Westphal et al. in Chapter 7.

Finally, Part IV covers some novel applications of pattern recognition techniques. In Chapter 8, a wavelet-based statistical method is used by He et al. for automated writer identification in Chinese handwritings. The book is concluded by Chapter 9, where Gimel'farb and Zhou apply a Generic Markov–Gibbs Model for structural analysis and synthesis of stochastic and periodic image textures.

We believe that the chapters included in our volume will provide a useful background for researchers and practitioners in pattern recognition and related areas. Our additional goal is to encourage more applications of pattern recognition techniques to novel and yet unexplored tasks.

October 2007

Horst Bunke
Abraham Kandel
Mark Last

Contents

Part I Face Recognition Applications

Skin-based Face Detection-Extraction and Recognition of Facial Expressions
N. Bourbakis and P. Kakumanu 3
1 Introduction 3
2 Neural color constancy based skin detection 4
3 Image segmentation 5
4 Local region graph (Facial feature region graph) 6
4.1 Facial feature matching using local graph 7
5 Skin region synthesis 9
5.1 Neighbor region searching and region synthesis 9
6 Matching multiple regions with Local Global (LG) Graph method 11
6.1 Image Representation with Local Global (LG) Graph 12
6.2 LG Graph matching 13
7 Experimental results 18
7.1 Effect of feature shape and spatial deformations on LG graph matching 18
7.2 Effect of the missing facial features (nodes) and partial LG graph matching 19
8 Extending LG Graph method for recognizing facial expressions 20
9 Conclusions 22
References 23

Facial Image Processing
Xiaoyi Jiang and Yung-Fu Chen 29
1 Introduction 29
2 Removal of glasses 31
3 Facial expression synthesis 34
4 Eye synthesis 36
5 Redeye removal 40

6 Restoration of facial images 42
7 Artistic processing of facial images 42
8 Facial weight change 43
9 Conclusion 43
References 45

Face Recognition and Pose Estimation with Parametric Linear Subspaces
Kazunori Okada and Christoph von der Malsburg 49
1 Introduction 49
2 Problem Overview and Definitions 52
2.1 Statistical Models of Pose Variation 52
2.2 Pose-Insensitive Face Recognition 53
3 Parametric Linear Subspace Model 56
3.1 Linear PCMAP Model 56
3.2 Parametric Piecewise Linear Subspace Model 61
4 Interpersonalized Pose Estimation 64
5 Experiments 65
5.1 Data Set 65
5.2 Personalized Pose Estimation and View Synthesis 66
5.3 Pose-Insensitive Face Recognition 69
5.4 Interpersonalized Pose Estimation 70
6 Conclusion 71
References 72

Part II Spatio-Temporal Patterns

4D Segmentation of Cardiac Data Using Active Surfaces with Spatiotemporal Shape Priors
Amer Abufadel, Tony Yezzi and Ronald W. Schafer 77
1 Introduction 78
2 Method 81
2.1 Combining Space and Time 81
2.2 Adding time to shape–based segmentation methods 82
2.3 Generating Training Data 84
2.4 Segmentation 88
3 Periodic data 90
4 Results 90
4.1 Consistency Tests 91
4.2 Accuracy Tests 96
5 Conclusion 99
References 99

Measuring Similarity Between Trajectories of Mobile Objects
Sigal Elnekave, Mark Last and Oded Maimon 101
1 Introduction 101
2 Related Work 103
2.1 Spatio-temporal data collection 103
2.2 Representing spatio-temporal data 104
2.3 Spatio-temporal data summarization 106
2.4 Querying spatio-temporal data 107
2.5 Indexing trajectories of moving objects 108
2.6 Clustering moving objects and trajectories 110
2.7 Spatio-temporal group patterns mining 111
2.8 Incremental maintenance of mobile patterns 113
2.9 Predicting spatio-temporal data 113
2.10 Spatio-temporal similarity measures 114
2.11 Spatio-temporal data generation 115
2.12 Summary of related work 116
3 Specific methods 116
3.1 An Algorithm for MBB-Based Trajectory Representation 116
3.2 Defining a new similarity measure 119
3.3 Clustering trajectories with the K-Means Algorithm 120
3.4 Using incremental approach for clustering 121
4 Evaluation Methods 121
5 Evaluation Experiments 122
5.1 Generating spatio-temporal data 122
5.2 Detailed Results 123
6 Conclusions 125
References 126

Part III Graph-Based Methods

Matching of Hypergraphs — Algorithms, Applications, and Experiments
Horst Bunke, Peter Dickinson, Miro Kraetzl, Michel Neuhaus and Marc Stettler 131
1 Introduction 131
2 Preliminaries 133
3 Hypergraph Matching 137
4 Algorithms for Hypergraph Matching 140
5 Experimental Results 144
5.1 Experiments on Synthetic Data 144
5.2 Experiments on Pseudo-real and Real Data 147
6 Conclusions 152
References 153

Feature-Driven Emergence of Model Graphs for Object Recognition and Categorization
Günter Westphal, Christoph von der Malsburg and Rolf P. Würtz 155
1 Introduction 156
2 Learning Set, Partitionings, and Categories 158
3 Parquet Graphs 162
3.1 Similarity Function 164
3.2 Local Feature Detectors 164
4 Learning a Visual Dictionary 165
4.1 Feature Calculators 165
4.2 Feature Vectors 167
5 Preselection Network 168
5.1 Neural Model 170
5.2 Position-Invariant Feature Detectors 170
5.3 Weighting of Feature Detectors 171
5.4 Neurons, Connectivity, and Synaptic Weights 172
5.5 Saliencies 175
5.6 Selection of Salient Categories and Model Candidates 175
6 Verification of Model Candidates 177
6.1 Construction of Graphs 177
6.2 Matching 181
6.3 Model Selection 182
7 Experiments 182
7.1 Object Recognition 183
7.2 Object Categorization 191
8 Summary and Future Work 196
References 197

Part IV Special Applications

A Wavelet-based Statistical Method for Chinese Writer Identification
Zhenyu He and Yuan Yan Tang 203
1 Introduction 203
2 A Classic Method for Writer Identification: Two-Dimensional Gabor Model 206
3 Our Algorithm for Writer Identification 207
3.1 Preprocessing 208
3.2 Feature Extraction Based on Wavelet 209
3.3 Similarity Measurement 214
4 Experiments 215
4.1 Identification performance evaluation 1 215
4.2 Identification performance evaluation 2 217

5 Conclusions 219
References 219

Texture Analysis by Accurate Identification of a Generic Markov–Gibbs Model
Georgy Gimel'farb and Dongxiao Zhou 221
1 Introduction 221
2 Identification of a generic Markov-Gibbs model 222
2.1 Basic notation 222
2.2 Generic MGRF with pairwise interaction 223
2.3 Accurate first approximation of potentials 224
2.4 Model-based interaction maps (MBIM) 228
3 Characteristic Structure and Texels 232
4 Texture Synthesis by Bunch Sampling 236
5 Comparisons and conclusions 237
References 244

Part I

Face Recognition Applications

Skin-based Face Detection-Extraction and Recognition of Facial Expressions

N. Bourbakis and P. Kakumanu

Information Technology Research Institute
Wright State University, Dayton Ohio 45435 USA
nikolaos.bourbakis@wright.edu

Summary. Face detection is the foremost task in building vision-based human-computer interaction systems and in particular in applications such as face recognition, face identification, face tracking, expression recognition and content based image retrieval. A robust face detection system must be able to detect faces irrespective of illuminations, shadows, cluttered backgrounds, facial pose, orientation and facial expressions. Many approaches for face detection have been proposed. However, as revealed by FRVT 2002 tests, face detection in outdoor images with uncontrolled illumination and in images with varied pose (non-frontal profile views) is still a serious problem. In this chapter, we describe a Local-Global Graph (LGG) based method for detecting faces and for recognizing facial expressions accurately in real world image capturing conditions both indoor and outdoor, and with a variety of illuminations (shadows, high-lights, non-white lights) and in cluttered backgrounds. The LG Graph embeds both the local information (the shape of facial feature is stored within the local graph at each node) and the global information (the topology of the face). The LGG approach for detecting faces with maximum confidence from skin segmented images is described. The LGG approach presented here emulates the human visual perception for face detection. In general, humans first extract the most important facial features such as eyes, nose, mouth, etc. and then inter-relate them for face and facial expression representations. Facial expression recognition from the detected face images is obtained by comparing the LG Expression Graphs with the existing the Expression models present in the LGG database. The methodology is accurate for the expression models present in the database.

1 Introduction

For face detection, a number of methods have been previously proposed [1, 3, 4, 11, 13–87]. To achieve a good performance, many of these methods assume that the face is either segmented or surrounded by a simple background and the images are well-illuminated with frontal-facial pose, which is not always true. The robustness of these approaches is challenged by many factors such as changes in illumination across the scene, shadows, cluttered backgrounds,

N. Bourbakis and P. Kakumanu: *Skin-based Face Detection-Extraction and Recognition of Facial Expressions*, Studies in Computational Intelligence (SCI) **91**, 3–27 (2008)
www.springerlink.com

image scale, facial pose, orientation and facial expressions. In this chapter, we present a *Local-Global (LG) Graph* approach for detecting faces and for recognizing facial expressions. The overall approach to face and facial expression detection presented here - the LG graph approach combined with the skin detection procedure is robust against cluttered backgrounds, uncontrolled illuminations, shadows, and to a variety of facial poses and orientations.

Given an image, the neural color constancy based skin detection technique (section 2) provides skin-color similar regions in the image. These segmented skin regions form potential candidate face regions. However, all skin-color similar regions do not represent faces and non-face regions must be rejected. First, to detect facial feature regions, a fuzzy-like segmentation method is applied on the segmented skin regions (section 3). A curve fitting procedure is applied on the segmented regions that result in a set of line segments which are later used to define a *local graph* that represents the shape of the region (section 4). Due to the parameter settings of the segmentation method and the fact that different regions of the faces are illuminated differently, the segmentation procedure might produce a large number of regions on the face. To reduce the number of regions and to highlight the key facial features, a skin region synthesis procedure based on color information is applied (section 5). From the remaining facial regions, a *local-global* graph which incorporates the spatial geometry of facial features is constructed. To detect a face, the candidate LG graph thus formed is finally compared with the face-model LG graph present in the LG database using a computationally effective graph matching technique (section 6). The LG graph approach in general can be applied to any object detection provided that there is a corresponding object model in the LG database. For recognizing facial expressions, new candidate Expression LG graphs are formed for each facial feature region, and then compared with the existing Expression models present in the LG database (section 8).

2 Neural color constancy based skin detection

To detect skin from images, a two stage color constancy based approach is used [2]. The first stage - the color correction stage consists of estimating the image illuminant using a skin color adapted Neural Network (NN) and then color correcting the image based on the NN illuminant estimate. The Neural Network is trained so as to adapt to the skin color. The Neural Network is trained on randomly selected images from a database consisting of images collected under various illumination conditions both indoor and outdoor, and containing skin colors of different ethnic groups [3]. The color correction step assigns achromatic color (gray) to skin pixels. The second stage – skin detection stage classifies the skin and non-skin pixels using a simple thresholding technique in RGB space based on the achromatic value of the color corrected images. To remove the noise and to fill the holes created by the skin detection

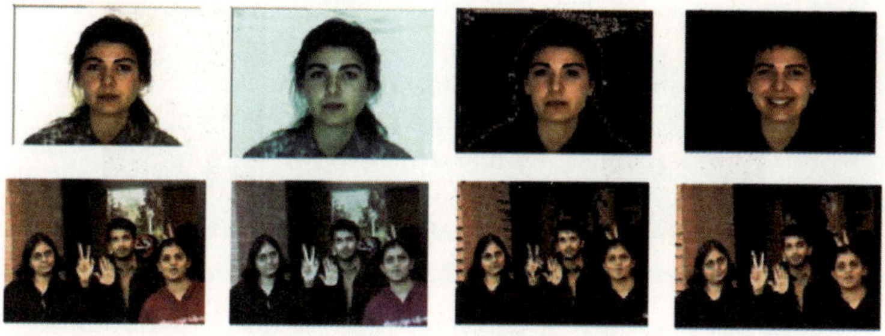

a) Original; b) NN color-corrected Images; c) Skin detection; d) Skin-similar regions left out

Fig. 1. Neural color constancy based skin detection results

procedure (for example, eye and eye-brow regions are not detected by the skin detection procedure), morphological operations such as erosion, dilation and hole-filling are applied on these images. Figure 1 shows the original image, NN color corrected image, detected skin regions and the skin-similar regions left out after applying the morphological operations. The image shown in row 1 is a part of AR Face Database [4] and the image shown in row 2 is a part of custom collected image database at ITRI/Wright State University.

The advantage of the NN method for color adaptation is that it does not have any inherent assumptions about the object surfaces in the image or the illumination sources as the input to the neural network is only the color from the image. The overall approach for skin detection is computationally inexpensive and is feasible for real-time applications. However, if the goal is to detect faces, skin detection alone is not sufficient. The skin detection step provides face-like regions in the image and serves as a primary step for the LG graph procedure described below.

3 Image segmentation

Segmentation is one of the most commonly used preprocessing steps in image analysis. The goal of segmentation is to partition the image into connected regions such that each region is homogeneous in respect to one or more characteristics. Each segment is composed of a continuous collection of neighboring pixels. When a segmentation algorithm terminates, each pixel in the image is assigned to a particular segment. More formally, segmentation divides the entire image (I) into n disjoint continuous regions (R_i) such that the union of all these regions results in the overall image.

$$I = R_1 \cup R_2 \cup \ldots . \cup R_i \ldots . \cup R_n; \qquad R_i \cap R_j = \phi \tag{1}$$

The Fuzzy Region Growing (FRG) segmentation method used in this research is a computationally efficient technique which uses smoothing, edge information, homogeneity criteria and degree of farness to segment image regions [5].

(a) Skin segmented image

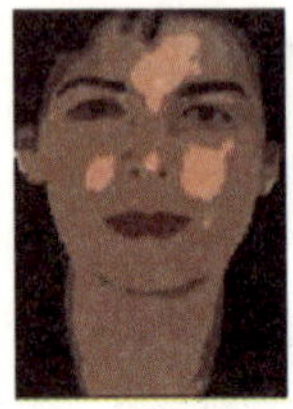

(b) FRG segmented image

Fig. 2. Fuzzy Region Growing (FRG) segmentation results

The algorithm first performs smoothing and edge operations to determine the interior pixels. A set of segments are then initialized by performing flood fill operations at the interior points. The decision as to whether a given adjacent (four-connected) pixel should be filled during the flood operation is based on its closeness in RGB color space to the original seed pixel of the segment. The pixels which have not been merged with any segment after the flood fill operation are merged through a region growing procedure. In order to merge the remaining pixels that are not assigned to any particular segment through region growing, the edge pixels of the existing set of segments are propagated outward or grown. Then, as unassigned pixels are encountered, they are merged with the closest segment of most similar color. This specific condition is calculated using a least squares difference of the RGB color components as well as a distance proportional to the distance from the original propagating edge pixel. Figure 2 shows the skin segmented image from Figure 1 and the corresponding FRG segmented image. Clearly, the FRG method segments important facial feature regions such as eyes, eye-brows, nose and mouth.

4 Local region graph (Facial feature region graph)

The application of FRG segmentation method generates a set of color regions. On these segmented regions, a curve fitting procedure is applied which results in a set of connected line segments that defines the shape of each region. The shape of the region is used in generation of a local region graph and later in the synthesis of neighboring regions (section 5). The representation of a region by a set of line segments is not sufficient as it does not hold all the connectivity relationships, which is very important for accurate description of the geometrical shape of the region. An accurate description of the shape of facial features is critical for robust face detection/recognition. A good solution to this problem is to build a local graph which encodes the spatial relationships of these line segments. Thus the shape S of the region is represented as:

$$S = L_1 R^c_{12} L_2 R^c_{23} L_3 R^c_{34} \dots L_{n-1} R^c_{n-1n} L_n \dots \tag{2}$$

where L_i represents the line segment and R_{ij} represents the connectivity relationship between line segments L_i and L_j of the region shape. R_{ij} is

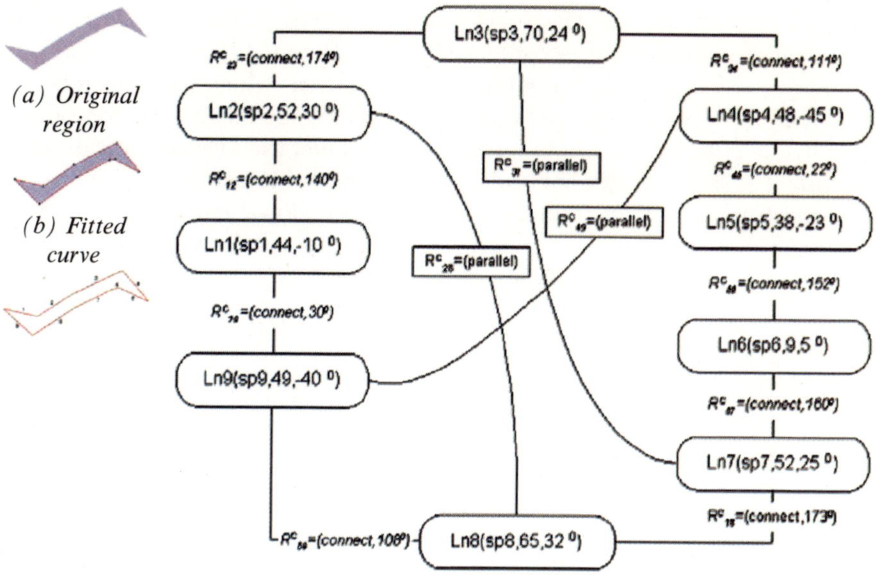

(a) Original region *(b) Fitted curve* *(c) Line fitted region* *(d) Region Local Graph*

Fig. 3. Fitted curve around an object and the corresponding region local graph [6]

characterized by connectivity (c), parallelism (p), symmetry (s), relative magnitude (rm), and relative distance (rd).

$$R_{ij} -> \{c, p, s, rm, rd\} \tag{3}$$

Each line segment (L_i) is characterized by its properties (P_i) such as starting point (sp), length (l), orientation (d), and curvature (cu).

$$L_i -> \{sp, l, d, cu\} \tag{4}$$

Figure 3 shows a segmented object (a), the fitted curve on this region (b, c) and the corresponding local graph (d). At this stage, we have defined the shape of the region with a local graph, with the line segments as the nodes and the connectivity between these line segments is described by the edges [6].

4.1 Facial feature matching using local graph

To detect a face, we first need to match the key facial regions. In the LG graph method, each facial region is represented by a local graph. Hence, to identify the segmented regions as facial regions, we need to match the local graph of the segmented regions to the stored model local graphs in the database.

Assume that the model facial feature region and the candidate region are represented by the curves $f(t)$ and $g(t)$. To match a region, we need to find the geometric transformation that best maps the curve $g(t)$ onto $f(t)$. The basic geometric transformation is composed of translation (T), rotation (M) and scaling (S) [7]. In the case of affine transformation, which includes reflection and shearing, there are six free parameters. These model the two components of the translation of the origin on the image plane, the overall rotation of the coordinate system, the global scale, together with the parameters of shearing and reflection. The transformation between two closed curves, $f(t)$ and $g(t)$ is defined:

$$g(t) = S \cdot R \cdot f(t) + T \tag{5}$$

Ignoring the parameters of shearing and reflection, the transformation in matrix notation is:

$$\begin{bmatrix} x'(t) \\ y'(t) \\ 1 \end{bmatrix} = \begin{bmatrix} 1 & 0 & s_x \\ 0 & 1 & s_y \\ 0 & 0 & 1 \end{bmatrix} \bullet \begin{bmatrix} \cos\theta & -\sin\theta & 0 \\ \sin\theta & \cos\theta & 0 \\ 0 & 0 & 1 \end{bmatrix} \bullet \begin{bmatrix} x(t) \\ y(t) \\ 1 \end{bmatrix} + \begin{bmatrix} 1 & 0 & t_x \\ 0 & 1 & t_y \\ 0 & 0 & 1 \end{bmatrix} \tag{6}$$

Most of the imaging equipments have the same scale factor to both x and y directions. Hence, we assume that the scale factors in both x and y directions the same ($s \equiv s_x \equiv s_y$). Since, in geometric transformation, rotation and scaling does not change the centroid, the translation is defined as:

$$T = F_{cen}(g(t)) - F_{cen}(f(t)) \tag{7}$$

where $\boldsymbol{F}_{cen}(f(\cdot))$ is a function which computes the centroid of the curve $f(\cdot)$. We compute the centroid (x_0, y_0) of each region and move the centroid's to the origin. Figure 4 (b) shows an example of translation. To find the scale parameter, we compute the momentum of the object. The momentum is a measure of the mass distribution of an object and is defined as:

$$Mom = \frac{1}{N} \sum_{i=1}^{N} m_i \cdot \|p_i - p_{centroid}\|^2 \tag{8}$$

where N is the number of curve points, m_i is the mass weight at point p_i, $p_{centroid}$ is the centroid point of the closed curve. $\|\bullet\|$ is a kind of norm, such as Euclidean norm. The translation and rotation do not change the shape of an object. So, Mom is identical before and after translation and rotation. Thus the scaling factor is defined as:

$$s = \left(\frac{Mom'}{Mom}\right)^{1/2} \tag{9}$$

To find the rotation parameter, we suppose that the curve, $f'(t)$ is generated by rotating the curve $f(t)$ by a certain angle θ. The rotation is obtained on the centroid of the curve $f(t)$. However, the problem associated with equation (6) is that we do not know the point-to-point correspondence to calculate the rotation angle θ. To by-pass the point-to-point correspondence, wavelet coefficient of the region borders are used. Figure 4 shows a complete example of single region matching.

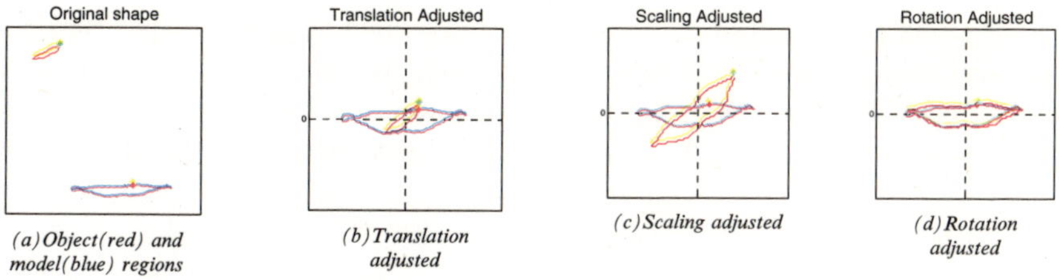

Fig. 4. Single region matching and the matching results at each stage (b, c, d)

5 Skin region synthesis

Different regions of the face are illuminated differently and hence the segmentation procedure might produce a large number of regions on the face. Also, due to the parameter settings of the segmentation procedure (such as region-filling thresholds) and the relative facial image size, the segmentation procedure might produce a number of regions on the face (Figure 2(b)). Not all regions generated by the segmentation procedure are needed. It should be noted that the image region used for segmentation was identified by the skin detection procedure previously as skin. To identify these skin regions as a face, it is sufficient if we can identify the distinctive facial features such as eyes, eyebrows, nostrils and mouth. Hence, to reduce the number of regions (nodes) and to simplify the successive graph matching procedure, we first apply a skin region synthesis procedure to merge neighborhood regions based on relative *RGB color similarity* as described below.

5.1 Neighbor region searching and region synthesis

Two regions are defined as neighbors, if their borders have common shape and the common parts are very close in Euclidean space. If the correspondent common parts have distance greater than zero, then these two regions are pseudo-neighbors. For skin region synthesis we consider only those regions whose RGB color value is close to average skin color value. The skin region synthesis procedure is described as below:

1) *Initialize the first region which is closer to average skin color as the active skin region. The degree of closeness is calculated as the color difference in RGB space.*
2) *Select the next region which is closer to average skin color. Find the common edge between this region and the active region.*
3) *If a common edge is found, synthesize the current region and the active region. Assign new region to active region.*
4) *If all regions have been processed, region synthesis completes; otherwise go to step 2.*

If two regions are neighbors, they have at least one common border. Since, the local graph represents the shape in high-level, we use the local graph to find common edge between two regions. Let L_1 and L_2 be the border line local graphs of two regions with n and m lines respectively and defined as follows:

$$\begin{aligned} L_1 &= Ln_1 R_{12}^c Ln_2 R_{23}^c Ln_3 R_{34}^c \ldots Ln_{n-1} R_{n-1n}^c Ln_n \\ L_2 &= L'n_1 R_{12}^c L'n_2 R_{23}^c L'n_3 R_{34}^c \ldots L'n_{m-1} R_{m-1m}^c L'n_m \end{aligned} \tag{10}$$

If L_1 and L_2 have a common edge, a part in L_1 must match with a part in L_2. We need to find, if there is a partly matching and where it is in the two graphs. For this, we treat every element in the line and the relationship as a character. Thus L_1 and L_2 can be treated as two strings. The problem of finding the common edge thus reduces to finding a common sub-string in these two strings.

Once the common edge is detected, a region synthesis process is started. This process merges all the neighbor regions and retains distinctive facial features as the color of these features is far from average skin color value. Given two regions, R_1 and R_2, the rules to synthesize a new shape R_{12} are defined as

$$shape(R_{12}) = \begin{cases} shape(R_1)shape(R_2), & if\ REL(R_1, R_2) = contiguous \\ shape(R_1), & if\ REL(R_1, R_2) = contain \\ shape(R_2), & if\ REL(R_1, R_2) = contained \\ \phi, & if\ REL(R_1, R_2) = separate \end{cases} \tag{11}$$

where the relationship between two regions (nodes) is as shown in Figure 5. In order to improve the performance of the region synthesis process, we build a table to accelerate the finding common-edge process. We do not need to compute the relationship between each node pair. Some of the relationships can be deduced from other relationships. Figure 6 shows the region synthesis procedure. The application of this step retains important facial features such as eyes, eyebrows, nostrils and mouth since the color value of these facial features is relatively different from the average skin color. The facial features can now be effectively used to represent a face and are further used for generating image LG graph.

(a) Contiguous *(b) Contain* *(c) Contained* *(d) Separate*

Fig. 5. Four relationships between two nodes (regions)

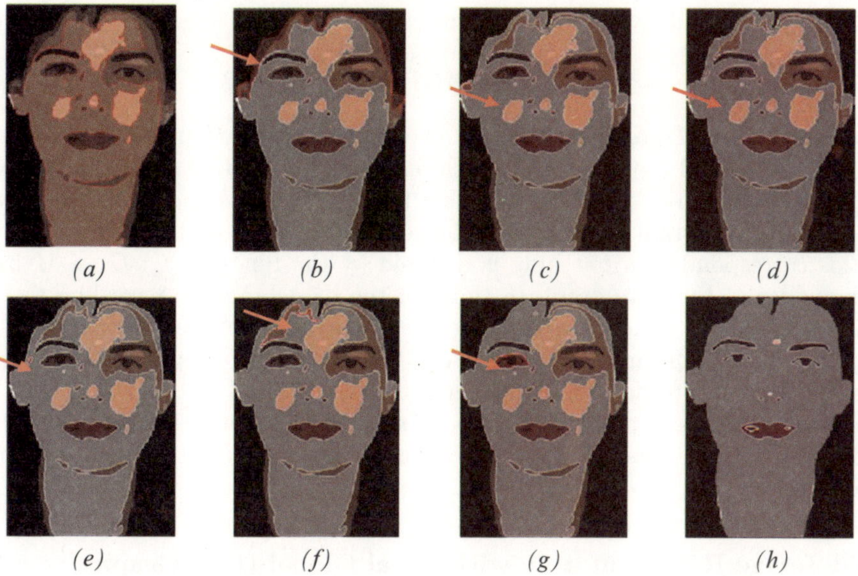

Fig. 6. Figure showing skin region synthesis procedure (a-h). At each stage, the next region considered is shown with a red border. Only the first few steps (a-g) and the last step are shown. All the regions which are merged by the skin synthesis procedure are shown with achromatic color

6 Matching multiple regions with Local Global (LG) Graph method

To perceive the group of facial features as a face, the spatial relationships between the corresponding facial features is an important constraint. It is just not enough to identify the facial features alone but their geometrical positions and placements are also very important. The location of spatial features serves as a natural choice (as landmarks) in relating multiple views of faces. Rather than using the shape constraints to establish similarity correspondence, we use the constraints provided by the spatial adjacency of the regions. These constraints are relaxed by separately triangulating the data and model regions. We use the neighborhood consistency of the correspondences in the triangulations to weight the contributions to the similarity function. In this section, we describe how the relational consistency is used in the matching process. In particular, we abstract the representation of correspondences using a bipartite graph. Because of its well-documented robustness to noise and change of viewpoint [8–10], we use the Voronoi Tessellation method, the Delaunay triangulation and the Local-Global graphs as our basic representation of the image structure.

The process of Delaunay triangulation generates relational graphs from the two sets of point-features. More formally, the point-sets are the nodes of a data graph and a model graph are:

$$G_D = \{D, E_D\}, G_M = \{M, E_M\} \tag{12}$$

where D and M are the node sets of data (image) and model respectively. $E_D \subseteq D \times D$ and $E_M \subseteq M \times M$ are the edge-sets of the data and model graphs. The key to the matching process is that it uses the edge-structure of Delaunay-Graphs to constrain the correspondence matches between the two point-sets. This correspondence matching is denoted by the function $f : D \rightarrow M$ from the nodes of the data-graph to those of the model-graph. According to this notation, $f(i) = j$ indicates that there is a matching between the nodes $i \in D$ of the data-graph and the node $j \in M$ of the model-graph. The Delaunay graph is robust to noise and geometry transformation. In other words, if the node set undergoes any kinds of transformations, the new Delaunay graph is the transformed version of model Delaunay graph, with the same transform parameters.

6.1 Image Representation with Local Global (LG) Graph

In the LG graph scheme, each node in the graph represents a facial region, not just a facial point and hence contains information about the region. Each node is defined with its spatial location, same as the center of gravity (x, y), color of the region *(color)*, local graph associated with this node (L), the number of pixels in the region *(size)* and the contour pixel set *(border)*.

$$node = \{(x, y), color/texture, L, size, border\} \tag{13}$$

After introducing the local graph information into the global graph scheme, the new method can now handle both local (region) and global (object) information in the matching process. We combine the segmented skin regions as described in section 2 with the LG graph scheme. The skin region synthesis procedure retains only few facial regions. Every region is a characteristic of the face. As previously defined, the object and model LG graphs are represented as:

$$G_D = \{D, E_D\}, G_M = \{M, E_M\} \tag{14}$$

Where D and M are node sets, E_D and E_M are edge sets. The node-set *(NS)* and edge-set (E) are defined as:

$$NS = \{node_i, i = 1, 2, 3 \ldots n\} \quad n \ is \ the \ total \ number \ of \ nodes \tag{15}$$

$$E(i, j) = \begin{cases} 1 & if \ node \ i \ connects \ with \ node \ j \\ 0 & otherwise \end{cases} \tag{16}$$

Figure 7 shows the model LG graph. As shown in the figure, for representing faces, we consider only eyes, eyebrows, nostrils and mouth regions.

(a) Model face image (b) Skin synthesized image (c) Selected facial regions (d) Delaunay graph

Fig. 7. Model face image and the corresponding Delaunay graph

6.2 LG Graph matching

Matching two graphs comprises of establishing point correspondences between two graph node sets that maximizes the likelihood between two graphs, given the spatial constraints [11]. Given only the geometry position of the nodes, the matching process in general requires a permutation or a recursive method. In the proposed LG scheme, the node correspondences is solved by using the local graph similarity constraint (color and shape similarity), thus bypassing the computationally expensive recursive or permutation methods. Thus the matching complexity can be reduced to a reasonable extent.

Establishing node point correspondences

In the LG graph scenario, each node corresponds to a region. Here, every region is a meaningful part or a characteristic of the face. We define that only regions with similar characteristics can be considered as Potential Correspondent Region Pair (*PCRP*). Random matching of one node in the graph with nodes in another graph is not reasonably acceptable. In the human perception system, color plays an important role in recognizing objects. It is reasonable to link node pairs that have similar color. The method proposed here does not allow one node to correspond to nodes with very different color. Suppose a model set M has N nodes. A threshold th_c is chosen to filter out dissimilar nodes in the data graph. This process is described below:

1) *Initialize PCRP table, which has n entries, the region count in the model graph.*
2) *For each region in the model graph, do the following*
   a. *Compute color distances between all regions in the data graph.*
   b. *For any region with distance* $< th_c$, *add it to current PCRP table entry.*
   c. *If no PCRP is found for a current region, return an ERROR.*

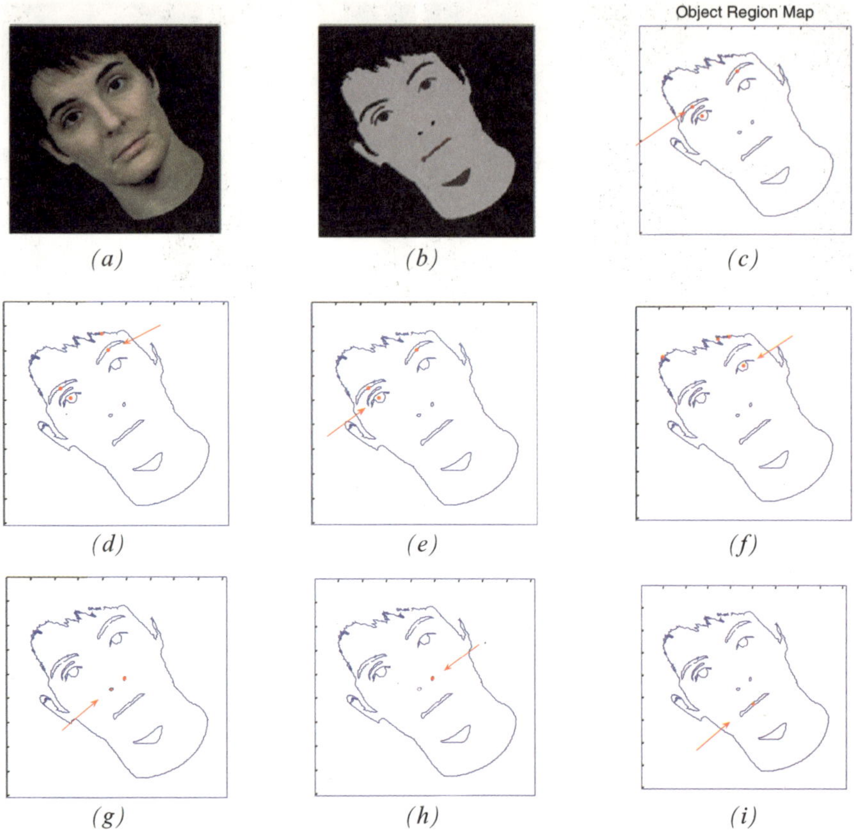

Fig. 8. Skin segmented data Image (a) and the corresponding segmented image obtained after applying skin region synthesis (b). (c-i) show the PCRP selection of every region. The arrows indicate the model regions that are compared with

Figure 8 shows an example of the skin segmented data image (image taken from FERET Database [11]) and the regions retained after the skin synthesis procedure. Figure 8 shows the process of selecting the PCRP regions by comparing the date image with that of the model Face LG graph in Figure 7(d).

Graph Similarity

Once the node correspondences are established, the next step is to compute the similarity between two graphs. The graph similarity can be determined by the relative spatial connectivity between the nodes. The translation, rotation and scaling do not change the graph spatial structure. The spatial structure of the graph is mainly represented by the angle, i.e., if the correspondent angles between arcs are similar, the graphs are similar too. We define the angle similarity function $S_{ANGSIM}(\Delta\theta)$ by using two thresholds. One is $\boldsymbol{\theta_{th1}}$,

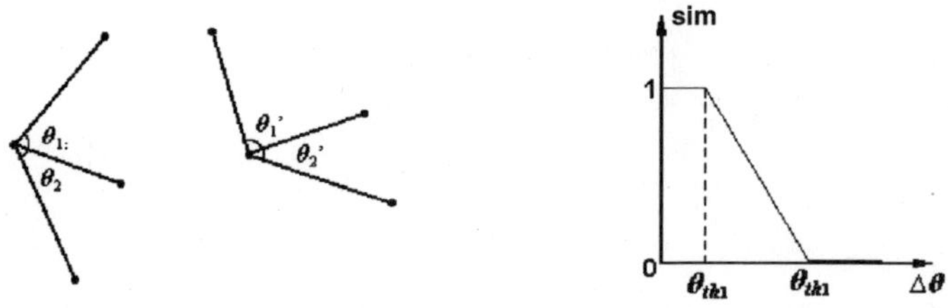

(a) (b)

Fig. 9. Compared nodes (a) and Angle similarity function (b)

lower bound of angle difference; another is $\boldsymbol{\theta_{th2}}$, the upper bound of angle difference. The angle similarity function, $S_{ANGSIM}(\Delta\theta)$ defined as follows:

$$S_{ANGSIM}(\Delta\theta) = \begin{cases} 1, & \Delta\theta < \theta_1 \\ \dfrac{\theta_2 - \Delta\theta}{\theta_2 - \theta_1}, & \theta_2 \geq \Delta\theta \geq \theta_1 \\ 0, & \Delta\theta > \theta_2 \end{cases} \tag{17}$$

where θ_1 is the lower bound of the angle difference and θ_2 is the upper bound of the angle difference as shown in Figure 9. The total similarity between two graphs, SIM_G is defined as

$$SIM_{LG} = \frac{1}{N}\sum_{i=1}^{N}\sum_{j=1}^{N} E(i,j)\,.\,S_{ANGSIM}(\theta_{ij} - \theta_{i0}) \tag{18}$$

where N represents the total number of nodes, θ_{ij} is the angle of the *edge(i, j)* and θ_{i0} is the base angle for every node. The base angle can be selected as the angle of the first arc associated with node i.

The PCRP graphs built as shown in Figure 10 are compared with the face model graph shown in Figure 7. Table 1 lists the similarity values computes between the PCRP graphs and the model graph using the above similarity equation.

The LG Graph Relation Checking

The PCRPs described in section 6.2 are selected only based on their color similarity. The combinations of all PCRPs have 13 possible graphs. Figure 10 shows all the *PCRP* graphs. However, only one graph among these can be a right match. In extreme case, we may not have the right match in the selected graphs. This indicates that only graph spatial constraints cannot ensure finding the right match. Thus we need to examine the validity further.

In equation 18, the graph similarity is measured based on the assumption that every node pair has the same weight. However, if the shapes of two nodes as defined by local graphs are different, it is very likely that it is a

Fig. 10. PCRP graphs for the data image in Figure 8(a)

Table 1. Graph similarity comparison result. All the possible 13 PCRP graphs are compared with the model graph

Graph Index	***1***	***2***	***3***	***4***	***5***	***6***	***7***
Graph Sim. Value	*0.46*	*0.71*	*0.47*	*0.46*	*0.76*	*0.32*	*0.33*
Graph Index	***8***	***9***	***10***	***11***	***12***	***13***	
Graph Sim. Value	*0.47*	*0.61*	*0.60*	*0.63*	1	*0.4*	

wrong PCRP. In this case, even though the PCRP has high node similarity with its counterpart in the model graph, we reduce its contribution to the graph similarity by introducing a shape similarity weight factor to the graph similarity measure. We weigh the graph similarity measure with the

Table 2. Relationship checking table

Data Image Relationship \ Model Relationship	Contiguous	Contain	Contained	Separate
Contiguous	*1*	*0*	*0*	*0*
Contain	*0*	*1*	*0*	*0*
Contained	*0*	*0*	*1*	*0*
Separate	*0*	*0*	*0*	*1*

corresponding node pair shape similarity measure calculated as described in section 4.1. High weight is given to the node pair with high shape similarity while low weight is given to the node pair with low shape similarity. After introducing the shape similarity factor, the graph similarity is defined as

$$SIM_{LG} = \frac{1}{N}\sum_{i=1}^{N}\sum_{j=1}^{N} E(i,j)\,.\,S_{ANGSIM}(\theta_{ij} - \theta_{i0})\,.\,Weight(i,j) \tag{19}$$

In other methods, graph nodes are only points in 2D or 3D space and have only geometry relationships among them [12]. In the LG Graph scenario, every node represents a region, i.e., a characteristic region of the face. These nodes have certain pre-determined constraints. We define four relationships between two regions – contiguous, contain, contained and separate as shown in section 5.1. The relationship checking table is shown in Table 2. For any two regions R1 and R2, if their relationship in the image is r_{12} and in the model graph is r_{12}', then the similarity relationship between these two regions is

$$sim_{REL} = T_{RE}(r_{12}, r_{12}') \tag{20}$$

The relationship between two graphs is determined by every relation checking between any two connected graph nodes.

$$SIM_{REL} = \prod_{i=1}^{N}\prod_{j=1}^{N} T_{RE}(r_{i,j}, r_{i,j}') \tag{21}$$

LG Graph Similarity

The overall LG graph error is defined as:

$$ERR_{LG} = (1 - ERR_{rel}) \bullet (ERR_G + ERR_{shape}) \tag{22}$$

where ERR_{LG} is the total matching error between object and model, ERR_{rel} - the matching error of relationships among regions, ERR_G - the matching error between global graphs, and ERR_{shape} - the matching error between two objects' shapes. The overall matching scheme is defined as:

1) *First compare the relationships between the object and the model. If all the region(node) relationships comply, then ERR_{rel} returns 0; otherwise return 1 and the process ends.*
2) *Build the global graphs for the object and model graphs. The global graph error ERR_G is obtained by comparing the object and model graphs. To cope up with differences in various individuals, a relative large threshold Th_{graph} is set for ERR_G. If the $ERR_G > Th_{graph}$, it means that the current object has a very different structure from the model. We will not consider them as similar and quit current step.*
3) *Finally, objects are merged using the region synthesis procedure described in section 5.4. This is in particular applied for recognizing expressions described in section 8. After synthesizing, we compute shape similarity between the object and the model. If the shape matching error is less than predetermined threshold Th_{shape}, i.e., $ERR_{shape} < Th_{shape}$, then the object is similar to the model.*

If all the three terms are computed and fall in the acceptable range, the face is located. The total similarity measure is as computed by equation 22. The lower the ERR_{LG}, the more similar the object and model are.

7 Experimental results

Figure 11 shows the application of LG graph based face detection on sample images from the AR face database [4]. The AR face database includes frontal view images with different facial expressions, illumination conditions and occlusion with by sunglasses and scarf. It should be noted that, at present the images with sunglasses are not used in evaluating the proposed method. Also, the results were shown on realistic images collected at ITRI/WSU (Figure 11-e) and on images collected from web (Figure 11-f). From the results, it is clear that the proposed method accurately detects faces from images.

7.1 Effect of feature shape and spatial deformations on LG graph matching

One of the major problems with detecting faces is that, face is dynamic in nature and the shape of the facial features deform with the facial expressions. The facial features also appear deformed depending on the non-frontal views of the face. Added to this, there are differences between the faces from one person to another. If the target is to detect faces, all these deformations must be accounted for. In the above described LG graph method for face detection, the facial feature information is stored at the local-level (within the local graph at each feature node) and the spatial geometry of the facial features is stored at the global level (within the local-global graph with Delaunay Triangulation). The PCRP selection method described in section 6 allows us first

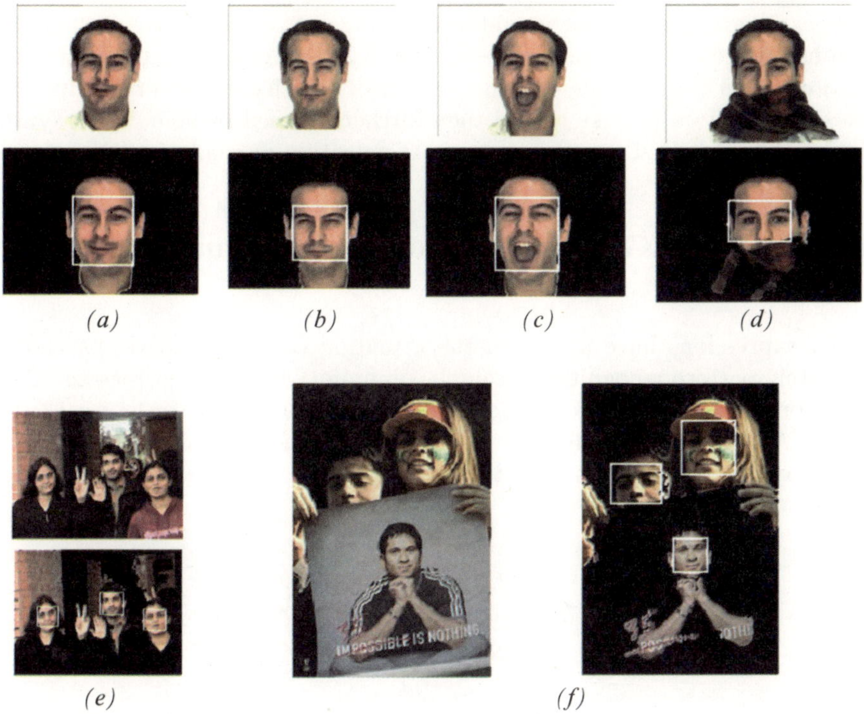

Fig. 11. Example of face detection on sample images. Shown in figure both the original images, the skin segmented images and the detected faces

to select different candidate facial features depending on the color, but not on the shape similarity. The validity of the selected facial feature's relationships is then constrained by the graph and shape similarity relationships. Selecting the nodes based on color similarity and then selecting a big threshold for graph similarity, Th_{graph} (section 6.2) allows us to cope up with the deformations in facial features and the differences between the individuals from the model graph. Figure 11 (a-c) shows the example of faces detected from images containing various facial expressions.

7.2 Effect of the missing facial features (nodes) and partial LG graph matching

Another problem associated with face detection is that due to partial occlusions, all the facial features used to represent the model graph might not be present in the data image. To detect partially occluded faces, we use partial LG graph matching. At the present stage, though we consider all the possible node permutations by eliminating each node at a time, since the node correspondences are solved using shape and color similarity, the overall computationally complexity is kept within an acceptable level. Node elimination and partial graph matching should be computationally expensive, if the PCRP regions selected in the data image (section 6.2) are relatively huge in number. However, this is rarely the case as the overall search space is first reduced by the skin detection procedure and then further reduced by skin region synthesis. Figure 11(d, f) shows the example of detecting partially occluded faces.

8 Extending LG Graph method for recognizing facial expressions

Facial expressions have attracted the attention of the cognitive, psychology and lately pattern recognition research community for different reasons. There are approximately 70 different facial expressions as shown in figure 12 (Unknown Author (UA) 1996). For recognizing facial expressions, the series of tasks to be performed are face detection, facial feature recognition and then

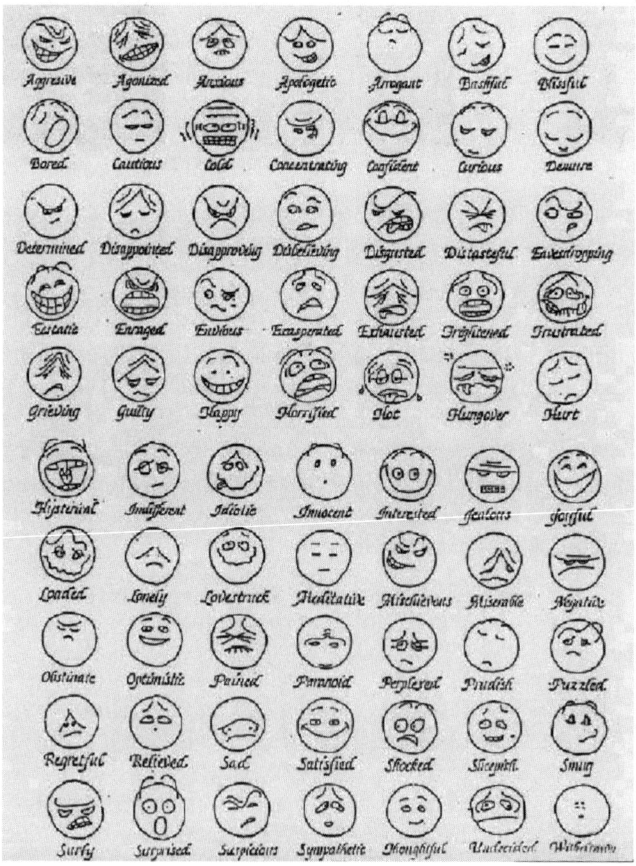

Fig. 12. 70 different cartoon-like facial expressions (UA 1996)

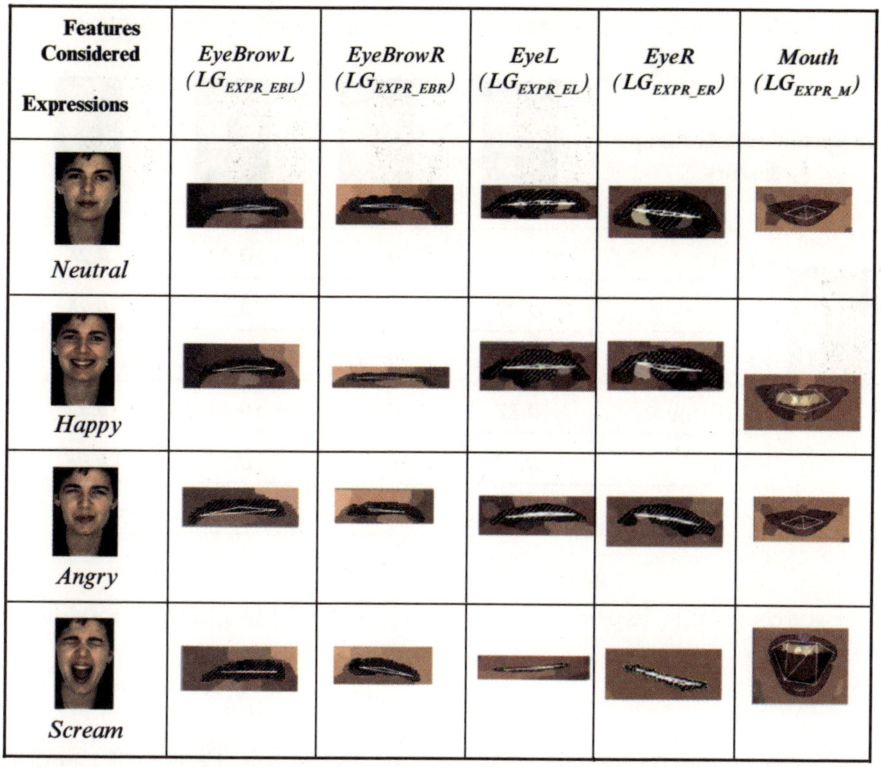

Fig. 13. Figure showing the LG Expression Graphs for the features considered for neutral, happy, angry and scream expressions respectively

facial expression recognition [13]. In the proposed LG Graph method, after the face detection step, the various facial features from the image can be retrieved by the local-global graph node correspondences. The extra step needed for recognizing facial expressions is that we need to compare the image LG graph with the existing Expression LG graphs. To recognize facial expressions, the LG database also consists of the corresponding Expression LG graphs. It should be noted that each facial expression is characterized by a specific configuration of the facial features. This particular configuration of each facial feature is represented by the Expression LG graphs, similar to the Face LG graph defined previously. The overall methodology of describing the facial feature configurations for each expression is in accordance with the human way of representing and perceiving expressions. To recognize expressions, humans first extract the most important facial features such as eyes, eye-brows, nose, mouth, etc and then inter-relate their specific configurations as one makes a transition from neutral to another expression. These specific expression configurations are represented by the Expression LG graphs.

Table 3. Table showing the Expression LG graph errors for each expression considered

Expression LG Graph Errors		 Neutral	 Happy	 Angry	 Scream
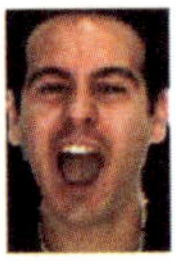	EyeL	*0.72*	*0.53*	*0.22*	***0.26***
	EyeR	*0.78*	*0.52*	*0.27*	***0.28***
	EyeBrowL	*0.33*	*0.34*	*0.40*	***0.42***
	EyeBrowR	*0.37*	*0.37*	*0.39*	***0.42***
	Mouth	*0.85*	*0.67*	*0.82*	***0.08***
	Avg. Error	*0.61*	*0.49*	*0.42*	***0.29***

For recognizing expressions, each node in the LG graph is modified as shown below:

$$node = \{(x, y), color/texture, L, size, border, LG_{EXPR1}, ..LG_{EXPRi}\} \quad (23)$$

where $\boldsymbol{LG_{EXPRi}}$ represents an Expression LG graph. Figure 13 shows the corresponding Expression LG graphs for neutral, happy, angry and scream expressions present in the AR Face Database.

In the LG graph scheme, after the face detection step, an approximate position of the facial features in the data image can be found from the node correspondences established during LG graph matching step. A part of the image region around the facial feature in the data image is considered. This region is segmented using the FRG segmentation method and an image Expression LG graph is built similar to described in the previous sections. To detect an expression, we compare each image facial feature to the existing LG Expression graphs corresponding to that facial feature (node). The combination of the LG Expression graphs for a particular expression for which we obtain the minimum average error as shown in Table 3 is selected as the corresponding expression for that image.

9 Conclusions

In this chapter, we proposed a novel face detection and facial expression recognition method based on Local-Global Graphs. A face is described by the Face – Local Global Graph model. To detect faces, the model Face-LG graph and the corresponding test image Face-LG graph are constructed and then compared with an efficient LG graph matching technique. The corresponding matching error is evaluated as the similarity between the two graphs. For representing facial expressions, the specific configurations of facial features corresponding to a particular expression are represented by the Expression LG graphs. Since, the LG graph embeds both the structure of the local facial features (at the nodes of the graph) and the whole geometric structure of the face; it is a more accurate way of representing the face and is in accordance with the psychological way of perceiving human faces and facial expressions. In general, humans first extract the most important facial features such as eyes, nose, mouth, etc. and then inter-relate them for facial and facial expression representations.

The proposed LG graph method does not require any training as in other methods and is feasible in situations where we do not have many training samples. The graph method is invariant to scale, rotations, to a certain extent to pose and is shown to perform robustly under various illumination conditions. The methodology is also shown to perform accurately for recognizing facial expressions corresponding to the expression models present in the database.

References

1. M.H. Yang, D.J. Kriegman, and N. Ahuja. Detecting faces in images: A survey. *IEEE Pattern Analysis and Machine Intelligence*, 24(1), 2002
2. P. Kakumanu, S. Makrogiannis, R. Bryll, S. Panchanathan, and N. Bourbakis. Image chromatic adaptation using ANNs for skin color adaptation. In *Proceedings of the 16th IEEE International Conference on Tools with Artificial Intelligence, ICTAI04*, 2004
3. P. Kakumanu. A face and facial expression detection method for visually impaired. Ph.D. Dissertation, Wright State University, Dayton, OH, 2006
4. A.M. Martinez and R. Benavente. The AR Face Database, CVC Technical Report #24, 1998
5. X. Yuan and D. Goldman, A. Moghaddamzadeh and N. Bourbakis. Segmentation of colour images with highlights and shadows using fuzzy-like reasoning. *Pattern Analysis and Applications*, 4(4):272–282, 2001
6. N. Bourbakis, P. Yuan, and S. Makrogiannis. Object recognition using local - global graphs. In *Proceedings of the 15th IEEE International Conference on Tools with Artificial Intelligence, ICTAI0,* 2003
7. D. Hearn and M.P. Baker. *Computer Graphics, C version.* Prentice-Hall, NJ, 1997
8. N. Ahuja. Dot processing using Voronoi neighborhoods. *IEEE Pattern Analysis and Machine Intelligence*, 4(3), 1982
9. N. Ahuja, B. An, and B. Schachter. Image representation using Vornoi tesseletation. *Computer Vision, Graphics and Image Processing*, 29, 1985
10. K. Arbter, W. E. Snyder, H. Burkhardt and G. Hirzinger. Application of affine invariant Fourier descriptors to recognition of 3D object. *IEEE Pattern Analysis and Machine Intelligence*, 12(7), 1990
11. P.J. Phillips, H. Moon, S. A. Rizvi, and P. J. Rauss. The Feret evaluation methodology for face-recognition Algorithms. *IEEE Pattern Analysis and Machine Intelligence*, 22(10), 2000

12. A.D.J. Cross and E.R. Hancock. Graph matching with a dual step EM algorithm. *IEEE Transactions on Pattern Analysis and Machine Intelligence*, 20(11), 1998
13. B. Fasel and J. Luettin. Automatic facial expression analysis: A survey. *Pattern Recognition*, 36:259–275, 2003
14. D. Chai and K.N. Ngan. Locating facial region of a head-and-shoulders color image. In *ICFGR98*, 1998
15. R. Chellappa, C. Wilson, and S. Sirohey. Human and machine recognition of faces: A survey. *Proceedings of IEEE*, 83(5):705–740, 1995
16. A.J. Colmenarez and T.S. Huang. Face detection with information-based maximum discrimination. In *Proceedings of CVPR*, 1997
17. T.F. Cootes and C.J. Taylor. Locating faces using statistical feature detectors. In *Proceedings of AFGR*, pages 204–209, 1996
18. I. Craw, H. Ellis, and J. Lishman, Automatic extraction of face features. *Pattern Recognition Letters*, 5:183–187, 1987
19. I. Craw, D. Tock, and A. Bennett. Finding face features. In *Proceedings of the Second European Conference on Computer Vision*, pages 92–96, 1992
20. Y. Dai and Y. Nakano. Face-texture model based on SGLD and its application in face detection in a color scene. *Pattern Recognition*, 29(6):1007–1017, 1996
21. J.J. de Dios and N. Garcia. Face detection based on a new color space YCgCr. In *ICIP03*, 2003
22. B.A. Draper, K. Baek, M.S. Bartlett, and J.R. Beveridge. Recognizing faces with PCA and ICA. *Computer Vision Image Understanding*, 91(1–2):115–137, 2003
23. G.J. Edwards, C.J. Taylor, and T. Cootes. Learning to Identify and Track Faces in Image Sequences. In *Proceedings of ICCV*, pages 317–322, 1998
24. P. Ekman and W. Frisen. *Facial Action Coding System*, Palo Alto, CA. Consulting Psychologists Press, 1978
25. R. Fe'raud, O.J. Bernier, J.-E. Villet, and M. Collobert. A fast and accurate face detector based on neural networks. *Pattern Analysis and Machine Intelligence*, 22(1): 42–53, 2001
26. C. Garcia and G. Tziritas. Face detection using quantized skin color regions merging and wavelet packet analysis. *IEEE Transactions on Multimedia*, 1(3):264–277, 1999
27. V. Govindaraju. Locating human faces in photographs. *International Journal of Computer Vision*, 19(2):129–146, 1996
28. V. Govindaraju, S.N. Srihari, and D.B. Sher. A computational model for face location. In *Proceedings of the International Conference on Computer Vision*, pages 718–721, 1990
29. E. Hjelmas and B.K. Low. Face detection: A survey. *Journal of Computer Vision and Image Understanding*, 83:236–274, 2001
30. R.L. Hsu, M. Abdel-Mottaleb, and A.K. Jain. Face detection in color images. *IEEE Transactions on Pattern Analysis and Machine Intelligence*, 24(5):696–706, 2002
31. S.H. Kim, N.K. Kim, S.C. Ahn, and H.G. Kim. Object oriented face detection using range and color information. In *AFGR98*, 1998
32. M. Kirby and L. Sirovich. Application of the Karhunen–Loe've procedure for the characterization of human faces. *Pattern Analysis and Machine Intelligence*, 12(1), 1990

33. Y.H. Kwon and N. da Vitoria Lobo. Face detection using templates. In *Proceedings of ICPR*, pages 764–767, 1994
34. S.G. Kong, J. Heo, B.R. Abidi, J. Paik, and M.A. Abidi. Recent advances in visual and infrared face recognition – a review. *Computer Vision and Image Understanding*, 97, 2005
35. J. Kovac, P. Peer, and F. Solina. Human skin color clustering for face detection. In *EUROCON2003,* 2003
36. C. Kotropoulos, A. Tefas, and I. Pitas, Frontal face authentication using morphological elastic graph matching. *IEEE Transactions on Image Processing*, 9(4):555–560, 2000
37. P. Kuchi, P. Gabbur, S. Bhat, and S. David. Human face detection and tracking using skin color modeling and connected component operators. *IETE Journal of Research, Special issue on Visual Media Processing*, 2002
38. V. Kumar and T. Poggio. Learning-based approach to real time tracking and analysis of faces. In *Proceedings of AFGR, 2000*
39. A. Lanitis, C.J. Taylor, and T.F. Cootes. An automatic face identification system using flexible appearance models. *Image and Vision Computing* 13(5):393–401, 1995
40. M.S. Lew and N. Huijsmans. Information theory and face detection. In *Proceedings of ICPR*, 1996
41. C. Liu and H. Wechsler. Comparative assessment of independent component analysis (ICA) for face recognition. In *Proceedings of the Second International Conference on Audio- and Video-based Biometric Person Authentication*, Washington, DC, 1999
42. S. Mann. Wearable, tetherless computer-mediated reality: wearcam as a wearable face-recognizer, and other applications for the disabled. *Technical Report TR 361*, MIT Media Lab Perceptual Computing Section, Cambridge, MA, 1996. Also, available at http://www.eyetap.org/
43. F. Marqués and V. Vilaplana. A morphological approach for segmentation and tracking of human face. In *ICPR 2000*, 2000
44. S. McKenna, S. Gong, and Y. Raja. Modeling facial colour and identity with Gaussian mixtures. *Pattern Recognition*, 31(12):1883–1892, 1998
45. L. Meng and T. Nguyen. Two subspace methods to discriminate faces and clutters. In *Proceedings of ICIP*, 2000
46. J. Miao, B. Yin, K. Wang, L. Shen, and X. Chen. A hierarchical multiscale and multiangle system for human face detection in a complex background using gravity-center template. *Pattern Recognition*, 32(7):1237–1248, 1999
47. A.V. Nefian and M. H. Hayes III. Face detection and recognition using hidden Markov models. In *Proceedings of ICIP*, 1:141–145, 1998
48. K. Okada, J. Steffens, T. Maurer, H. Hong, E. Elagin, H. Neven, and C. von der Malsburg. The Bochum/USC face recognition system and how it fared in the Feret phase III test. In *Face Recognition: From Theory to Applications*. Springer, Berlin Heidelberg New York, 1998
49. N. Oliver, A. Pentland, and F. Berard. Lafter: Lips and face real time tracker. In *CVPR97*, 1997
50. E. Osuna, R. Freund, and F. Girosi. Training support vector machines: An application to face detection. In *Proceedings of CVPR,* pages 130–136, 1997
51. Z. Pan, G. Healey, M. Prasad, and B. Tromberg. Face recognition in hyperspectral images. *IEEE Transactions on Pattern Analysis and Machine Intelligence*, 25(12), 2003

52. A. Pentland, B. Moghaddam, and T. Starner. View-based and modular eigenspaces for face recognition. In *Proceedings of IEEE International Conference CVPR*, pages 84–91, 1994
53. P.J. Phillips, P. Grother, R.J. Micheals, D.M. Blackburn, E. Tabassi, M. Bone. Face recognition vendor test: Evaluation report, 2003
54. S.L. Phung, A. Bouzerdoum, and D. Chai. A novel skin color model in YCBCR color space and its application to human face detection. In *ICIP02*, 2002
55. M. Propp and A. Samal. Artificial neural network architecture for human face detection. *Intelligent Engineering Systems Through Artificial Neural Networks,* 2:535–540, 1992
56. D. Roth, M.-H. Yang, and N. Ahuja, A SNoW-based face detector. In *NIPS,* volume 12. MIT, Cambridge, MA, 2000
57. H. Rowley, S. Baluja, and T. Kanade, Neural network-based face detection. In *CVPR*, pages 203–208, 1996
58. H. Rowley, S. Baluja, and T. Kanade, Neural network-based face detection. *Pattern Analysis and Machine Intelligence*, 20(1):23–38, 1998
59. H. Rowley, S. Baluja, and T. Kanade, Rotation invariant neural network-based face detection. In *Proceedings of CVPR*, pages 38–44, 1998
60. E. Saber and A.M. Tekalp, Frontal-view face detection and facial feature extraction using color, shape and symmetry based cost functions. *Pattern Recognition Letters*, 17(8), 1998
61. H. Sahbi and N. Boujemaa. Coarse to fine face detection based on skin color adaptation. In *Workshop on Biometric Authentication, 2002,* volume 2359, LNCS, pages 112–120, 2002
62. F. Samaria and S. Young. HMM based architecture for face identification. *Image and Vision Computing,* 12:537–583, 1994
63. A. Samal and P.A. Iyengar. Human face detection using silhouettes. *International Journal of Pattern Recognition and Artificial Intelligence* 9(6), 1995
64. H. Schneiderman and T. Kanade. A statistical method for 3D object detection applied to faces and cars. In *Proceedings of CVPR*, volume 1, pages 746–751, 2000
65. K. Schwerdt and J.L. Crowely. Robust face tracking using color. In *AFGR00*, 2000
66. D.A. Socolinsky, A. Selinger, and J.D. Neuheisel. Face recognition with visible and thermal infrared imagery. *Computer Vision and Image Understanding*, 91(1–2):72–114, 2003
67. K. Sobottka and I. Pitas, Extraction of facial regions and features using color and shape information. In *ICPR96*, 1996
68. K. Sobottka and I. Pitas. A novel method for automatic face segmentation, facial feature extraction and tracking. *Signal Processing: Image Communication*, 12:263–281, 1998
69. F. Soulie, E. Viennet, and B. Lamy. Multi-modular neural network architectures: Pattern recognition applications in optical character recognition and human face recognition. *International Journal of Pattern Recognition and Artificial Intelligence*, 7(4):721–755, 1993
70. S. Srisuk and W. Kurutach. New robust face detection in color images. In *AFGR02*, pages 291–296, 2002
71. K.-K. Sung and T. Poggio. Example-based learning for view-based human face detection. *Pattern Analysis and Machine Intelligence*, 20, 1998

72. A. Tefas, C. Kotropoulos, and I. Pitas. Using support vector machines to enhance the performance of elastic graph matching for frontal face authentication. *Pattern Analysis and Machine Intelligence*, 23(7):735–746, 2001
73. J. Terrillon, M. Shirazi, M. Sadek, H. Fukamachi, and S. Akamatsu. Invariant face detection with support vector machines. In *Proceedings of ICPR*, 2000
74. M. Turk and A. Pentland. Face recognition using eigenfaces. In *Proceedings of CVPR*, pages 586–591, 1991
75. J.G. Wang and E. Sung. Frontal-view face detection and facial feature extraction using color and morphological operations.*Pattern Recognition Letters*, 20:1053–1068, 1999
76. Y. Wang and B. Yuan. A novel approach for human face detection from color images under complex background. *Pattern Recognition*, 34(10):1983–1992, 2001
77. L. Wiskott and C. von der Malsburg. Recognizing faces by dynamic link matching. *Neuroimage*, 4(3):S14–S18, 1996
78. K.W. Wong, K.M. Lam, and W.C. Siu. A robust scheme for live detection of human faces in color images. *Signal Processing: Image Communication*, 18(2):103–114, 2003
79. M.H. Yang and N. Ahuja. Detecting human faces in color images. In *ICIP98,* 1998
80. M.-H. Yang, N. Ahuja, and D. Kriegman. Face detection using mixtures of linear subspaces. In *Proceedings Fourth IEEE International Conference on Automatic Face and Gesture Recognition*, 2000
81. T.W. Yoo and I.S. Oh. A fast algorithm for tracking human faces based on chromatic histograms. *Pattern Recognition Letters*, 20(10):967–978, 1999
82. A. Yuille, P. Hallinan, and D. Cohen, Feature extraction from faces using deformable templates. *International Journal of Computer Vision*, 8(2):9–111, 1992
83. W. Zhao, R. Chellappa, P.J. Philips, and A. Rosenfeld, Face recognition: A literature survey, *ACM Computing Surveys*, 85(4):299–458, 2003
84. X. Zhang, Y. Jia, A linear discriminant analysis framework based on random subspace for face recognition. *Pattern Recognition*, 40:2585–2591, 2007
85. J. Meyneta, V. Popovicib, and J.-P. Thirana, Face detection with boosted Gaussian features, *Pattern Recognition* 40, 2007
86. Q. Chen.,W.-k. Cham, and K.-k. Lee, Extracting eyebrowcontour and chin contour for face recognition, *Pattern Recognition*, 40, 2007
87. S.-I. Choi, C. Kim, and C.-H. Choi. Shadowcompensation in 2D images for face recognition. *Pattern Recognition*, 40, 2007

Facial Image Processing

Xiaoyi Jiang[1] and Yung-Fu Chen[2]

[1] Department of Mathematics and Computer Science
University of Münster, Germany
xjiang@math.uni-muenster.de

[2] Department of Health Services Administration
China Medical University, Taichung 404, Taiwan
yungfu@mail.cmu.edu.tw

Summary. Faces are among the most important classes of objects computers have to deal with. Consequently, automatic processing and recognition of facial images have attracted considerable attention in the last decades. In this chapter we focus on a strict view of facial image processing, i.e. transforming an input facial image into another and involving no high-level semantic classification like face recognition. A brief overview of facial image processing techniques is presented. Typical applications include removal of eyeglasses, facial expression synthesis, red-eye removal, strabismus simulation, facial weight-change simulation, caricature generation, and restoration of facial images.

1 Introduction

In communication between people, faces play one of the most important roles. Faces allow to recognize a person's identity but moreover they carry rich information about a person's emotional state. We are able to sense the smallest differences in the appearance of a face and easily sense irony, the slightest disagreements or understanding. Thus, faces serve as an information carrier in a much more subtle manner than identity or emotions and help us to adapt our behavior. Not surprisingly, faces have been long a research topic in psychology.

Automatic processing and interpretation of facial images has attracted much attention in the last decades. Without doubt the most prominent research topic in this context is face recognition and verification in still images and videos [8, 26, 30, 65]. A fundamental issue in any face analysis system is to detect the locations in images where faces are present. A lot of work has been done on face localization which is often a preprocessing step within recognition systems [19, 61]. Moreover, many methods have been developed to detect facial features such as eyes, nose, nostrils, eyebrows, mouth, lips, ears, etc. [65]. These features can either be used for a direct recognition approach or

X. Jiang and Y.-F. Chen: *Facial Image Processing*, Studies in Computational Intelligence (SCI) **91**, 29–48 (2008)
www.springerlink.com

to normalize facial images for holistic recognition approaches like eigenfaces and Fisherfaces.

In this chapter we will take a strict view of facial image processing only, i.e. transforming an input facial image into another. Several tasks fall into this category, including removal of eyeglasses, facial expression synthesis, red-eye removal, strabismus simulation, facial weight-change simulation, and caricature generation. They are all image processing tasks and only involve very limited high-level semantic reasoning.

The motivation for these facial image processing tasks is manifold:

- Improvement of facial image quality
  A typical representative of this class is red-eye removal.
- Preprocessing for face recognition
  One crucial requirement on successful face recognition is robustness to variations arising from different lighting conditions, poses, scales, or occlusion by other objects. Glasses belong to the most common occluding objects and have a significant effect on the recognition performance. While it is possible to identify and use non-occluded local regions only, we may want to estimate the non-occluded facial image by removing the occluding objects, e.g. glasses.
- Simulation of effects
  Given a facial image, the task is to simulate expressions, strabismus, or to draw caricatures.

Facial image processing operations can be found as a part of various application scenarios. For instance, facial expression synthesis may serve to enrich face databases for improved recognition performance.

Compared to other images, processing of facial images is particularly delicate. We see and interpret many faces everyday and thus are continuously trained to successfully distinguish between a large number of different faces from the childhood on. As a consequence, we are very sensitive to small distortions and changes in the appearance of faces. The so-called Thatcher illusion [53] is an excellent example to illustrate this point, see Figure 1. In this illusion, the eyes and mouth of a face are inverted. The result looks grotesque in an upright face. However, when shown inverted, the face looks fairly normal. Although we do observe some distortions between the two facial images, the perceptual divergence is much smaller than seen upright. This and other related observations indicate that our brain tends to perceptually amplify the quantitatively measurable differences when seeing faces. In fact, it is the wonderful ability of the human's visual system in face perception that makes the automatic facial image processing so difficult.

While considerable survey work has been done in the past for face detection and face recognition, the goal of this chapter is to give a brief overview of facial image processing techniques. It is not our intention to provide a thorough review. Instead, we focus on the tasks mentioned before. For these tasks we

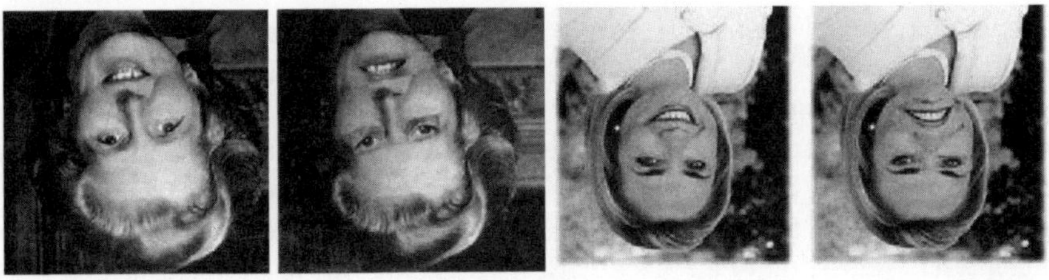

Fig. 1. Thatcher illusion. A normal face and a manipulated version of Thatcher (left); the same for Clinton

give a motivation, present the most recent and important methods, summarize the current state of research, and discuss the directions of future research.

2 Removal of glasses

The first publications concerning glasses in facial images are limited to the existence decision and position detection of glasses only. Jiang et al. [22] define six measures for classifying the presence of glasses in a facial image without suggesting a classifier. Wu et al. [58] go a step further and devise a sophisticated classifier for the existence of glasses based on support vector machines. For this classifier a recognition performance of 90 percent is reported.

Besides the existence decision similar to [22], Jing and Mariani [24] also extract the contour of glasses using a deformable model combining edge features and geometric properties. In [60] the extraction is done by decomposing face shape using the Delaunay triangulation. A 3D technique is reported in [59]. The authors perform a 3D Hough transform in trinocular stereo facial images based on the determination of a 3D plane passing through the rims of the glasses, without any assumption of the facial pose or the shape of glasses. Certainly, 3D information can make a substantial contribution to glasses localization. However, we pay a price of multiple cameras and the computational costs for 3D reconstruction.

The work [46] is among the first ones on the removal of glasses. The fundamental assumption is the availability of a set of M glasses-free images. Implicitly, it is further assumed that these prototype images are normalized to the same size and properly aligned. Let F_0 be the average of all glasses-free images and $F_1, F_2, \ldots, F_{M-1}$ the eigenfaces from a principal component analysis (PCA). In some sense the $M-1$ eigenfaces span the space of glasses-free images. Given an input image I_g with glasses, then the corresponding glasses-free image I_f is computed by a PCA reconstruction method

$$I_f = F_0 + \sum_{i=1}^{M-1} ((I_g - F_0) \cdot F_i) \cdot F_i \qquad (1)$$

It is not untypical that this simple PCA reconstruction cannot remove all traces of glasses.

The problem is partly caused by the fact that the eigenfaces represent a space, which the input image I_g does not belong to. This gives rise to some modeling inaccuracy which remains visible after the reconstruction. An obvious solution lies in improving the modeling accuracy by using M prototype images with glasses instead. Denoting the average of these prototypes by G_0 and their eigenfaces by $G_1, G_2, \ldots, G_{M-1}$, an input image I_g can be well represented by:

$$I_g = G_0 + \sum_{i=1}^{M-1} c_i \cdot G_i \tag{2}$$

in the space of facial images with glasses, where $c_i = (I_g - G_0) \cdot G_i$ holds. Assume that a set of prototype pairs be available, i.e. there exist one glasses-free image and one glasses image for the same person in the example set. Then, we could construct an image I_f by retaining the coefficients c_i and replacing G_0 and G_i in Eq. (4) by F_0 and F_i respectively.

The example-based reconstruction method [42] follows this line. However, the PCA computation is done on M prototype images, each resulting from a concatenation of a glasses image and a glasses-free image of the same person. The single average image of these images can be decomposed into two separate average images G_0 and F_0 again. Similarly, the resulting eigenfaces can be decomposed into two separate sets of images $G = \{G_1^*, G_2^*, \ldots, G_{M-1}^*\}$ and $F = \{F_1^*, F_2^*, \ldots, F_{M-1}^*\}$. Note that while G_0 and F_0 are identical to those before, the other images G_i^* and F_i^* differ from the eigenfaces of glasses images and glasses-free images, respectively, due to the different ways of PCA computation. Given an input image I_g, we look for its optimal representation in the space of facial images with glasses spanned by set G and then apply the representation coefficients to set F for constructing a glasses-free image I_f. In this case, however, the images G_i^* are no more orthogonal to each other and thus a least square minimization is required to obtain the optimal representation of I_g in terms of G.

A refined algorithm by using recursive error compensation is presented in [42]. It is based on the same idea of PCA reconstruction (in terms of glasses-free prototype images). An iterative scheme is defined, where in each iteration, the pixel values are adaptively computed as a linear combination of the input image and the previously reconstructed pixel values.

Some results for the three removal methods (simple PCA reconstruction [46], example-based reconstruction [42], and recursive error compensation [42]) are shown in Figure 2. Visually, the error compensation method seems to produce the best results; the images in Figure 2(d) have no traces of glasses and look seamless and natural. This impression is also confirmed by a quantitative error measure. The number on each facial image indicates the average pixel-wise distance to the corresponding original facial image without glasses. The numbers in the last column represent the mean of pixel-wise

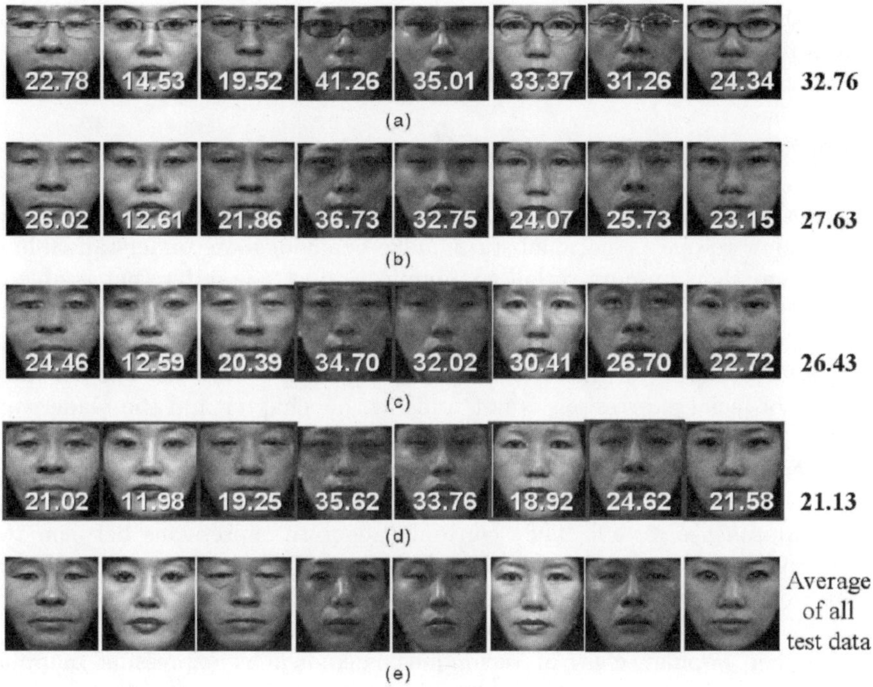

Fig. 2. Examples of glasses removal: (a) input images with glasses; (b) simple PCA reconstruction method; (c) example-based reconstruction method; (d) recursive error compensation; (e) original faces without glasses. (Courtesy of J.-S. Park, © 2005 IEEE)

distances of a total of 100 test images. The recursive error compensation method tends to have the smallest error measure.

Another advanced glasses removal technique is presented in [59]. Given an image I_g with glasses, the glasses-free reconstruction is based on the maximum a posteriori (MAP) criterion:

$$I_f^* = \arg\max_{I_f} p(I_f|I_g) = \arg\max_{I_f} p(I_g, I_f)$$

The joint distribution $p(I_g, I_f)$ is estimated by means of a set of aligned pairs of glasses and glasses-free images of the same person.

An assessment of glasses removal quality depends on applications. For image editing purpose the only criterion is simply to which degree the reconstructed image is seamless and natural looking. In the context of face recognition, however, the same question must be answered in a task-based evaluation manner. That is, we conduct face recognition experiments using both original glasses-free and reconstructed glasses-free images. The difference in recognition rate is the ultimate performance measure for glasses removal. Currently, very few work has been done following this line; only [42] reports a small-scale study. More work is needed to demonstrate the potential of this preprocessing step in real-world face recognition systems.

3 Facial expression synthesis

Facial expressions play a major role in how people communicate. They serve as a window to one's emotional state, make behavior more understandable to others, and they support verbal communication. A computer that is able to interact with humans through facial expressions (in addition to other modalities) would greatly advance human-computer interfaces. This ability includes both understanding of facial expressions and their synthesis. The semantic analysis of facial expressions is not topic of this chapter and the readers are referred to [14, 15, 28].

Several algorithmic paradigms for facial expression synthesis can be found in the literature [28, 43]. One class of methods are variants of the morph-based approaches [6, 49]. They can only generate expressions between two given images of the same person and their ability of generating arbitrary expressions is thus more than limited. If merely one image of a person is available, these approaches are not applicable at all.

Another popular class of techniques is known as expression mapping (performance-driven animation) [32, 43]. Its principle is quite simple: Given an image A of a person's neutral face and another image A' of the same person's face with a desired expression, the movement of facial features from A to A' is geometrically performed on a second person's neutral image B to produce its facial image B' with the expression. The major algorithmic steps are:

- Find the facial features (eyes, eyebrows, mouth, etc.) in A, A', and B, either manually or by some automatic method.
- Compute the difference vectors between the feature positions of A and A'.
- Move the features of B along the difference vector of corresponding feature of A and warp the image to a new one B' accordingly.

This technique is a geometry-driven feature mapping combined with image warping. As such, its applicability includes animation of 2D drawings and images far beyond facial expression synthesis.

The image B' produced by this expression mapping technique usually looks reasonable, since the transformation from B to B' captures all necessary geometric changes that are needed to copy the facial expression as exemplified by A and A'. However, the algorithm totally ignores the photometric changes and thus the result image lacks details such as the wrinkles on the forehead. An attempt of including photometric changes is made in [34]. Assume that all images A, A', and B are geometrically aligned. Using the Lambertian reflectance model, the authors show that the relationship:

$$B'(u,v) = B(u,v) \cdot \frac{A'(u,v)}{A(u,v)}$$

holds for each image position (u, v). It tells us how B' is photometrically related to B in a way so that the same geometric changes between A and A' are done between B and B'. This relationship can be used to extend the traditional geometric warping method to the following algorithm:

- Perform traditional geometric warping on images A, A', and B to compute an image B'.
- Align A and A' with B' through image warping and denote the warped images by $\tilde{A}$ and $\tilde{A}'$, respectively.
- Compute the photometrically corrected final result $B^*(u,v) = B'(u,v) \cdot \frac{\tilde{A}'(u,v)}{\tilde{A}(u,v)}$ for each pixel (u, v).

The result image B^* has exactly the same geometry as the initial result B'. It also respects the photometric changes from A to A' in addition to the geometric changes. Among others, it is very effective in generating wrinkles on the forehead.

In the recent work [63] quite different assumptions are made. The authors assume that a set of images of the same person is available. If these images cover enough different facial expressions, then they span in some sense the space of expressions (of that particular person) and any other expression can be represented as their convex combination [45]. Let G_i, $i = 1, 2, \ldots, m$, denote the geometry of the i-th example expression image I_i (vector of positions of feature points). Then, the geometry G of an arbitrary new expression image is represented as a convex combination:

$$G = \sum_{i=1}^{m} c_i \cdot G_i$$

where the coefficients c_i satisfy the condition $\sum_{i=1}^{m} c_i = 1$. The optimal coefficients are determined by solving the optimization problem:

$$\text{minimize: } ||G - \sum_{i=1}^{m} c_i \cdot G_i||^2, \quad \text{subject to: } \sum_{i=1}^{m} c_i = 1, \; c_i \geq 0$$

which is a quadratic programming problem. If the m example images themselves are I_i, $i = 1, 2, \ldots, m$, then an image I corresponding to the desired expression geometry G can be composed by:

$$I = \sum_{i=1}^{m} c_i \cdot I_i$$

Since the example images are assumed to be aligned, this step is simply a pixel-wise color blending. This algorithm can be used for expression editing,

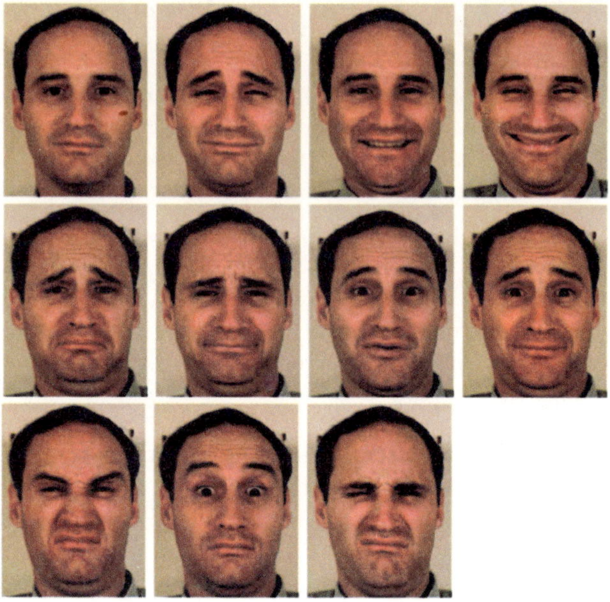

Fig. 3. Eleven example images of a male person. (Courtesy of Z. Liu, © 2006 IEEE)

where the geometry of the desired expression is, for instance, constructed using some editing tool. Figure 3 shows the set of example images. The six expression editing results in Figure 4 indicate the algorithm's ability of generating expression images of a different geometry than any of the example images. A second application scenario is expression mapping, where another person's facial image serves as the source of the geometry G and the algorithm mimics the expression of that person. In Figure 4 the right column shows image pairs of a female and a male face. The male faces are synthetic images generated by taking over the geometry of the female image to synthesize a similar expression. Interestingly, this technique has been extended in [63] to synthesize facial expressions of 3D head models.

Facial expression synthesis has attracted many researchers in the last years [1, 2, 7, 13, 18, 62, 66]. It is expected to have substantial impact on facial image editing and human-computer interfaces.

4 Eye synthesis

The most important perceptual feature of a face is probably the eye as the eye appearance plays a central role in human communication. Consequently, eye synthesis is of interest in several contexts.

Face recognition is confronted by significant natural variations, e.g. lighting conditions, image size, etc. Although some factors like image size can be

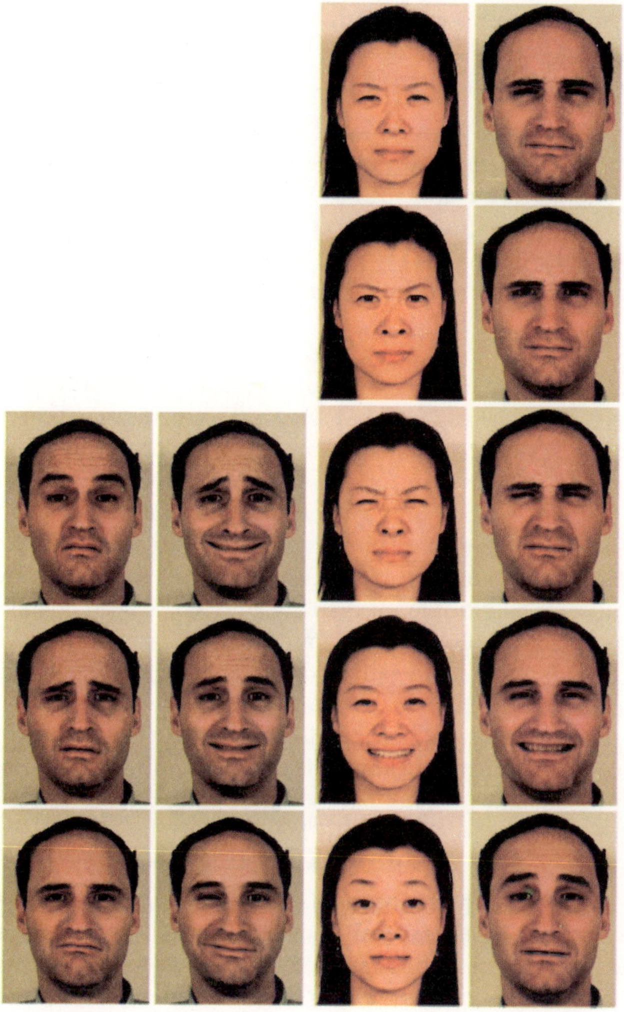

Fig. 4. Expression editing: Six synthetic images based on editing the geometry (left). Expression mapping: The male's synthetic image is generated by taking over the geometry of the female (right). (Courtesy of Z. Liu, © 2006 IEEE)

alleviated by geometric and photometric normalization, the availability of a collection of training images covering other variations and appearances is of great help. Since it is hardly possible to capture all possible variations, one solution lies in their synthesis. In [25] the authors investigated a number of operators for changing the eye shape such as:

- Pull down or push up of the eyebrow (without any change to eye location)
- Rotation of the entire eyebrow around its middle

- Pull down or push up of the upper eyelid so that the eye appears less or more open
- Pull down or push up of the lower eyelid so that the eye appears more or less open

Although these operations do not produce precise anatomical changes, they seem to be adequate for appearance-based systems. The synthesized images artificially enlarge the training set of a face recognition system in order to represent the entire space of possible variations more completely. In [25] the authors show that this enrichment indeed improves the recognition performance.

Gaze redirection is another important application of eye synthesis. In video-conferencing and other visual communication scenarios, the participant watches the display rather than the camera. This turns the gaze direction away from the conversation partner and impairs the desired eye contact. Several hardware solutions have been proposed to alleviate the problem. Sellen [48] places the display/camera unit sufficiently far away from the user, so that gazing at the screen becomes indistinguishable from gazing at the camera. A more sophisticated solution is suggested in [40], where the camera is positioned behind a semi-transparent display. Due to the need of special hardware, however, this kind of solutions has found limited acceptance so far. The authors of [56] propose to artificially redirect the gaze direction. The eye-white and the iris are detected in a facial image and replaced by corresponding parts from a real eye image at desired position.

Synthesizing strabismic face images, each of a different angle, based on a normal frontal face image is needed for conducting studies in psychosocial and vocational implications of strabismus and strabismus surgery. Strabismus can have a negative impact on an individual's employment opportunity, school performance, and self-perception. A recent study [11], for instance, indicates that large-angle horizontal strabismus appeared to be vocationally significant particularly for female applicants, reducing their ability to obtain employment. One possibility of synthesizing strabismic face images lies in manual image editing using standard image processing packages [11]. Alternatively, one may ask a person to simulate different strabismic angles. Both methods are tedious and time-consuming, thus not applicable if many face images and various strabismic angles are needed. The latter approach has an additional problem that it is hardly possible for a person to precisely simulate a particular desired angle. The work [23] proposes an algorithm for synthesizing strabismic face images of an arbitrary angle. It consists of the following main steps (see Figure 5):

- Detection of the contour of the iris and the reflection point
- Removal of the iris
- Detection of the contour of the eye
- Rotation of the eye, i.e. re-insertion of the iris
- Placement of the reflection point

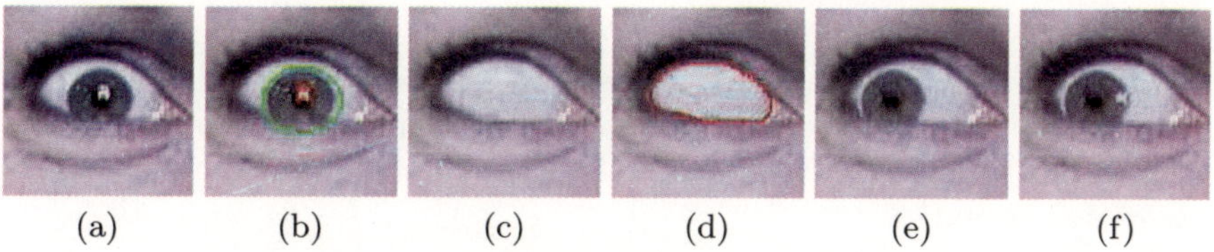

Fig. 5. Main steps of strabismus synthesis algorithm: (a) (part of) input image; (b) detected contour of iris and reflection point; (c) removal of iris; (d) detected eye contour; (e) rotated eye; (f) embedded reflection point. The final result is given in (f)

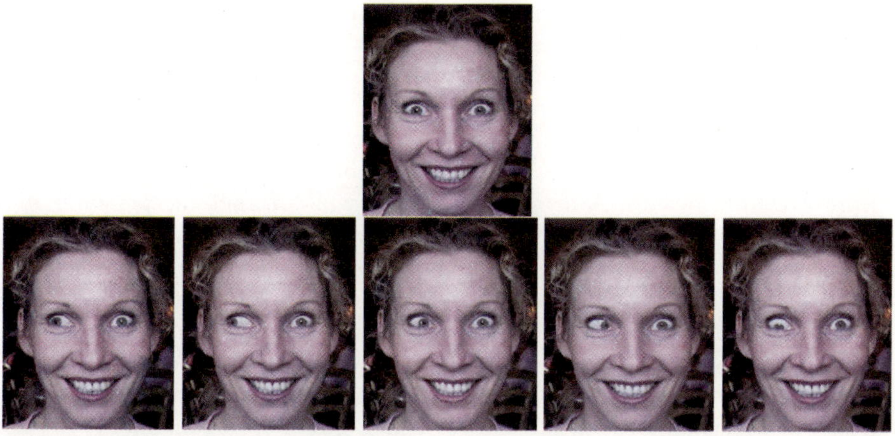

Fig. 6. Results of strabismus simulation. Top: input image. Bottom: 20° and 40° to the right, 20° and 40° to the left, and strabismus with a vertical angle (from left to right). Only the right eye of the person is processed

All three contour detection subtasks (iris, reflection point, and eye) are solved by a dynamic programming technique. The iris removal should be done with care. Typically, the eyelashes partially interlay with the iris so that a straightforward removal and subsequently filling this area by the eye background would produce unnatural appearance in the eyelash. Instead, one has to fill the missing background and continue the missing eyelashes in a natural way simultaneously. In [23] this step is performed by means of an image inpainting algorithm. Despite of strabismus the reflection point should remain unchanged. Therefore, as the last operation the detected reflection point is embedded exactly at the same coordinates as in the input image. Accordingly, the reflection point has to be removed from the source image before it is re-inserted into the inpainted background without iris. Figure 6 shows some results of strabismus simulation.

The reversed direction of correction of strabismic face images is of great interest in plastic surgery applications. This gives the patient an approximate post-operation look of strabismus surgery in the pre-operation phase. In fact, both strabismus simulation and correction are closely related to gaze redirection. The main difference lies in the required image quality. Psychosocial studies tend to need static pictures of higher quality than typically in communication scenarios.

5 Redeye removal

Redeye is a common problem in consumer photography. When a flash is needed to illuminate the scene, the ambient illumination is usually low and a person's pupils will be dilated. Light from the flash can thus reflect off the blood vessels in the person's retina. In this case it appears red in color and this reddish light is recorded by the camera.

One possible hardware technique to avoid redeye is to increase the distance between the flash unit and the camera lens. Another popular solution is the use of pre-exposure flashes. A pre-exposure flash will contract the person's pupil and thus reduce the chance that light reflected off the retina will reach the lens. The drawback of this approach is that people will sometimes close their eyes by reflex and the substantial need of power; additional flashes further lower the battery life. In addition the red-eye artifacts are reduced, but not completely eliminated.

With the advent of digital photography, software solutions become popular. While several photo editing software packages allow for manual correction of redeye, they tend to be semiautomatic and do not always give satisfactory results. Today, there exist several patents for detecting and removing redeyes (see listings in [38, 47, 51]).

All research work published in the literature has a modular structure of redeye detection followed by a removal operation. As for redeye detection, the two papers [16, 17] are based on a face detection. This narrows the search area for redeye artifacts to facial regions only. In addition the face information can be used to infer properties like the location and size of possible redeye artifacts. Within a face region, features such as redness and changes in luminance and redness are then applied to find redeye parts.

In [55] an active appearance model is trained using a set of typical redeye subimages. Given an input image, color cues are first used to locate potential redeye regions. Then, a matching is performed to minimize the difference between a candidate image part and one synthesized by the appearance model.

Several algorithms do not perform a complete face detection prior to the redeye detection. Although the prior face detection could provide extremely useful information it is avoided as it is a challenging task itself [19, 61]. Instead, the redeye detection is performed by a framework for pattern classification where image subparts which serve as candidates for redeye artifacts are found by relatively easy methods. Then, a classifier trained by a set of typical redeye subimages throws away many false positives. The references [37, 64] follow this line. For instance, red oval regions are found by simple image processing operations and regarded as redeye candidates [37]. In contrast, [36] trains a classifier to decide whether a region of fixed size 11×11 is a redeye subimage. This classifier is applied to image patches of the same size at all possible positions. To handle different sizes of redeyes, the same operation is also carried out on scaled versions of the image.

The approach in [51] starts with a skin detection, followed by a morphological postprocessing of the skin-like regions. Then, the color is converted to a greylevel image, in which redeye artifacts in skin-like regions are highlighted as bright spots. Finally, convolutions are done to find pupils of circular shape.

After the detection of redeye parts, the correction can be performed in many different ways. The simplest solution of correction is to replace the red value of a pixel in a redeye part by its other color channel values such as the average of green and blue [51]. The reference [54] discusses a means to correct redeyes in a perceptually pleasing manner. Figure 7 shows two examples of redeye removal.

An unconventional hard/software solution is suggested in [38]. For the same scene two images are acquired, one with flash and one without flash. This conceptual idea could be implemented on today's digital cameras since many of them support "continuous shooting". If two images of the same scene are shot within a very short time interval, it can be assumed that no motion occurred and the images are well aligned. A flash and a non-flash image should have similar chromatic properties except in the redeye artifacts. This fact gives the fundamental of redeye detection in [38]. Furthermore, the chromatic information in the non-flash image is also used to restore the original iris color for redeye correction.

For the redeye detection part, a quantitative performance evaluation is straightforward. Some of the publications [16, 36, 37, 55, 64] mention this kind of performance measures, typically on hundreds of test images, while others do not. All the experiments were done on the authors' own images. Today, there exist still no common public test image databases with a well

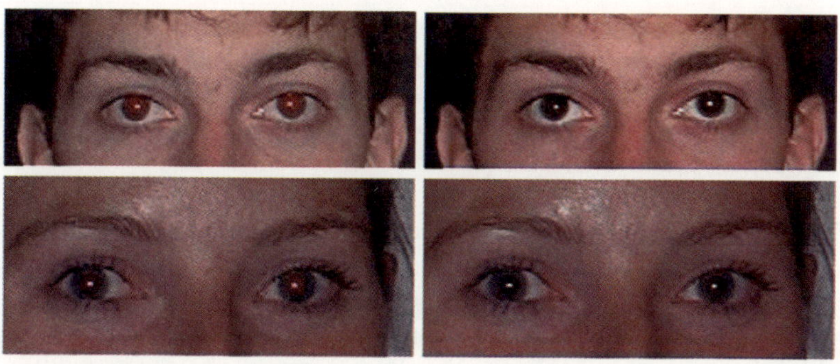

Fig. 7. Examples of redeye removal. (Courtesy of S. Jirka and S. Rademacher)

accepted performance evaluation protocol like in other fields (face recognition [44], range image segmentation [20], etc.). In contrast, a (quantitative) evaluation of the redeye correction is much more subtle. Despite of the difficulties, performance evaluation on common data would certainly advance the research on redeye detection and correction.

6 Restoration of facial images

If some parts of a facial image, for instance the eyes and mouth, are damaged or missing, then we are faced with the task of facial image reconstruction. This problem is different from what image inpainting techniques [50] intend to solve, namely removing large objects and replacing them with visually plausible backgrounds. In contrast, face reconstruction recovers in some sense the "foreground" object (face) and as such some face modeling work is certainly needed.

Hwang and Lee [21] present an approach to face reconstruction. Face modeling is done by PCA on a set of prototype facial images so that the face space is spanned by the computed eigenfaces. A damaged facial image is considered as a point in a subspace spanned by reduced eigenfaces which only contain the pixels not damaged in the input image. Its optimal representation in terms of a linear combination of the reduced eigenfaces can be computed by a least square method. Then, using the same representation coefficients to linearly combine the original eigenfaces delivers a reconstructed facial image. In [21] this fundamental idea is separately applied to reconstruct the geometry and texture of a face. Afterwards they are fused to synthesize a facial image without damages. The same principle has also been used to enhance the resolution of facial images [41].

7 Artistic processing of facial images

A caricature is a humorous illustration that exaggerates the basic essence of a person to create an easily identifiable visual likeness. Artists have the amazing ability to capture distinguished facial features that make a subject different from others and to exaggerate these features. The central issue here is the question which facial features are significant from others and how to exaggerate.

There is only very few work on caricature generation [10, 27, 31]. In [10] the authors define a set of 119 nodes based on the MPEG-4 face definition parameters and face animation parameters to control the geometric shape of a face. They use 100 pictures of Asian female with manually labeled nodes to build an average shape. Given a new facial image, the corresponding mesh representation is computed by a chain of processing steps including locating facial features like mouth and iris, and an iterative mesh optimization procedure. This mesh is then compared with the average face representation. Those nodes far away from the average are selected for exaggeration. The caricature generation itself is an image warping.

Related to caricature generation is the task of generating a line drawing from a facial image [9]. A further artistic processing of facial images is automatic generation of sketches [33, 52], which however has an important real application in low enforcement. An essential task there is automatic retrieval of suspects from a face database using a mug-shot only. In [52] facial images of the database are all transformed into a sketch, which are compared with the mug-shot. Based on a set of example photo-sketch pairs the photo-to-sketch transformation is performed by a technique similar to those for glasses removal using pairs of glasses and glasses-free example images.

8 Facial weight change

The task of weight change is to make a person's face to look fatter or thinner while maintaining the natural appearance of the face. Potential applications include beauty industry, security and police work. In [12] the authors report another use of weight change as a diagnostic tool in medicine. To diagnose, evaluate and treat eating disorders, especially Anorexia Nervosa, weight-changed facial images are needed for testing purpose.

The weight change simulator built in [12] is a straightforward image warping. The user selects thirteen landmarks to specify the shape of the neck and the cheeks. Their positions after the weight change are defined by a transformation controlled by a factor w. Then, the thin plate spline warping [5] is applied to transform other image points.

9 Conclusion

In this chapter we have given a brief overview of facial image processing tasks. The tasks considered include glasses removal, facial expression synthesis, eye synthesis, redeye removal, restoration of facial images, artistic processing of facial images, and facial weight change.

In addition to studying the algorithms for particular problems, it should be emphasized that several techniques we have seen in the last sections are of general nature and may find applications in other situations. In particular, the image mapping approach based on a set of pairs of images is despite of its simplicity powerful enough for challenging tasks like glasses removal [42, 46] and sketch generation [52].

A few topics are not discussed in this chapter. For instance, simulation of aging effects [29, 35] (making somebody look younger) is both a fascinating cosmetic image operation and a helpful tool in forensic medicines and law

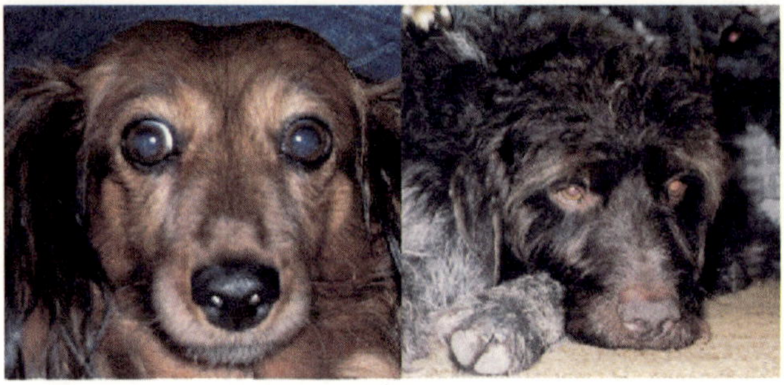

Fig. 8. Redeye effect with dogs. (Courtesy of S. Jirka and S. Rademacher)

enforcement. In the latter case we may need to recognize a person, for which we only have older pictures in the database showing the person at a younger age. Handling varying imaging conditions like lighting is essential for robust image analysis systems. Face relighting [57] intends to re-render a facial image to arbitrary lighting conditions. Finally, the characteristic of facial images requires particular consideration in watermarking [39].

The facial image processing tasks considered in this chapter are mostly restricted to frontal images; the exception is redeye removal. Manipulating images of faces from arbitrary viewpoints clearly causes increased complexity. Some work based on 3D face modeling can be found in [3, 4].

Some of the discussed facial image processing operations could be extended to deal with animals. For instance, photographs of animals have similar effects as redeye, see Figure 8. Although in this case the artifacts may be green or blue, an automatic detection and correction still makes sense.

Tasks of facial image processing are mostly very challenging. As discussed in the introduction section, our brain with his intelligent visual processing ability in general and the very sensitive perception of faces increases the complexity. Nevertheless some powerful techniques have been developed so far and further development can be expected to produce powerful tools for facial image editing, communication, and recognition in a variety of application scenarios.

Acknowledgments

The work was done when the first author was visiting Da-Yeh University, Taiwan. The financial support from both the Ministry of Education of Taiwan and Da-Yeh University is greatly appreciated. The authors want to thank S. Wachenfeld for proofreading the manuscript.

References

1. B. Abboud, F. Davoine, and M. Dang. Facial expression recognition and synthesis based on an appearance model. *Signal Processing: Image Communication*, 19(8):723–740, 2004
2. B. Abboud and F. Davoine. Bilinear factorisation for facial expression analysis and synthesis. *IEE Proceedings of the Vision Image and Signal Processing*, 152(3):327–333, 2005
3. V. Blanz, C. Basso, T. Poggio, and T. Vetter. Reanimating faces in images and video. *Computer Graphics Forum*, 22(3):641–650, 2003
4. V. Blanz, K. Scherbaum, T. Vetter, and H.-P. Seidel. Exchanging faces in images. *Computer Graphics Forum*, 23(3):669–676, 2004
5. F. Bookstein. Principal warps: Thin-plate splines and the decomposition of deformations. *IEEE Transactions on Pattern Analysis and Machine Intelligence*, 11(6):567–585, 1989
6. C. Bregler, M. Covell, and M. Slaney. Video rewrite: Driving visual speech with audio. In *Proceedings of SIGGRAPH*, pages 353–360, 1997
7. N.P. Chandrasiri, T. Naemura, and H. Harashima. Interactive analysis and synthesis of facial expressions based on personal facial expression space. In *Proceedings of IEEE International Conference on Automatic Face and Gesture Recognition*, pages 105–110, 2004
8. K. Chang, K. Bowyer, and P. Flynn. A survey of approaches and challenges in 3D and multi-modal 2D+3D face recognition. *Computer Vision and Image Understanding*, 101(1):1–15, 2006
9. H. Chen, Y.-Q. Xu, H.-Y. Shum, S.-C. Zhu, and N.-N. Zheng. Example-based facial sketch generation with non-parametric sampling. In *Proceedings of International Conference on Computer Vision*, volume 2, pages 433–438, 2001
10. P.-Y. Chiang, W.-H. Liao, and T.-Y. Li. Automatic caricature generation by analyzing facial features. In *Proceedings of Asian Conference on Computer Vision*, 2004
11. D.K. Coats, E.A. Paysse, A.J. Towler, and R.L. Dipboye. Impact of large angle horizontal strabismus on ability to obtain employment. *Ophthalmology*, 107(2):402–405, 2000
12. U. Danino, N. Kiryati, and M. Furst. Algorithm for facial weight-change. *Proceedings of International Conference on Electronics, Circuits and Systems*, pages 318–321, 2004
13. Y. Du and Y. Lin. Emotional facial expression model building. *Pattern Recognition Letters*, 24(16):2923–2934, 2003
14. P. Eisert. Analysis and synthesis of facial expressions. In N. Sarris and G.M. Strintzis, editors, *3D Modeling and Animation: Synthesis and Analysis Techniques for the Human Body*, pages 235–265, Idea Group Inc., 2004
15. B. Fasel and J. Luettin. Automatic facial expression analysis: A survey. *Pattern Recognition*, 36(1):259–275, 2003
16. F. Gasparini, R. Schettini. Automatic redeye removal for smart enhancement of photos of unknown origin. In *Proceedings of the International Conference on Visual Information Systems*, 2005
17. M. Gaubatz and R. Ulichney. Automatic red-eye detection and correction. In *Proceedings of International Conference on Image Processing*, volume I, pages 804–807, 2002

18. J. Ghent and J. McDonald. Photo-realistic facial expression synthesis. *Image and Vision Computing*, 23(12):1041–1050, 2005
19. E. Hjelmas and B.K. Low. Face detection: A survey. *Computer Vision and Image Understanding*, 83:237–274, 2001
20. A. Hoover, G. Jean-Baptiste, X. Jiang, P.J. Flynn, H. Bunke, D. Goldgof, K. Bowyer, D. Eggert, A. Fitzgibbon, and R. Fisher. An experimental comparison of range image segmentation algorithms. *IEEE Transactions on Pattern Analysis and Machine Intelligence*, 18(7):673–689, 1996
21. B.-W. Hwang and S.-W. Lee. Reconstruction of partially damaged face images based on a morphable face model. *IEEE Transactions on Pattern Analysis and Machine Intelligence*, 25(3):365–372, 2003
22. X. Jiang, M. Binkert, B. Achermann, and H. Bunke. Towards detection of glasses in facial images. *Pattern Analysis and Applications*, 3(1):9–18, 2000
23. X. Jiang, S. Rothaus, K. Rothaus, and D. Mojon. Synthesizing face images by iris replacement: Strabismus simulation. In *Proceedings of first International Conference on Computer Vision Theory and Applications*, pages 41–47, 2006
24. Z. Jing and R. Mariani. Glasses detection and extraction by deformable contour. In *Proceedings of International Conference on Pattern Recognition*, 933–936, 2000
25. B. Kamgar-Parsi and A.K. Jain. Synthetic eyes. In J. Kittler and M.S. Nixon, editors, *Audio- and Video-Based Biometric Person Authentication*, pages 412–420. Springer, Berlin Heidelberg New York, 2003
26. S. Kong, J. Heo, B. Abidi, J. Paik, and M. Abidi. Recent advances in visual and infrared face recognition – A review. *Computer Vision and Image Understanding*, 97(1), 103–135, 2005
27. H. Koshimizu, M. Tominaga, T. Fujiwara, and K. Murakami. On KANSEI facial image processing for computerized facialcaricaturing system PICASSO. In *Proceedings of Conference on Systems, Man, and Cybernetics*, volume 6, pages 294–299, 1999
28. S. Krinidis, I. Buciu, and I. Pitas. Facial expression analysis and synthesis: A survey. In *Proceedings of International Conference on Human–Machine Interaction*, pages 1432–1436, 2003
29. A. Lanitis, C.J. Taylor, and T.F. Cootes. Toward automatic simulation of aging effects on face images. *IEEE Transactions on Pattern Analysis and Machine Intelligence*, 24(4):442–455, 2002
30. S.Z. Li and A.K. Jain, editors. *Handbook of Face Recognition.* Springer, Berlin Heidelberg New York, 2005
31. L. Liang, H. Chen, Y.Q. Xu, and H.Y. Shum. Example-based caricature generation with exaggeration. In *Proceedings of 10th Pacific Conference on Computer Graphics and Applications*, 2002
32. P. Litwinowicz and L. Williams. Animating images with drawings. In *Proceedings of SIGGRAPH*, pages 409–412, 1994
33. Q. Liu, X. Tang, H. Jin, H. Lu, and S. Ma. A nonlinear approach for face sketch synthesis and recognition. In *Proceedings of Conference on Computer Vision and Pattern Recognition*, volume 1, pages 1005–1010, 2005
34. Z. Liu, Y. Shan, and Z. Zhang. Expressive expression mapping with ratio images. In *Proceedings of SIGGRAPH*, pages 271–276, 2001
35. Z. Liu, Z. Zhang, and Y. Shan. Image-based surface detail transfer. *IEEE Computer Graphics and Applications*, 24(3):30–35, 2004

36. S. Loffe. Red eye detection with machine learning. In *Proceedings of International Conference on Image Processing*, volume II, pages 871–874, 2003
37. H. Luo, J. Yen, and D. Tretter. An efficient automatic redeye detection and correction algorithm. In *Proceedings of International Conference on Pattern Recognition*, volume 2, pages 883–886, 2004
38. X.P. Miao and T. Sim. Automatic red-eye detection and removal. In *Proceedings of International Conference on Multimedia and Expo*, pages 1195–1198, 2004
39. A. Nikolaidis and I. Pitas. Robust watermarking of facial images based on salient geometric pattern matching. *IEEE Transactions on Multimedia*, 2(3):172–184, 2000
40. K.-I. Okada, F. Maeda, Y. Ichikawaa, and Y. Matsushita. Multiparty videoconferencing at virtual social distance: MAJIC design. In *Proceedings of ACM Conference on Computer Supported Cooperative Work*, pages 385–393, 1994
41. J.-S. Park and S.-W. Lee. Reconstruction of high-resolution facial images for visual surveillance. In C.H. Chen and P.S.P. Wang, editors, *Handbook of Pattern Recognition and Computer Vision*, 3rd edn., pages 445–459. World Scientific, Singapore, 2005
42. J.-S. Park, Y.-H. Oh, S.-C. Ahn, and S.-W. Lee. Glasses removal from face image using recursive error compensation. *IEEE Transactions Pattern Analysis and Machine Intelligence*, 27(5):805–811, 2005
43. F.I. Parke and K. Waters. *Computer Facial Animation*. A.K. Peters Ltd., 1996
44. P.J. Phillips, H. Moon, S.A. Rizvi, and P.J. Rauss. The FERET evaluation methodology for face-recognition algorithms. *IEEE Transactions on Pattern Analysis and Machine Intelligence*, 22(10):1090–1104, 2000
45. F. Pighin, J. Hecker, D. Lischinski, R. Szeliski, and D. Salesin. Synthesizing realistic facial expressions from photographs. In *Proceedings of SIGGRAPH*, pages 75–84, 1998
46. Y. Saito, Y. Kenmochi, and K. Kotani. Estimation of eyeglassless facial images using principal component analysis. In *Proceedings of International Conference on Image Processing*, volume 4, pages 197–201, 1999
47. J.S. Schildkraut and R.T. Gray. A fully automatic redeye detection and correction algorithm. In *Proceedings of International Conference on Image Processing*, volume I, pages 801–803, 2002
48. A. Sellen. Remote conversations: The effects of mediating talk with technology. *Human Computer Interaction*, 10(4):401–444, 1995
49. S.M. Seitz and C.R. Dyer. View morphing. In *Proceedings of SIGGRAPH*, pages 21–30, 1996
50. T.K. Shih and R.-C. Chang. Digital inpainting – survey and multilayer image inpainting algorithms. In *Proceedings of International Conference on Information Technology and Applications*, volume 1, pages 15–24, 2005
51. B. Smolka, K. Czubin, J.Y. Hardeberg, K.N. Plataniotis, M. Szczepanski, and K. Wojciechowski. Towards automatic redeye effect removal. *Pattern Recognition Letters*, 24:1767–1785, 2003
52. X. Tang and X. Wang. Face sketch recognition. *IEEE Transactions Circuits and Systems for Video Technology*, 14(1):50–57, 2004
53. P. Thompson. Margaret Thatcher: A new illusion. *Perception*, 9:483–484, 1980
54. R. Ulichney and M. Gaubatz. Perceptual-based correction of photo red-eye. In *Proceedings of IASTED International Conference on Signal and Image Processing*, 2005

55. J. Wan and X.P. Ren. Automatic red-eyes detection based on AAM. In *Proceedings of International Conference on Systems, Man and Cybernetics*, volume 7, pages 6337–6341, 2004
56. D. Weiner and N. Kiryati. Virtual gaze redirection in face images. In *Proceedings of International Conference on Image Analysis and Processing*, pages 76–81, 2003
57. Z. Wen, Z. Liu, and T.S. Huang. Face relighting with radiance environment maps. In *Proceedings of Conference on Computer Vision and Pattern Recognition*, volume 2, pages 158–165, 2003
58. C. Wu, C. Liu, and J. Zhou. Eyeglasses verification by support vector machine. In *Proceedings of IEEE Pacific Rim Conference on Multimedia*, pages 1126–1131, 2001
59. H. Wu, G. Yoshikawa, T. Shioyama, T. Lao, and T. Kawade. Glasses frame detection with 3D Hough transform. In *Proceedings of International Conference on Pattern Recognition*, volume 2, pages 346–349, 2002
60. Y. Xiao and H. Yan. Extraction of glasses in human face images. In *Proceedings of International Conference on Biometric Authentication*, pages 214–220, 2004
61. M.-H. Yang, D. Kriegman, and N. Ahuja. Detecting faces in images: A survey. *IEEE Transactions on Pattern Analysis and Machine Intelligence*, 24(1):34–58, 2002
62. L. Yin and A. Basu. Generating realistic facial expressions with wrinkles for model-based coding. *Computer Vision and Image Understanding*, 84(2):201–240, 2001
63. Q. Zhang, Z. Liu, B. Guo, D. Terzopoulos, and H.-Y. Shum. Geometry-driven photorealistic facial expression synthesis. *IEEE Transactions Visualization and Computer Graphics*, 12(1):48–60, 2006
64. L. Zhang, Y. Sun, M. Li, and H. Zhang. Automated red-eye detection and correction in digital photographs. In *Proceedings of International Conference on Image Processing*, volume IV, pages 2363–2366, 2004
65. W. Zhao, R. Chellappa, J. Phillips, and A. Rosenfeld. Face recognition: A literature survey. *ACM Computing Surveys*, 35(4):399–458, 2003
66. C. Zhou and X. Lin. Facial expressional image synthesis controlled by emotional parameters. *Pattern Recognition Letters*, 26(16):2611–2627, 2005

Face Recognition and Pose Estimation with Parametric Linear Subspaces

Kazunori Okada[1] and Christoph von der Malsburg[2,3]

[1] Department of Computer Science
San Francisco State University
San Francisco, CA 94132-4163 USA
kazokada@sfsu.edu

[2] Frankfurt Institute of Advanced Studies
Science Campus Riedberg
60438 Frankfurt, Germany
malsburg@fias.uni-frankfurt.de

[3] Computer Science Department
University of Southern California
Los Angeles, CA90089-2520 USA

Summary. We present a general statistical framework for modeling and processing head pose information in 2D grayscale images: analyzing, synthesizing, and identifying facial images with arbitrary 3D head poses. The proposed framework offers a compact view-based data-driven model which provides bidirectional mappings between facial views and their corresponding parameters of 3D head angle. Such a mapping-based model implicitly captures 3D geometric nature of the problem without explicitly reconstructing a 3D structural model from data. The proposed model consists of a hierarchy of local linear models that cover a range of parameters by piecing together a set of localized models. This piecewise design allows us to accurately cover a wide parameter range, while the linear design, using efficient principal component analysis and singular value decomposition algorithms, facilitates generalizability to unfamiliar cases by avoiding overfitting. We apply the model to realize robust pose estimation using the view-to-pose mapping and pose-invariant face recognition using the proposed model to represent a known face. Quantitative experiments are conducted using a database of Cyberware-scanned 3D face models. The results demonstrate high accuracy for pose estimation and high recognition rate for previously unseen individuals and views for a wide range of 3D head rotation.

1 Introduction

Face recognition is one of the most interesting and challenging problems in computer vision and pattern recognition. In the past many aspects of this problem have been rigorously investigated because of its importance for realizing various applications and understanding our cognitive processes. For

K. Okada and C. von der Malsburg: *Face Recognition and Pose Estimation with Parametric Linear Subspaces*, Studies in Computational Intelligence (SCI) **91**, 49–74 (2008)
www.springerlink.com

reviews, we refer the reader to [6, 37, 42]. Past studies in this field have revealed that our utmost challenge is to reliably recognize people in the presence of image/object variations that occur naturally in our daily life [32]. As head pose variation is one of the most common variations,it is an extremely important factor for many application scenarios. There have been a number of studies which specifically addressed the issue of pose invariance in face recognition [1–3, 5, 11–13, 15, 16, 20, 23, 30, 31, 38, 40, 41]. Despite the accumulation of studies and the relative maturity of the problem, however, performance of the state-of-the-art has unfortunately remained inferior to human ability and sub-optimal for practical use when there is no control over subjects and when one must deal with an unlimited range of full 3D pose variation.

Our main goal is to develop a simple and generalizable framework which can be readily extended beyond the specific focus on head pose (e.g., illumination and expression), while improving the pose processing accuracy of the state-of-the-art. For this purpose, we propose a general statistical framework for compactly modeling and accurately processing head pose information in 2D images. The framework offers means for analyzing, synthesizing, and identifying facial images with arbitrary head pose. The model is data-driven in the sense that a set of labeled training samples are used to learn how facial views change as a function of head pose. For realizing the compactness, the model must be able to learn from only a few samples by generalizing to unseen head poses and individuals. Linearity is emphasized in our model design, which simplifies the learning process and facilitates generalization by avoiding typical pitfalls of non-linearity, such as overfitting [4].

For pose-insensitive face recognition, previous work can roughly be categorized into two types: single-view and multi-view approaches. Single-view approaches are based on person-independent transformation of test images [16, 20, 23]. Pose invariance is achieved by representing each known person by a facial image with a fixed canonical head pose and by transforming each test image to the canonical pose. This forces head pose of test and gallery entries to always be aligned when they are compared for identification. An advantage of this approach is the small size of the known-person gallery. However, recognition performance tends to be low due to the difficulty of constructing an accurate person-independent transformation. The multi-view approach [2, 3, 12, 41], on the other hand, is based on a gallery that consists of views of multiple poses for each known person. Pose-invariance is achieved by assuming that for each input face there exists a view with the same head pose as the input for each known person in the gallery. These studies have reported generally better recognition performance than the single-view approach. The large size of the gallery is, however, a disadvantage of this approach. The recognition performance and the gallery size have a trade-off relationship; better performance requires denser sampling of poses, increasing the gallery size. Larger gallery size makes it difficult to scale the recognition system to larger sets of known people and makes the recognition process more time-consuming.

One solution to the trade-off problem is to represent each known person by a compact model. Given the multi-view gallery, each set of views of a known person can be used as training samples to learn such a personalized model, reducing the gallery size while maintaining high recognition performance. The parametric eigenspace method of Murase and Nayar [25] and the virtual eigen-signature method of Graham and Allinson [13] are successful examples of this approach. These methods represent each known person by compact manifolds in the embedded subspace of the eigenspace. Despite their good recognition performance, generalization capability is their shortcoming. Both systems utilized non-linear methods (cubic-spline for the former and radial basis function network for the latter) for parameterizing/modeling the manifolds. Such methods have a tendency to overfit peculiarities in training samples, compromising capability to generalize over unseen head poses. The solution proposed here emphasizes linearity in model design, facilitating such generalization, as well as model compactness.

Our investigation explores the model-based solution of pose estimation, pose animation, and pose-insensitive face recognition using parametric linear subspace models. As a basic component, we exploit local principal component mapping (LPCMAP) [27], which offers a compact view-based model with bidirectional mappings between face views and 3D head angles. A mapping from face to pose we call *analysis mapping* and that from pose to face *synthesis mapping*. Concatenation of the two mappings creates an analysis-synthesis chain for model matching. Availability of such mappings avoids the necessity of an exhaustive search in the parameter space. Its parametric nature also provides an intuitive interface that permits clear interpretation of image variations and enables the model to continuously cover the pose variation thereby improving accuracy of the previous systems. Such a mapping-based model implicitly captures the three-dimensional geometric nature of the problem without explicitly reconstructing 3D facial structure from data. The model is learned by using efficient principal component analysis (PCA) and singular value decomposition (SVD) algorithms resulting in a linear form of the functions. They are however only locally valid due to their linearity. Therefore this local model is further extended to mitigate this limitation. The parametric piecewise linear subspace (PPLS) model [29] extends the LPCMAP for covering a wider pose range by piecing together a set of LPCMAP models. Our multiple-PPLS model further extends the PPLS in the sense of generalization over different individuals [26]. The discrete local models are continuously interpolated, improving the structurally discrete methods such as the view-based eigenface by Pentland et al. [31].

These proposed models are successfully applied to solve pose estimation and animation by using analysis and synthesis mappings, respectively. A novel pose-insensitive face recognition framework is also proposed by exploiting the PPLS model to represent each known person. In our recognition framework, the model matching with the PPLS models provides a flexible pose alignment of model views and input faces with arbitrary head poses, making the recognition invariant against pose variation. As a pure head pose estimation application, the analysis mapping can also be made to generalize inter-personally by using the multiple-PPLS model that linearly combines a set of personalized PPLS models.

The rest of this article is organized as follows. In Section 2, we give an overview of our framework and introduce some basic terminologies. Section 3 describes the LPCMAP and PPLS models in details. Section 4 shows how we can extend the PPLS model inter-personally. In Section 5, we empirically evaluate effectiveness of the proposed models. In Section 6, we conclude this article by summarizing our contributions and discussing future work.

2 Problem Overview and Definitions

Our problem here is to learn a data-driven statistical model of how facial appearance changes as a function of corresponding head angles and to apply such learned models for building a face recognition system that is insensitive to pose variation. The following introduces formal descriptions of our problem, as well as terminology used throughout this paper.

2.1 Statistical Models of Pose Variation

Analysis and Synthesis Mappings

We employ the parametric linear subspace (PLS) model [27, 29] for representing pose-modulated statistics. A PLS model consists of bidirectional, continuous, multivariate, mapping functions between a vectorized facial image $\mathbf{v}$ and 3D head angles $\boldsymbol{\theta}$. We call a mapping $\mathcal{A}_\Omega$ from the image to angles *analysis mapping*, and its inverse $\mathcal{S}_\Omega$ *synthesis mapping*:

$$\begin{aligned} \mathcal{A}_\Omega : \mathbf{v} &\stackrel{\Omega}{\longrightarrow} \boldsymbol{\theta} \\ \mathcal{S}_\Omega : \boldsymbol{\theta} &\stackrel{\Omega}{\longrightarrow} \mathbf{v}(\Omega), \end{aligned} \tag{1}$$

where Ω denotes a model instance that is learned from a set of training samples. We suppose that a set of M training samples, denoted by M pairs $\{(\mathbf{v}_m, \boldsymbol{\theta}_m)|m = 1, .., M\}$, is given where a single labeled training sample is denoted by a pair of vectors $(\mathbf{v}_m, \boldsymbol{\theta}_m)$, $\mathbf{v}_m$ being the m-th vectorized facial image, and $\boldsymbol{\theta}_m = (\theta_{m1}, \theta_{m2}, \theta_{m3})$ the corresponding 3D head angles of the face presented in $\mathbf{v}_m$.

An application of the analysis mapping can be considered as *pose estimation*. Given an arbitrary facial image $\mathbf{v} \notin \{\mathbf{v}_1, .., \mathbf{v}_M\}$, $\mathcal{A}_\Omega$ provides a 3D head angle estimate $\hat{\boldsymbol{\theta}} = \mathcal{A}_\Omega(\mathbf{v})$ of a face in $\mathbf{v}$. On the other hand, an application of the synthesis mapping provides a means of *pose transformation* or *facial animation*. Given an arbitrary 3D head angle $\boldsymbol{\theta} \notin \{\boldsymbol{\theta}_1, .., \boldsymbol{\theta}_M\}$, $\mathcal{S}_\Omega$ provides a synthesized sample or model view $\hat{\mathbf{v}} = \mathcal{S}_\Omega(\boldsymbol{\theta})$ whose head is rotated according to the given angle but its appearance is due to the learned model Ω.

Personalized and Interpersonalized Models

The type of training samples used for learning a model determines the nature of a specific PLS model instance. A model is called *personalized* when it is learned with pose-varying samples from a single individual. In this case, both analysis and synthesis mappings become specific to the person presented in the training set. Therefore the synthesis mapping output $\mathbf{v}(\Omega)$ exhibits personal appearance that solely depends on Ω encoding specificities of the person, while its head pose is given by an input to this mapping. On the other hand, a model is called *interpersonalized* when the training set contains multiple individuals. For pose estimation, this provides a more natural setting where the analysis mapping $\mathcal{A}_\Omega$ continuously covers not only head pose variations but also variations over different individuals.

2.2 Pose-Insensitive Face Recognition

Model Matching

Given an arbitrary person's face as input, a learned PLS model can be fit against it by concatenating the analysis and synthesis mappings. We call this model matching process the *analysis-synthesis chain* (ASC) process:

$$\mathcal{M}_\Omega : \mathbf{v} \xrightarrow{\Omega} \boldsymbol{\theta} \xrightarrow{\Omega} \mathbf{v}(\Omega). \tag{2}$$

The output of ASC is called the *model view* $\mathbf{v}(\Omega)$. It provides a facial view $\mathbf{v}(\Omega)$ of the person learned in Ω whose head pose is aligned to the input face in $\mathbf{v}$. This process not only fits a learned model to the input but also gives simultaneously a 3D pose estimate as a byproduct that can be used for other application purposes.

Note that, when matching a personalized PLS model to a different person's face, the resulting model view can be erroneous due to the pose estimation errors caused by the identity mismatch. To overcome this problem, the $\mathcal{A}_\Omega$ of an interpersonalized model [26] can be exploited for reducing pose estimation errors. For the purpose of face recognition, however, these errors actually serve as an advantage because it makes model views of mismatched individuals less similar to the input helping to single out the correct face. Moreover such errors are typically small due to the geometrical proximity of different faces.

Overview of the Proposed Recognition Framework

Figure 1 illustrates our framework for pose-insensitive face recognition. The framework employs the PLS model as the representation of a known person. For each known person, a personalized model is learned from the pose-varying samples of the person. We call a database of P known people, as a set of learned personalized models $\{\Omega_p | p = 1, .., P\}$, the *known-person gallery*. Given

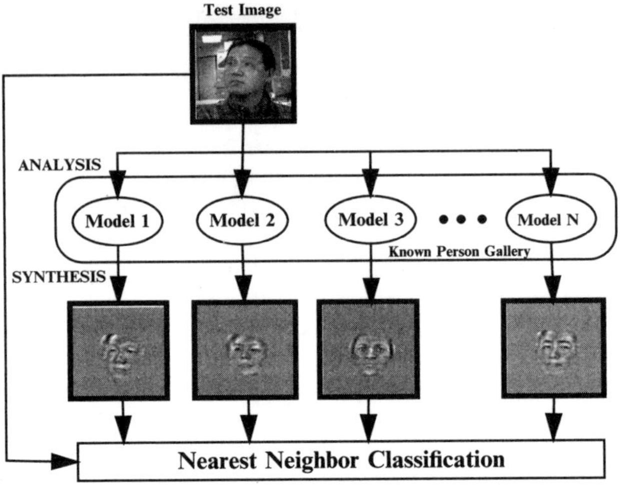

Fig. 1. Pose-insensitive face recognition framework with parametric linear subspace models used to represent each known person

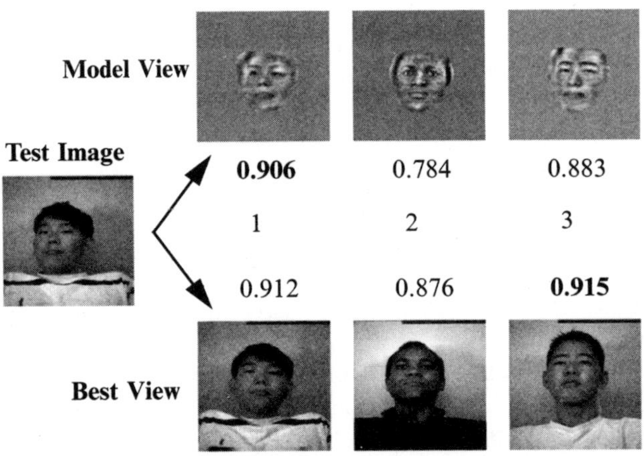

Fig. 2. An illustrative example of face recognition with pose variation using the model-based and the example-based systems

a test image of an arbitrary person with an arbitrary head pose, each model in the gallery is matched against the image by using its ASC process. The process results in pose-aligned model views of all known persons. After this pose alignment, the test image is compared against the model views in nearest neighbor classification fashion. In this scheme, the view-based comparison only occurs between views of the same pose improving the recognition performance.

Figure 2 illustrates the advantage of the proposed model-based method over an example-based multi-view method using a gallery of three known persons. A set of training samples for each person is used to construct the multi-view gallery entry. The top row displays model views of three learned models; the bottom row displays the best views of each known person that are most similar to the test image. Decimal numbers shown adjacent to the images denote their similarity to the test. There were no views in the gallery whose head pose was the same as the test image shown in the left. Therefore head pose of the test and the best matched views are always different. This results in a mis-identification by the multi-view system. On the other hand, the model-based solution, constructed by using the exactly same samples as in the multi-view system, identifies the test image correctly. This is realized by the model's ability to generalize to unseen views, resulting in model views whose head pose is better aligned to the test.

The proposed framework flexibly aligns head pose of the inputs and model views at an arbitrary pose, exploiting the PLS's continuous generalization capability to unseen views. Figure 3 and Table 1 illustrate this advantage in comparison with two other recognition frameworks: the multi-view (MVS) and single-view (SVS) systems. Given facial images with arbitrary head poses shown in the first row, a PLS, learned for this face, can provide model views

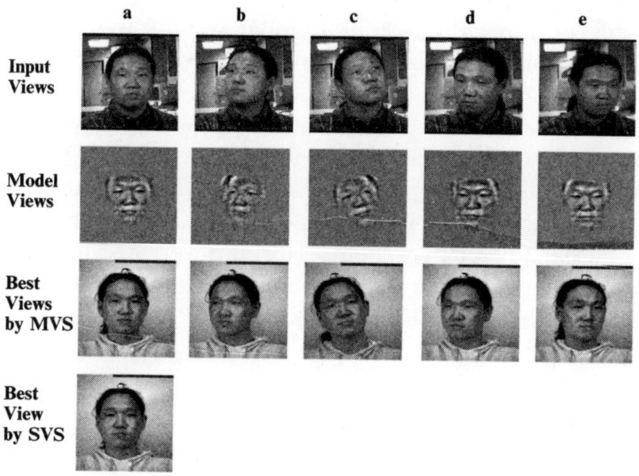

Fig. 3. Comparison of three recognition frameworks in terms of pose-alignment ability. Model views shown in the second row are given by the proposed method. MVS: example-based multi-view system; SVS: example-based single-view system. See texts for their description

Table 1. Similarity scores between the test input and different types of model and best views in Figure 3.

Model Type	a	b	c	d	e	std.dev.
Model Views	0.915	0.871	0.862	0.891	0.878	0.0184
Best Views by MVS	0.930	0.872	0.876	0.913	0.897	0.0220
Best View by SVS	0.926	0.852	0.816	0.878	0.862	0.0359

whose head pose is well aligned to the inputs. MVS provides the most similar view (best view) to the input among the training samples used to learn the model, while SVS employs always the same frontal view that acts as sole representation of the person. The figure shows that the proposed model appears to provide better pose-alignment than the two other systems. Table 1 shows actual facial similarity values between the inputs and the three different types of model or best view. The standard deviation shown in the right column indicates the degree of invariance against the pose variations by each framework. The parametric linear model provided the smallest standard deviation among the three, demonstrating the model's favorable characteristics toward pose-insensitivity.

3 Parametric Linear Subspace Model

This section describes two instances of the PLS model: the linear principal component-mapping (LPCMAP) model [27] and the parametric piecewise linear subspace (PPLS) model [29]. The PPLS model employs a set of LPCMAP models, each of which realizes continuous analysis and synthesis mappings. For maintaining the continuous nature in a global system, we consider that local mapping functions cover the whole parameter space, without imposing a rigid parameter window. Due to the linearity, however, the range over which each local mapping is accurate is limited. In order to cover a wide range of continuous pose variation, a PPLS model pieces together a number of local models distributed over the 3D angle space of head poses. In order to account for the local model's parameter-range limitation, each model is paired with a radial-basis weight function. The PPLS then performs a weighted linear combination of the outputs of the local models, realizing a continuous global function. The following introduces details of these models.

3.1 Linear PCMAP Model

The LPCMAP is a PLS model that realizes continuous, but only locally valid, bidirectional mapping functions. It consists of a combination of two linear systems: 1) *linear subspaces* spanned by principal components (PCs) of training samples and 2) *linear transfer matrices*, which associate projection coefficients of training samples onto the subspaces and their corresponding 3D head angles. It linearly approximates the entire parameter space of head poses by a single model.

Shape-Texture Decomposition and Image Reconstruction

The LPCMAP model treats shape and texture information separately in order to utilize them for different purposes. It has also been shown in the literature

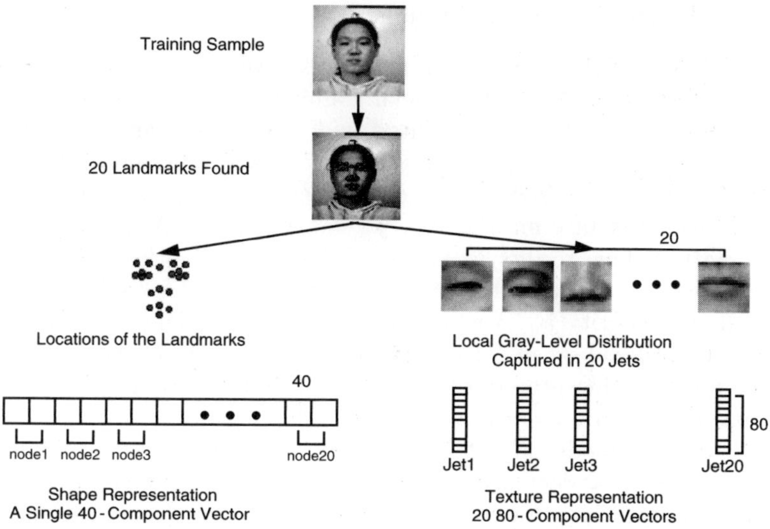

Fig. 4. Shape and texture decomposition process, illustrating parameter settings used for our experiments in Section 5. The number of landmarks $N = 20$ and the length of a texture vector $L = 80$ with a bank of 5-level and 8-orientation 2D complex Gabor filters

that combining shape with shape-free texture information improves recognition performance [7, 21, 38]. Figure 4 illustrates the process of decomposing shape and texture information in facial images. First, N predefined landmarks are located in each facial image $\mathbf{v}_m$ by a landmark finder or other means. Using this location information, shape and texture representations $(\mathbf{x}_m, \{\mathbf{j}_{m,n}\})$ are extracted from the image. The shape is represented by an array $\mathbf{x}_m \in \mathbf{R}^{2N}$ of object-centered 2D coordinates of the N landmarks. Texture information, on the other hand, is represented by a set of spatially sparse local feature vectors sampled at the N landmark points. A multi-orientation and multi-scale Gabor wavelet transformation [8] is used to define local features. The texture representation $\{\mathbf{j}_{m,n} \in \mathbf{R}^L | n = 1, .., N\}$ stands for a set of Gabor jets (L-component complex coefficient vector of the Gabor transform) sampled at the N landmarks [19, 28, 39]. Let $\mathcal{D}_x$ and $\mathcal{D}_j$ denote operations of the shape and texture decomposition, respectively:

$$\mathbf{x}_m = \mathcal{D}_x(\mathbf{v}_m) \quad \mathbf{j}_{m,1}, .., \mathbf{j}_{m,N} = \mathcal{D}_j(\mathbf{v}_m). \tag{3}$$

As an inverse operation, a gray-level facial image can be reconstructed approximately from a pair of shape and texture representations $(\mathbf{x}_m, \{\mathbf{j}_{m,n}\})$, following the work by Poetzsch et al. [33]. $\mathcal{R}$ denotes this reconstruction operation:

$$\mathbf{v}_m = \mathcal{R}(\mathbf{x}_m, \mathbf{j}_{m,1}, .., \mathbf{j}_{m,N}). \tag{4}$$

Learning Linear Subspace Models

As the first step of the model learning process, we extract a small number of significant statistical modes from facial training images using Principal Component Analysis (PCA), as illustrated in Figure 5. Given training samples $\{(\mathbf{v}_m, \boldsymbol{\theta}_m)|m = 1,..,M\}$, a set of extracted shape representations $\{\mathbf{x}_m\}$ is subjected to PCA [34], solving the eigen decomposition problem of the centered sample covariance matrix, $XX^t\mathbf{y}_p = \lambda_p\mathbf{y}_p$, where X is a $2NM$ column sample matrix. This results in an ordered set of $2N$ principal components (PCs) $\{\mathbf{y}_p|p = 1,..,2N\}$ of the shape ensemble. We call such PCs *shape PCs.* The local texture set $\{\mathbf{j}_{m,n}\}$ at a landmark n is also subjected to PCA, resulting in an ordered set of L PCs $\{\mathbf{b}_{s,n}|s = 1,..,L\}$. We cal such PCs *texture PCs.* Performing this procedure for all the N landmarks results in a set of local texture PCs $\{\mathbf{b}_{s,n}|s = 1,..,L; n = 1,..,N\}$.

The subspace model [31, 36] is defined by a vector space spanned by a subset of the PCs in decreasing order of their corresponding variances as illustrated in Figure 5(a). An image $\mathbf{v}$ is approximated as the sum of the average image $E(\mathbf{v})$ and the PCs $(\mathbf{e}_1,..,\mathbf{e}_p)$. The weight vector $(w_1,..,w_p)$ is defined by orthogonal projection onto the subspace and serves as a compact representation of the image $\mathbf{v}$. Due to the orthonormality of PCs, a linear combination of the PCs with the above mixing weights provides the best approximation of an original representation which minimizes the L_2 reconstruction error.

As illustrated in Figure 5(b), a *shape model* Y is constructed by the first $P_0 \leq 2N$ shape PCs, $Y = (\mathbf{y}_1,..,\mathbf{y}_{P_0})^t$. And a *texture model* $\{B_n\}$ is then constructed by the first $S_0 \leq L$ texture PCs at each landmark n, $\{B_n = (\mathbf{b}_{1,n},..,\mathbf{b}_{S_0,n})^t|n = 1,..,N\}$. These subspace models are used to parameterize and compress a centered input representation by orthogonally projecting it onto the subspace:

$$\mathbf{q}_m = Y(\mathbf{x}_m - \mathbf{u}_x) \quad \mathbf{u}_x = \frac{1}{M}\sum_{m=1}^{M}\mathbf{x}_m \tag{5}$$

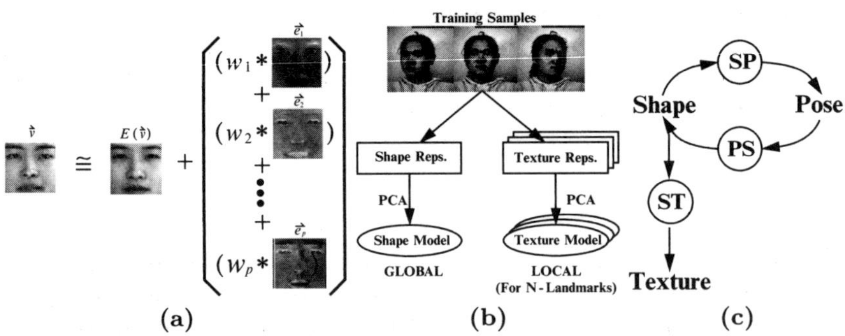

Fig. 5. Learning processes of the LPCMAP model: (a) PCA subspace model by Sirovich and Kirby [36], (b) shape and texture models using linear subspaces, and (c) linear transfer matrices relating different model parameters A rectangle in (b) denotes a set of training samples and an ellipse denotes a PCA subspace model

$$\mathbf{r}_{m,n} = B_n(\mathbf{j}_{m,n} - \mathbf{u}_{jn}) \quad \mathbf{u}_{jn} = \frac{1}{M}\sum_{m=1}^{M} \mathbf{j}_{m,n}. \tag{6}$$

We call the projection coefficient vectors for shape representation $\mathbf{q}_m \in \mathbf{R}^{P_0}$ *shape parameters* and those of texture representation $\mathbf{r}_{m,n} \in \mathbf{R}^{S_0}$ *texture parameters*, respectively. We also refer to these parameters (equivalent to the weight vector in Figure 5(a)) as *model parameters* collectively.

$$\mathbf{x}_m \approx \mathbf{u}_x + Y^t \mathbf{q}_m \tag{7}$$

$$\mathbf{j}_{m,n} \approx \mathbf{u}_{jn} + (B_n)^t \mathbf{r}_{m,n} \tag{8}$$

Learning Linear Transfer Matrices

As the second step of the learning process, model parameters are linearly associated with head pose parameters for realizing direct mappings between $\mathbf{v}$ and $\boldsymbol{\theta}$, as illustrated in Figure 5(c).

Clearly, the model parameters are non-linearly related to the 3D head angles, and therefore the intrinsic mapping between them is non-linear. In order to linearly approximate such non-linear mapping, we first transform the 3D head angles $\boldsymbol{\theta}_m = (\theta_{m,1}, \theta_{m,2}, \theta_{m,3})$ to *pose parameters* $\boldsymbol{\varphi}_m \in \mathbf{R}^{T \geq 3}$ such that the mapping between the pose and model parameters can be linearly approximated. We consider the following trigonometric function $\mathcal{K}$ for this purpose.

$$\boldsymbol{\varphi}_m = \mathcal{K}(\boldsymbol{\theta}_m) = (\cos(\tilde{\theta}_{m,1}), \sin(\tilde{\theta}_{m,1}), \cos(\tilde{\theta}_{m,2}), \sin(\tilde{\theta}_{m,2}), \cos(\tilde{\theta}_{m,3}), \sin(\tilde{\theta}_{m,3}))$$
$$\tilde{\theta}_{m,i} = \theta_{m,i} - u_{\theta i} \quad \mathbf{u}_\theta = (u_{\theta 1}, u_{\theta 2}, u_{\theta 3}) = \tfrac{1}{M}\textstyle\sum_{m=1}^{M} \boldsymbol{\theta}_m \tag{9}$$

There exists an inverse transformation $\mathcal{K}^{-1}$ such that

$$\boldsymbol{\theta}_m = \mathcal{K}^{-1}(\boldsymbol{\varphi}_m) = \mathbf{u}_\theta + (\arctan(\frac{\varphi_{m,2}}{\varphi_{m,1}}), \arctan(\frac{\varphi_{m,4}}{\varphi_{m,3}}), \arctan(\frac{\varphi_{m,6}}{\varphi_{m,5}})) \tag{10}$$

For both the analysis and synthesis mappings, the pose parameters $\boldsymbol{\varphi}_m$ are linearly related only with the shape parameters $\mathbf{q}_m$.

$$\boldsymbol{\varphi}_m = F\mathbf{q}_m \tag{11}$$

$$\mathbf{q}_m = G\boldsymbol{\varphi}_m \tag{12}$$

A $T \times P_0$ transfer matrix F (denoted as **SP** in Figure 5(c)) is learned by solving an overcomplete set of linear equations, $FQ = \Phi$, $Q = (\mathbf{q}_1, .., \mathbf{q}_M)$, $\Phi = (\boldsymbol{\varphi}_1, .., \boldsymbol{\varphi}_M)$. The Singular Value Decomposition (SVD) [34] is used to solve this linear system. Moreover, a $P_0 \times T$ transfer matrix G (denoted as **PS** in Figure 5(c)) is also learned by solving, $G\Phi = Q$, in the same manner. For the synthesis mapping, the shape parameters $\mathbf{q}_m$ are linearly related with the texture parameters $\mathbf{r}_{m,n}$ at each landmark n.

$$\{\mathbf{r}_{m,n} = H_n \mathbf{q}_m | n = 1, .., N\} \tag{13}$$

A set of $S_0 \times P_0$ transfer matrices $\{H_n\}$ (denoted as **ST** in Figure 5(c)) is learned by solving, $H_n Q = R_n, R_n = (\mathbf{r}_{1,n}, .., \mathbf{r}_{M,n})$, by using SVD for all the N landmarks.

Model Definition

The above two learning steps generate a set of data entities that collectively capture facial appearance in a given set of training samples. A LPCMAP model LM is defined by such data entities that are learned from training samples

$$LM := \{\mathbf{u}_x, \{\mathbf{u}_{jn}\}, \mathbf{u}_\theta, Y, \{B_n\}, F, G, \{H_n\}\}, \tag{14}$$

where $\mathbf{u}_x$ and $\{\mathbf{u}_{jn}\}$ are average shape and texture representations, $\mathbf{u}_\theta$ is an average 3D head angle vector, Y and $\{B_n\}$ are shape and texture models, F and G and $\{H_n\}$ are shape-to-pose, pose-to-shape, and shape-to-texture transfer matrices.

Mapping and Chain Functions

The analysis and synthesis mappings are constructed as a function of the learned LPCMAP model LM, as illustrated in Figure 6. The analysis mapping

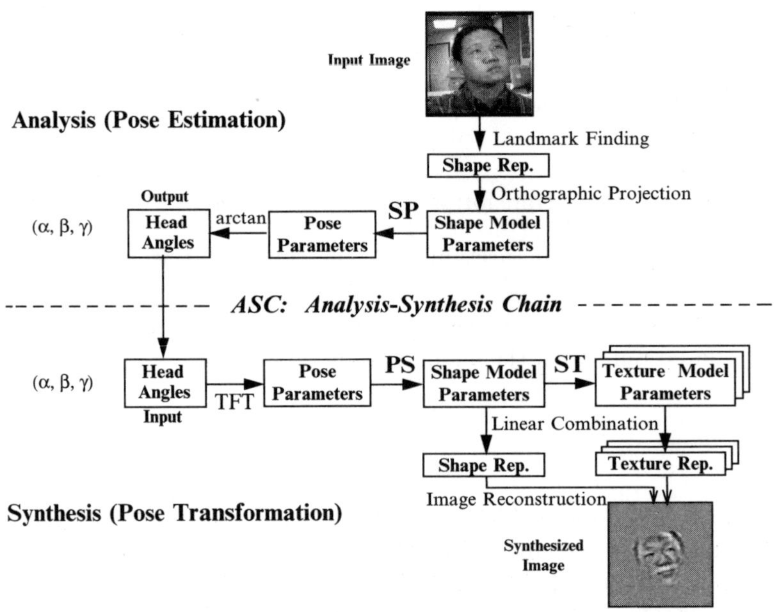

Fig. 6. Analysis and synthesis mapping and analysis-thesis-chain functions. Trigonometric transfer functions $\mathcal{K}$ and $\mathcal{K}^{-1}$ are denoted by **TFT** and **arctan**, respectively. **SP**, **PS** and **ST** denote the transfer matrices shown in Figure 5(c)

function $\mathcal{A}_{LM}(\mathbf{v})$ is given by combining formulae (3), (5), (11), and (10):

$$\hat{\boldsymbol{\theta}} = \mathcal{A}_{LM}(\mathbf{v}) = \mathbf{u}_\theta + \mathcal{K}^{-1}(F \cdot Y \cdot (\mathcal{D}_x(\mathbf{v}) - \mathbf{u}_x)). \tag{15}$$

The analysis function only utilizes the shape information of faces, following results of our preliminary experiments in which the head angles are better correlated with the shape representations than the texture representations [26]).

The synthesis mapping function $\mathcal{S}(\boldsymbol{\theta})$ is given by relating the 3D head angles to the shape coefficients and the shape coefficients to the texture coefficients. Because we have separate shape and texture decompositions, we address distinct synthesis processes for shape and texture. We refer to the shape and texture synthesis mapping functions as $\mathcal{SS}$ and $\mathcal{TS}$, respectively.

The shape synthesis mapping function $\mathcal{SS}_{LM}(\boldsymbol{\theta})$ is given by combining formulae (9), (12), and (7), using only the shape information similar to the analysis function. On the other hand, the texture synthesis mapping function $\mathcal{TS}_{LM}(\boldsymbol{\theta})$ is given by formulae (9), (12), (13), and (8), utilizing the correlation between shape and texture parameters. The synthesis mapping function $\mathcal{S}_{LM}(\boldsymbol{\theta})$ is then given by substituting the shape and texture synthesis functions to formula (4):

$$\begin{aligned} \hat{\mathbf{v}} &= \mathcal{S}_{LM}(\boldsymbol{\theta}) = \mathcal{R}(\mathcal{SS}_{LM}(\boldsymbol{\theta}), \mathcal{TS}_{LM}(\boldsymbol{\theta})) \\ \hat{\mathbf{x}} &= \mathcal{SS}_{LM}(\boldsymbol{\theta}) = \mathbf{u}_x + Y^t \cdot G \cdot \mathcal{K}(\boldsymbol{\theta} - \mathbf{u}_\theta) \\ \{\hat{\mathbf{j}}_n | n = 1, .., N\} &= \mathcal{TS}_{LM}(\boldsymbol{\theta}) = \{\mathbf{u}_{jn} + B_n \cdot H_n \cdot G \cdot \mathcal{K}(\boldsymbol{\theta} - \mathbf{u}_\theta) | n = 1, .., N\}. \end{aligned} \tag{16}$$

Finally, the ASC function $\mathcal{M}(\mathbf{v})$ is given by concatenating Eq. (15) and Eq. (16) as shown in Figure 6:

$$\hat{\mathbf{v}} = \mathcal{M}_{LM}(\mathbf{v}) = \mathcal{R}(\mathcal{SS}_{LM}(\mathcal{A}_{LM}(\mathbf{v})), \mathcal{TS}_{LM}(\mathcal{A}_{LM}(\mathbf{v}))). \tag{17}$$

3.2 Parametric Piecewise Linear Subspace Model

Model Definition

The parametric piecewise linear subspace (PPLS) model [29] extends the LPCMAP model by using the piecewise linear approach [35]. Due to the linear approximation, the LPCMAP model can only be accurate within a limited range of pose parameters. A piecewise linear approach approximates the nonlinear pose variation within a wider range by piecing together a number of locally valid models distributed over the pose parameter space. The PPLS model PM consists of a set of K local models in the form of the above-described LPCMAP model:

$$PM := \{LM_k | k = 1, .., K\}. \tag{18}$$

We assume that the local models are learned by training data sampled from appropriately distanced local regions of the *3D angle space*: the 3D finite

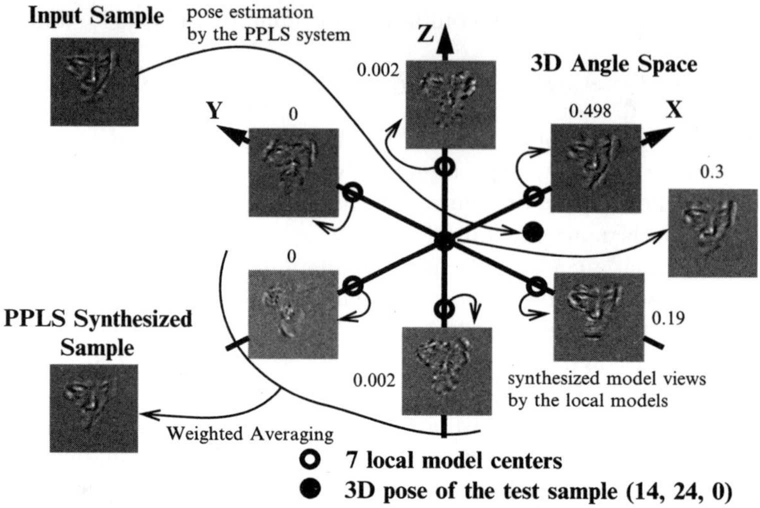

Fig. 7. A sketch of the PPLS model with seven LPCMAP models. An input image is shown at the top-left. Model centers of the LPCMAPs are denoted by circles. Pose estimation is performed by applying the analysis mapping $\mathcal{A}_{PM}$, giving the global estimate denoted by a filled circle. Pose transformation, on the other hand, is performed by applying the synthesis mapping $\mathcal{S}_{PM}$. Model views, shown next to the model centers, are linearly combined with Gaussian weights, resulting in the global synthesis shown at the bottom-left

parameter space spanned by the head angles. Each set of the local training samples is associated with a *model center*, the average 3D head angles $\mathbf{u}_{\theta}^{LM_k}$, which specifies the learned model's location in the 3D angle space. Figure 7 illustrates seven local models distributed in the 3D angle space. Model centers are denoted by circles and model views of the input are also shown next to them. Missing components of shape representations due to large head rotations are handled by the *mean-imputation method* [22], which fills in each missing component by a mean computed from all available data at the component dimension.

Mapping and Chain Functions

The analysis mapping function $\mathcal{A}_{PM}$ of the PPLS model is given by averaging K local pose estimates with appropriate weights as illustrated in Figure 7:

$$\hat{\boldsymbol{\theta}} = \mathcal{A}_{PM}(\mathbf{v}) = \sum_{k=1}^{K} w_k \mathcal{A}_{LM_k}(\mathbf{v}). \tag{19}$$

Similarly, the synthesis mapping function $\mathcal{S}_{PM}$ is given by averaging K locally synthesized shape and texture estimates with the same weights as illustrated in Figure 7:

$$\begin{aligned} \hat{\mathbf{v}} &= \mathcal{S}_{PM}(\boldsymbol{\theta}) = \mathcal{R}(\mathcal{SS}_{PM}(\boldsymbol{\theta}), \mathcal{TS}_{PM}(\boldsymbol{\theta})) \\ \hat{\mathbf{x}} &= \mathcal{SS}_{PM}(\boldsymbol{\theta}) = \sum_{k=1}^{K} w_k \mathcal{SS}_{LM_k}(\boldsymbol{\theta}) \\ \{\hat{\mathbf{j}_n}\} &= \mathcal{TS}_{PM}(\boldsymbol{\theta}) = \sum_{k=1}^{K} w_k \mathcal{TS}_{LM_k}(\boldsymbol{\theta}). \end{aligned} \tag{20}$$

A vector of the weights $\mathbf{w} = (w_1, .., w_K)$ in Eq. (19) and Eq. (20) is responsible for localizing the output space of the LPCMAP models, since their outputs themselves are continuous and unbounded. For this purpose, we defined the weights, as a function of the input pose, by using a normalized Gaussian function of distance between an input pose and each model center

$$w_k(\boldsymbol{\theta}) = \frac{\rho_k(\boldsymbol{\theta} - \mathbf{u}_\theta^{LM_k})}{\sum_{k=1}^{K} \rho_k(\boldsymbol{\theta} - \mathbf{u}_\theta^{LM_k})} \quad \rho_k(\boldsymbol{\theta}) = \frac{1}{\sqrt{2\pi}\sigma_k} \exp(-\frac{\|\boldsymbol{\theta}\|^2}{2\sigma_k^2}) \tag{21}$$

where σ_k denotes the k-th Gaussian width. It is set by the standard deviation of the 3D head angle vectors used for learning LM_k and determines the extent to which each local model influences the outputs $\hat{\boldsymbol{\theta}}$ and $\hat{\mathbf{v}}$. The weight value reaches maximum when the input pose coincides with one of the model centers; it decays as the distance increases. Outputs of local models that are located far from an input pose can become distorted because of the pose range limitation. However, these distorted local outputs do not strongly influence the global output because their contribution is suppressed by relatively low weight values.

The ASC function $\mathcal{M}(\mathbf{v})$ is again given by connecting an analysis output to a synthesis input.

$$\hat{\mathbf{v}} = \mathcal{M}_{PM}(\mathbf{v}) = \mathcal{R}(\mathcal{SS}_{PM}(\mathcal{A}_{PM}(\mathbf{v})), \mathcal{TS}_{PM}(\mathcal{A}_{PM}(\mathbf{v}))) \tag{22}$$

Gradient Descend-based Pose Estimation

Note that Eq. (19) cannot be solved in closed-form because its r.h.s. include the weights as a function of an unknown $\boldsymbol{\theta}$. To overcome this problem, a gradient descent-based iterative solution is formulated. Let a shape vector $\mathbf{x}$ be an input to the algorithm. Also let $\mathbf{x}_i$ and $\boldsymbol{\theta}_i$ denote the shape and angle estimates at the i-th iteration. The algorithm iterates the following formulae until the mean-square error $\|\Delta\mathbf{x}_i\|^2$ becomes sufficiently small:

$$\begin{aligned} \Delta\mathbf{x}_i &= \mathbf{x} - \mathbf{x}_i, \\ \Delta\boldsymbol{\theta}_i &= \sum_{k=1}^{K} w_k(\boldsymbol{\theta}_i)\mathcal{A}'_{LM_k}(\Delta\mathbf{x}_i), \\ \boldsymbol{\theta}_{i+1} &= \boldsymbol{\theta}_i + \eta\Delta\boldsymbol{\theta}_i, \\ \mathbf{x}_{i+1} &= \sum_{k=1}^{K} w_k(\boldsymbol{\theta}_{i+1})\mathcal{SS}_{LM_k}(\boldsymbol{\theta}_{i+1}), \end{aligned} \tag{23}$$

where η is a learning rate and $\mathcal{A}'$ is a slight modification of Eq. (15) that replaces a vectorized facial image $\mathbf{v}$ by a shape (difference) vector $\Delta\mathbf{x}_i$. The initial conditions $\mathbf{x}_0$ and $\boldsymbol{\theta}_0$ are given by the local model whose center shape $\mathbf{u}_x^{LM_k}$ is most similar to $\mathbf{x}$.

Note that the weighted sum of the analysis mappings in (23) is used as an approximation of the gradient of $\boldsymbol{\theta}$ with respect to $\mathbf{x}$ at the current shape estimate $\mathbf{x}_i$. In the PPLS model, such gradients are only available at the K discrete model centers. The second formula in (23), therefore, interpolates the K local gradient matrices for computing the gradients at an arbitrary point in the 3D angle space. The good local accuracy of the LPCMAP model shown in [27] supports the validity of this approximation. When a sufficient number of local models are allocated in the 3D angle space, the chance of being trapped at a local minimum should decrease. In our experimental setting described in the next section, the above initial condition settings resulted in no local minimum trappings significantly away from the global minimum. Note also that the algorithm performs pose estimation and shape synthesis simultaneously since it iterates between pose and shape in each loop. This gives an alternative for the shape synthesis, although the global synthesis mapping in (20) remains valid.

4 Interpersonalized Pose Estimation

As mentioned in Section 2.1, a single system should be able to solve the pose estimation task across different individuals by exploiting the geometrical similarity between faces. In the previous sections, we focused on how to model pose variations by using an example of personalized PLS models. This section discusses how to extend the PLS framework to capture variations due to head pose and individual differences simultaneously. The resulting *interpersonalized model* is applied to realize pose estimation across different people.

There are two approaches for realizing such an interpersonalized PLS model. The first is simply to train a LPCMAP or PPLS model by using a set of training samples that contain different-pose views from multiple people. The generic design of the proposed PLS models allows this straightforward extension, we must, however, empirically validate if the learned linear model adequately captures both variations correctly. After learning, both LM and PM can be used in the manner described in Section 3 for exploiting the corresponding analysis-synthesis mappings and ASC model matching. We refer to this type of model as *single-PLS model.*

The second approach is to linearly combine a set of personalized models similar to the way we constructed PPLS using a set of LPCMAPs. We refer to this type of model as *multiple-PPLS model.* A multiple-PPLS model MM consists of a set of P personalized models in the form of PPLS:

$$MM := \{PM_p | p = 1, .., P\}. \tag{24}$$

We assume that each individual PPLS model is personalized by learning it with pose-varying samples of a specific person and that the training samples cover an adequate range of head poses in the 3D angle space. The analysis mapping function $\mathcal{A}_{MM}$ of the multiple-PPLS model is then defined by a weighted linear combination of P pose estimates by the personalized models, realizing an interpersonalized pose estimation:

$$\hat{\boldsymbol{\theta}} = \mathcal{A}_{MM}(\mathbf{v}) = \sum_{p=1}^{P} w_p \mathcal{A}_{PM_p}(\mathbf{v}). \tag{25}$$

The weight vector $\mathbf{w} = (w_1, .., w_P)$ is responsible for choosing appropriate personalized models and ignoring models that encode faces very different from the input. We consider an error of shape reconstruction $err_p(\mathbf{x})$ by using a shape-only analysis-synthesis chain of the PPLS model p. In a way similar to (21) we then let a normalized Gaussian function of such errors indicate fidelity of the personalized models to the input:

$$\begin{aligned} w_p(\boldsymbol{\theta}) &= \frac{\rho_p(err_p(\mathbf{x}))}{\sum_{p=1}^{P} \rho_p(err_p(\mathbf{x}))} \\ err_p(\mathbf{x}) &= \mathbf{x} - \hat{\mathbf{x}_p} = \mathbf{x} - \mathcal{SS}_{PM_p}(\mathcal{A}_{PM_p}(\mathbf{x})) \\ \rho_p(\boldsymbol{\theta}) &= \frac{1}{\sqrt{2\pi}\sigma_p} \exp(-\frac{\|err_p(\mathbf{x})\|^2}{2\sigma_p^2}), \end{aligned} \tag{26}$$

where a shape synthesis mapping $\mathcal{SS}_{PM_p}$ of the multiple-PPLS model is defined similar to (25) and σ_p denotes the Gaussian width of model p.

5 Experiments

5.1 Data Set

For evaluating our system's performance we must collect a large number of samples with controlled head poses, which is not an easy task. For mitigating this difficulty, we use 3D face models pre-recorded by a Cyberware scanner.

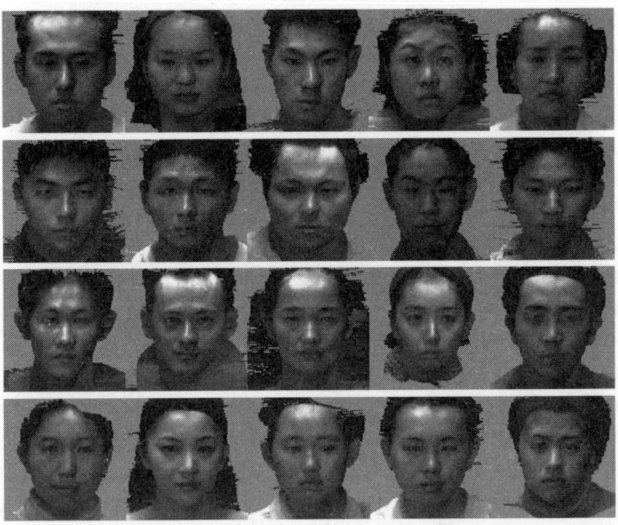

Fig. 8. 20 frontal views rendered from the 3D face models

Given such data, relatively faithful image samples with an arbitrary, but precise, head pose can easily be created by image rendering. We used 20 heads randomly picked from the ATR-Database [17], as shown in Figure 8.

For each head, we created 2821 training samples. They consist of 7 local sample sets each of which covers a pose range of ±15 degrees at one-degree intervals. These local sets are distributed over the 3D angle space such that they collectively cover a pose range of ±55 degrees along each axis of 3D rotations; their model centers are distanced by ±40 degrees from the frontal pose (origin of the angle space). We also created 804 test samples for each head. In order to test the model's generalization capability to unknown head poses, we prepared test samples whose head poses were not included in the training samples. Head angles of some test samples were in between the multiple local models and beyond their ±15 degree range. They cover a pose range of ±50 degrees. For more details of the data, see our previous reports [26, 29].

For each sample, the 2D locations of 20 inner facial landmarks, such as eyes, nose and mouth, are derived by rotating the 3D landmark coordinates, initialized manually, and by projecting them onto the image plane. The explicit rotation angles of the heads also provide 3D head angles of the samples. The rendering system provides the self-occlusion information. Up to 10% of the total landmarks were self-occluded for each head.

5.2 Personalized Pose Estimation and View Synthesis

We compare the PPLS and LPCMAP models learned using the training samples described above. The PPLS model consists of 7 local linear models, each of which is learned from one of the local training sets. On the other hand, a single LPCMAP model was learned from the total set of 2821 samples. The shape and texture representation are extracted using the specification $N = 20$ and $L = 80$ described in Figure 4. The PPLS model uses σ_k set to the sample standard deviation, gradient descent is done over 500 iterations, and η is set to 0.01. The learned models are tested with both 2821 training samples themselves and 804 separate test samples of unknown pose. We refer to the former by *accuracy test* and the latter by *generalization test.*

Figure 9(a) compares average pose estimation errors of the PPLS and LPCMAP models in both accuracy and generalization tests. In the accuracy test, the average angular error with the first 8 PCs was 0.8 ± 0.6 and 3.0 ± 2.4 degrees and the worst error was 5.6 and 18.9 degrees for the PPLS and LPCMAP models, respectively. In the generalization test, the average error was 0.9 ± 0.6 and 2.4 ± 1.4 degrees, and the worst error was 4.5 and 10.2 degrees for the two models. Figure 9(b) compares average shape synthesis errors of the two models in the two test cases. In the accuracy test, the average landmark position error with the first 8 PCs was 0.8 ± 0.4 and 2.2 ± 1.2 pixels, and the worst error was 3.0 and 7.6 pixels for the PPLS and LPCMAP models, respectively. In the generalization test, the average error was 0.9 ± 0.4 and 2.4 ± 0.7 pixels, and the worst error was 2.7 and 5.6 pixels for the two models.

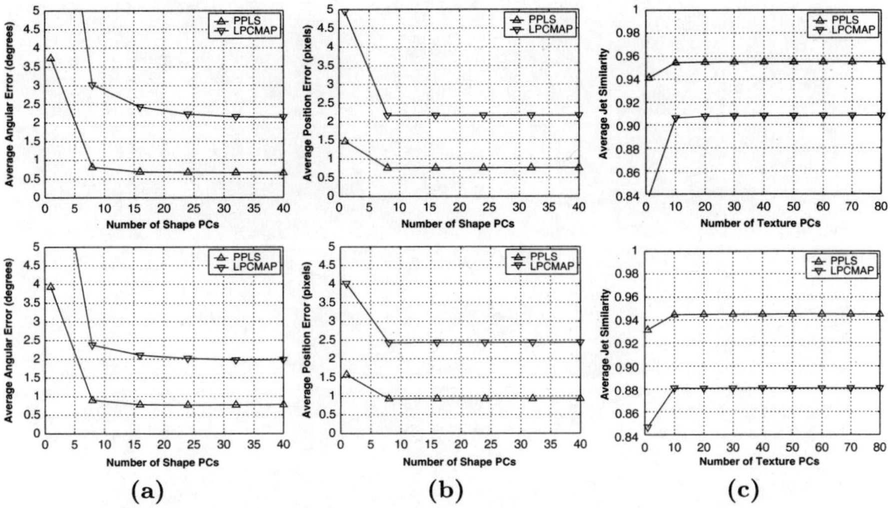

Fig. 9. Comparison of the PPLS and LPCMAP models in terms of pose estimation and transformation errors. The first and second rows show results of the accuracy and generalization tests, respectively. Errors (similarities) are plotted over the number of PCs used to construct a subspace. (a) pose estimation errors in degrees averaged over 3 rotation angles. (b) shape synthesis errors in pixels averaged over 20 landmarks. (c) texture synthesis error by Gabor jet similarity averaged over 20 landmarks

Figure 9(c) compares average similarities between synthesized texture vector $\hat{\mathbf{j}}$ and ground-truth texture vector $\mathbf{j}$ for the two models in the two test cases. Local texture similarity is computed as a normalized dot-product (cosine) of Gabor jet magnitudes, $JetSim := \frac{amp(\mathbf{j}) \cdot amp(\hat{\mathbf{j}})}{\|amp(\mathbf{j})\| \, \|amp(\hat{\mathbf{j}})\|}$, where amp extracts magnitudes of a Gabor jet in polar coordinates and $\| \cdot \|$ denotes $L2$ vector norm. The similarity values range from 0 to 1, where 1 denotes equality of two jets. In the accuracy test, the average similarity with the first 20 texture PCs was 0.955 ± 0.03 and 0.91 ± 0.04, and the worst similarity was 0.81 and 0.73 for the PPLS and LPCMAP models, respectively. In the generalization test, the average similarity was 0.945 ± 0.03 and 0.88 ± 0.03, and the worst similarity was 0.82 and 0.77 for the two models.

For all three tasks, the PPLS model greatly improved performance over the LPCMAP model in both test cases, resulting in *sub-degree* and *sub-pixel* accuracy. The results also show that the average errors between the two test cases were similar, indicating good generalization to unknown poses. As a reference for our texture similarity analysis, we computed average texture similarities over 450 people from the FERET database [32]. The average similarity was 0.94 ± 0.03 for the same person pairs and 0.86 ± 0.02 for the most similar, but different, person pairs. The average similarity of the PPLS model was higher

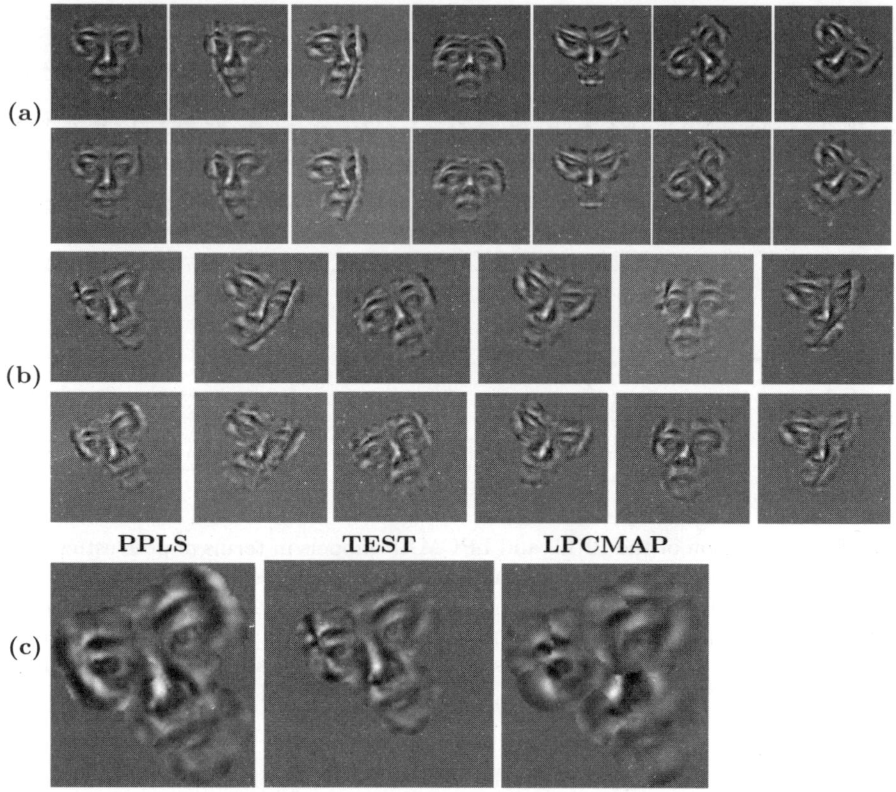

Fig. 10. Examples of synthesized model views by the PPLS model. In (a) and (b), model views in the first and second rows are reconstructed from ground-truth and synthesized pose-aligned samples, respectively. (a): training samples with known head pose (accuracy test case); (b): test samples with unknown head poses (generalization test case); (c): illustrative comparison of model views synthesized by the PPLS and LPCMAP models

than that of the large FERET database, which validates the results of our texture similarity analysis.

Figure 10 illustrates model views: images reconstructed from samples synthesized by formula (20) of the PPLS model. Note that facial images reconstructed by the Pötzsch algorithm [33] do not retain original picture quality. This is because a transformation $\mathcal{D}_j$ from images to the Gabor jet representations is lossy due to coarse sampling in both the spatial and frequency domains. Nonetheless, these images still capture characteristics of faces fairly well. Figure 10(a) compares reconstructed images of original and synthesized training samples. The left-most column shows frontal views while the rest of columns show views with ±45 degree rotation along one axis. Figure 10(b) compares original and synthesized test samples. For all three cases, the original and synthesized model views were very similar, indicating good accuracy and successful generalization to unknown head poses. Figure 10(c) compares model views synthesized by the PPLS and LPCMAP models. The PPLS's model view was more similar to the original than the LPCMAP's model view. This agrees with the results of our error and similarity analyses.

5.3 Pose-Insensitive Face Recognition

For comparison, we constructed four recognition systems with 20 known persons: 1) the single-view system (SVS), which represents each known person by a single frontal view, 2) the LPCMAP system with a gallery of LPCMAP models, 3) the PPLS system with a gallery of PPLS models, and 4) the multi-view system (MVS), which represents each person by various raw views of the person. The LPCMAP, PPLS and MVS are constructed by using the same 2821 training samples per person; the SVS serves as a base-line. For both models, P_0 and S_0 are set to 8 and 20, respectively. The PPLS models consist of 7 local models and perform 500 iterations with η set to 0.01 for each test sample. Each pair of views is compared by an average of normalized dot-product similarities between the corresponding Gabor jet's magnitudes.

Table 2 summarizes the results of our recognition experiments. Identification rates in the table are averaged over the 20 persons; the compression rates represent the size of the known-person gallery as a fraction of the MVS. The results show that recognition performance of the PPLS system was more robust than the LPCMAP system (7% higher rate). Performance of our model-based systems was much better than the base-line SVS. Identification rates of the PPLS and MVS were almost the same while the former compressed the data by a factor of 20.

In some application scenarios, head pose information can be independently measured by the other means prior to identification. In such a case, the proposed recognition system can be realized by using only the synthesis mapping instead of model matching. Table 3 compares average identification rates of the two cases: with and without the knowledge of head pose. The results show that the knowledge of head pose gave a slight increase in recognition performance, however the increase was minimal.

Table 2. Average correct-identification and relative compression rates for four different systems.

Test Samples	Identification	Compression
SVS	59.9±10.6%	0.035%
LPCMAP	91.6±5.0%	0.74%
PPLS	98.7±1.0%	5%
MVS	99.9±0.2%	—

5.4 Interpersonalized Pose Estimation

For comparison, we tested both single-PPLS and multiple-PPLS models for two test cases: interpolation and extrapolation tests, using the data described in Section 5.1. For the interpolation (known persons) test, both models are trained with all the 56420 training samples (20 people × 2821 samples). A single-PPLS model with 7 LPCMAPs is trained with all the samples. On the other hand, a multiple-PPLS model is build by training each personalized model with 2821 samples for a specific person. These two models are then tested with the same 16080 test samples from the 20 individuals. For the extrapolation (unknown persons) test, each model is trained with 53599 training samples of 19 individuals, excluding training samples referring to the person being tested, so that the model does not contain knowledge of testing faces. The two models are trained in the same way as the interpolation test and tested with the same 16080 test samples. The same parameter settings of LPCMAP and PPLS models are used as described in Section 5.2.

Figure 11 compares the single-PPLS model and multiple-PPLS model in the two test settings. Down-triangles denote the average angular errors of the single-PPLS model and up-triangles denote those of the multiple-PPLS model with $\sigma_p = 7$ for all p. As reference, average pose estimation errors of the personalized model shown in Figure 9 are also included and denoted by solid

Table 3. Identification rates when head pose of tests is unknown or given as ground-truth.

	PPLS	**LPCMAP**
Unknown: $\mathcal{M}(\mathbf{v})$	98.7±1.0%	91.6±5.0%
Known: $\mathcal{S}(\boldsymbol{\theta})$	99.3±0.7%	92.4±4.0%

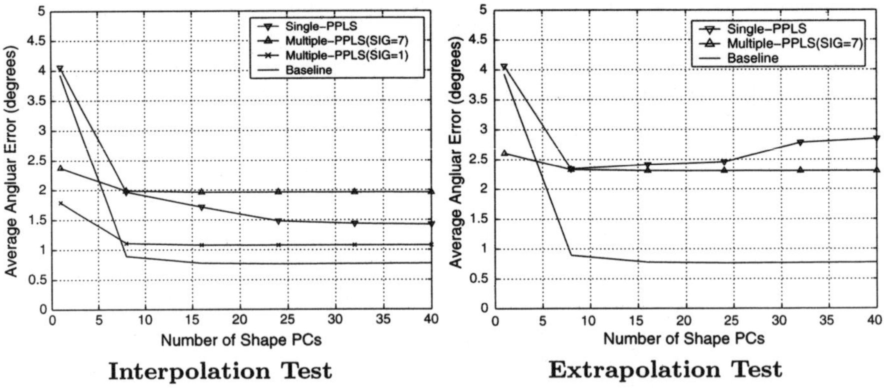

Fig. 11. Comparison of the single-PPLS and multiple-PPLS models for the interpolation and extrapolation tests in terms of interpersonalized pose estimation errors. Baseline plots indicate average pose estimation errors by the personalized models shown in Figure 9 for reference

lines without markers. Errors are plotted against 6 different sizes of the shape model. Our pilot study indicated that $\sigma_p = 7$ for all p is optimal for both interpolation and extrapolation cases. However $\sigma_p = 1$ for all p was optimal when only an interpolation test was considered. For this reason, errors with $\sigma_p = 1$ is also included for the interpolation test.

When σ_p is set optimally for both test cases, the average errors of the two models were very similar between two test cases. With the first 8 shape PCs, the errors of the two models were the same: 2.0 and 2.3 degrees for the interpolation and extrapolation tests, respectively. For the interpolation test, the standard deviation of the errors and the worst error were 0.9 and 5.5 degrees for the single-PPLS model and 0.8 and 5.1 degrees for the multiple-PPLS model. For the extrapolation test, the standard deviation and the worst error were 0.9 and 5.9 degrees for the former and 0.9 and 5.5 degrees for the latter. For both tests, the average errors of the two models are roughly 1 to 1.5 degrees larger than the baseline errors. When σ_p is set optimally for the interpolation condition, the multiple-PPLS model clearly outperformed the single-PPLS model, improving the average errors by roughly 1 degree and becoming similar to the baseline result (only 0.2 degree difference). These experimental results indicate that both models are fairly accurate, indicating the feasibility of the proposed approach to generalize over different persons.

6 Conclusion

This article presents a general statistical framework for modeling and processing head pose information in 2D grayscale images: analyzing, synthesizing, and identifying facial images with arbitrary 3D head poses. Three types of PLS model are introduced. The LPCMAP model offers a compact view-based model with bidirectional analysis and synthesis mapping functions. A learned model can be matched against an arbitrary input by using an analysis-synthesis chain function that concatenates the two. The PPLS model extends the LPCMAP for covering a wider pose range by combining a set of local models. Similarly the multiple-PPLS model extends the PPLS for generalizing over different people by linearly combining a set of PPLSs. A novel pose-insensitive face recognition framework is proposed by using the PPLS model to represent each known person. Our experimental results for 20 people covering a 3D rotation range as wide as ±50 degree demonstrated the proposed model's accuracy for solving pose estimation and pose animation tasks and robustness for generalizing to unseen head poses and individuals while compressing the data by a factor of 20 and more.

The proposed framework was evaluated by using accurate landmark locations and corresponding head angles computed by rotating 3D models explicitly. In reality, a stand-alone vision application based on this work will require a landmark detection system as a pre-process. A Gabor jet-based landmark tracking system [24] can be used to provide accurate landmark positions. It requires, however, the landmarks to be initialized by some other method. Pose-specific graph matching [18] provides an another solution but with much lower precision. In general, the landmark locations and head angles will contain measurement errors. Although our previous studies indicated robustness to such errors [26], more systematic investigation on this matter should be performed in future.

Our future goal must address other types of variation such as illuminations and expressions for realizing more robust systems. There has been progress with variation of both illumination [9, 12] and expression [10, 14]. However, the issue of combining these variation-specific solutions into a unified system robust against the different types of variation simultaneously has not fully been investigated. Our simple and general design approach may help to reach this goal.

Acknowledgments

The authors thank Shigeru Akamatsu and Katsunori Isono for making their 3D face database available for this study. This study was partially supported by ONR grant N00014-98-1-0242, by a grant by Google, Inc. and by the Hertie Foundation.

References

1. M.S. Bartlett and T.J. Sejnowski. Viewpoint invariant face recognition using independent component analysis and attractor networks. In *Neural Information Processing Systems: Natural and Synthetic*, volume 9, pages 817–823. MIT, Cambridge, MA, 1997
2. D. Beymer. Face recognition under varying pose. Technical Report A.I. Memo, No. 1461, Artificial Intelligence Laboratory, MIT, Cambridge, MA, 1993
3. D. Beymer and T. Poggio. Face recognition from one example view. Technical Report 1536, Artificial Intelligence Laboratory, MIT, Cambridge, MA 1995
4. C.M. Bishop. *Neural Networks for Pattern Recognition.* Oxford University Press, New York, 1995
5. X. Chai, L. Qing, S. Shan, X. Chen, and W. Gao. Pose invariant face recognition under arbitrary illumination based on 3d face reconstruction. In *Proceedings of AVBPA*, pages 956–965, 2005
6. R. Chellappa, C.L. Wilson, and S. Sirohey. Human and machine recognition of faces: A survey. *Proceedings of the IEEE*, 83(5):705–740, 1995
7. I. Craw, N. Costen, T. Kato, G. Robertson, and S. Akamatsu. Automatic face recognition: Combining configuration and texture. In *Proceedings of International Conference on Automatic Face and Gesture Recognition*, pages 53–58, Zurich, 1995
8. J.G. Daugman. Complete discrete 2-D Gabor transforms by neural networks for image analysis and compression. *IEEE Transactions on Acoustics, Speech, and Signal Processing*, 36:1169–1179, 1988

9. P. Debevec, T. Hawkins, H.P. Tchou, C. Duiker, W. Sarokin, and M. Sagar. Acquiring the reflectance field of a human face. In *Proceedings of Siggraph*, pages 145–156, 2000
10. G. Donato, M.S. Bartlett, J.C. Hager, P. Ekman, and T.J. Sejnowski. Classifying facial actions. *IEEE Transactions on Pattern Analysis and Machine Intelligence*, 21(10):974–988, 1999
11. S. Duvdevani-Bar, S. Edelman, A.J. Howell, and H. Buxton. A similarity-based method for the generalization of face recognition over pose and expression. In *Proceedings of the International Conference on Automatic Face and Gesture Recognition*, pages 118–123, Nara, 1998
12. A.S. Georghiades, P.N. Belhumeur, and D.J. Kriegman. From few to many: Generative models for recognition under variable pose and illumination. In *Proceedings of the International Conference on Automatic Face and Gesture Recognition*, pages 277–284, Grenoble, 2000
13. D.B. Graham and N.M. Allinson. Characterizing virtual eigensignatures for general purpose face recognition. In *Face Recognition: From Theory to Applications*, pages 446–456. Springer, Berlin Heidelberg New York, 1998
14. H. Hong. *Analysis, recognition and synthesis of facial gestures.* PhD thesis, University of Southern California, 2000
15. F. J. Huang, Z. Zhou, H. J. Zhang, and T. Chen. Pose invariant face recognition. In *Proceedings of the International Conference on Automatic Face and Gesture Recognition*, pages 245–250, Grenoble, France, 2000
16. H. Imaoka and S. Sakamoto. Pose-independent face recognition method. In *Proceedings of the IEICE Workshop of Pattern Recognition and Media Understanding*, pages 51–58, June 1999
17. K. Isono and S. Akamatsu. A representation for 3D faces with better feature correspondence for image generation using PCA. In *Proceedings of the IEICE Workshop on Human Information Processing*, pages HIP96–17, 1996
18. N. Krüger, M. Pötzsch, and C. von der Malsburg. Determination of face position and pose with a learned representation based on labeled graphs. Technical Report, Institut fur Neuroinformatik, Ruhr-Universität Bochum, 1996
19. M. Lades, J. C. Vorbrüggen, J. Buhmann, J. Lange, C. von der Malsburg, R. P. Würtz, and W. Konen. Distortion invariant object recognition in the dynamic link architecture. *IEEE Transactions on Computers*, 42:300–311, 1993
20. M. Lando and S. Edelman. Generalization from a single view in face recognition. In *Proceedings of the Inteernational Confeerence on Automatic Face and Gesture Recognition*, pages 80–85, Zurich, 1995
21. A. Lanitis, C.J. Taylor, and T.F. Cootes. Automatic interpretation and coding of face images using flexible models. *IEEE Transactions on Pattern Analysis and Machine Intelligence*, 19:743–755, 1997
22. R.J.A. Little and D.B. Rubin. *Statistical Analysis with Missing Data.* Wiley, New York, 1987
23. T. Maurer and C. von der Malsburg. Single-view based recognition of faces rotated in depth. In *Proceedings of the International Conference on Automatic Face and Gesture Recognition*, pages 248–253, Zurich, 1995
24. T. Maurer and C. von der Malsburg. Tracking and learning graphs and pose on image sequences. In *Proceedings of the International Conference on Automatic Face and Gesture Recognition*, pages 176–181, Vermont, 1996
25. H. Murase and S.K. Nayar. Visual learning and recognition of 3-D objects from appearance. *International Journal of Computer Vision*, 14:5–24, 1995

26. K. Okada. Analysis, Synthesis and Recognition of Human Faces with Pose Variations. PhD thesis, University of Southern California, 2001
27. K. Okada, S. Akamatsu, and C. von der Malsburg. Analysis and synthesis of pose variations of human faces by a linear PCMAP model. In *Proceedings of the International Conference on Automatic Face and Gesture Recognition*, pages 142–149, Grenoble, 2000
28. K. Okada, J. Steffens, T. Maurer, H. Hong, E. Elagin, H. Neven, and C. von der Malsburg. The Bochum/USC face recognition system: And how it fared in the FERET phase III test. In *Face Recognition: From Theory to Applications*, pages 186–205. Springer, Berlin Heidelberg New York, 1998
29. K. Okada and C. von der Malsburg. Analysis and synthesis of human faces with pose variations by a parametric piecewise linear subspace method. In *Proceedings of the IEEE Conference on Computer Vision and Pattern Recognition*, volume I, pages 761–768, Kauai, 2001
30. K. Okada and C. von der Malsburg. Pose-invariant face recognition with parametric linear subspaces. In *Proceeedings of the International Conference on Automatic Face and Gesture Recognition*, Washington, DC, 2002
31. A. Pentland, B. Moghaddam, and T. Starner. View-based and modular eigenspaces for face recognition. Technical Report, Media Laboratory, MIT, 1994
32. P.J. Phillips, H. Moon, S.A. Rizvi, and P.J. Rauss. The FERET evaluation methodology for face-recognition algorithms. *IEEE Transactions on Pattern Analysis and Machine Intelligence*, 22:1090–1104, 2000
33. M. Pötzsch, T. Maurer, L. Wiskott, and C. von der Malsburg. Reconstruction from graphs labeled with responses of Gabor filters. In *Proceedings of the International Conference Artificial Neural Networks*, pages 845–850, Bochum, 1996
34. W.H. Press, S.A. Teukolsky, W.T. Vetterling, and B.P. Flannery. *Numerical Recipes in C: The Art of Scientific Computing.* Cambridge University Press, New York, 1992
35. S. Schaal and C.G. Atkeson. Constructive incremental learning from only local information. *Neural Computing*, 10:2047–2084, 1998
36. L. Sirovich and M. Kirby. Low dimensional procedure for the characterisation of human faces. *Journal of the Optical Society of America*, 4:519–525, 1987
37. D. Valentin, H. Abdi, A. J. O'Toole, and G. W. Cottrell. Connectionist models of face processing: A survey. *Pattern Recognition*, 27:1209–1230, 1994
38. T. Vetter and N. Troje. A separated linear shape and texture space for modeling two-dimensional images of human faces. Technical Report TR15, Max-Plank-Institut fur Biologische Kybernetik, 1995
39. L. Wiskott, J.-M. Fellous, N. Krüger, and C. von der Malsburg. Face recognition by elastic bunch graph matching. *IEEE Transactions on Pattern Analysis and Machine Intelligence*, 19:775–779, 1997
40. L. Zhang and D. Samaras. Pose invariant face recognition under arbitrary unknown lighting using spherical harmonics. In *Proceedings of the Biometric Authentication Workshop*, 2004
41. W.Y. Zhao and R. Chellappa. SFS based view synthesis for robust face recognition. In *Proceedings of the International Conference on Automatic Face and Gesture Recognition*, pages 285–292, Grenoble, 2000
42. W.Y. Zhao, R. Chellappa, P.J. Phillips, and A. Rosenfeld. Face recognition: A literature survey. *ACM Computing Surveys*, 35:399–458, 2003

Part II

Spatio-Temporal Patterns

4D Segmentation of Cardiac Data Using Active Surfaces with Spatiotemporal Shape Priors

Amer Abufadel,[1] Tony Yezzi[1] and Ronald W. Schafer[2]

[1] School of Electrical and Computer Engineering, Georgia Institute of Technology, Atlanta, GA 30332

[2] Mobile and Media Systems Lab, Hewlett-Packard Laboratories, Palo Alto, CA 94304

Summary. We present a 4D spatiotemporal segmentation algorithm based on the Mumford-Shah functional coupled with shape priors. Our application is a fully automatic segmentation of cardiac MR sequences where the blood pool of the left ventricle is segmented and important diagnostic measurements extracted from the segmentation. When used in a clinical setting, our algorithm could greatly alleviate the time that clinicians must spend working with the acquired data to manually retrieve diagnostically meaningful measurements. This is due to the fact that most current approaches work only in 2D (sometimes 3D) and perform well for 2D slices away from the base and apex of the heart. The advantage of the 4D algorithm is that segmentation occurs in both space and time simultaneously, improving accuracy and robustness over existing 2D and 3D methods. The segmentation contour or hyper–surface is a zero level set function in 4D space that exploits the coherence within continuous regions not only between spatial slices, but between consecutive time samples as well. Shape priors are incorporated into the segmentation to limit the result to a known shape. Variations in shape are computed using principal component analysis (PCA), of a signed distance representation of the training data derived from manual segmentation of 18 carefully selected data sets. The automatic segmentation occurs by manipulating the parameters of this signed distance representation to minimize a predetermined energy functional. The training data sets have a large but equal number of slices and time samples covering the entire region of interest from the apex to the base of the myocardium. The spatial and temporal resolution of the training data is greater than what is usually measured in a clinical setting. However, typical patient data can be successfully segmented if it is upsampled appropriately. This will be demonstrated in numerous numerical experiments that will also show the improvement gained when using a 4D segmentation instead of 3D. Moreover, a comparison between an automatic 4D segmentation, a classical 3D segmentation and a manual segmentation carried out by an experienced clinician will also be presented.

A. Abufadel et al.: *4D Segmentation of Cardiac Data Using Active Surfaces with Spatiotemporal Shape Priors*, Studies in Computational Intelligence (SCI) **91**, 77–100 (2008)
www.springerlink.com

1 Introduction

In this chapter we improve upon the state of the art in shape-based methods by adding the time variable into the automatic segmentation process. Organs and structures in medical images are homogeneous regions with neighboring pixels or voxels having the same characteristics. Region–based segmentation methods are methods that are able to capture the coherence of continuous or similar regions, which makes them robust to noise and initial placement of the segmentation contour [15–18].

A special flavor of region–based methods are shape–based methods. These methods take advantage of the prior knowledge of the shape of the structure being segmented [11, 16]. In addition to shape, higher–level information such as appearance and relative geometry of anatomy may also be incorporated into the solution to complete the segmentation task [5, 7, 14, 20]. Hence, when using region–based and shape–based methods, labeling a voxel as part of the segmented region depends not only on the intensity value of the pixel itself, but on the values of the surrounding pixels as well. Consequently, increasing the number of neighbors will increase the amount of information available to the automatic segmentation procedure. The use of shape priors has also proven very helpful in overcoming problems of missing features or corrupted edges, a common characteristic of medical images [3, 5, 11, 13, 19]. However, current approaches produce good results in 2D and 3D only. These classical algorithms often fail to give adequate results, mostly because of the under–determined nature of the segmentation process [6, 15, 20].

Medical imaging modalities such as CT or MR collect data in slices that are stacked to form a 3D volume of points or voxels placed on a structured rectangular grid. The position of each voxel can be referenced by its cartesian (x, y, z) coordinates and has six immediate spatial neighbors. When circumstances permit, medical imaging systems are also able to acquire data that spans over a period of time. In cardiac MR, for example, data is collected over several time frames spanning the entire cardiac cycle. The acquired data is also gated or synchronized to the electrocardiogram (ECG) graph of the heart. For each location in the field of view (FOV) of the MR scanner, there are usually 10–20 measurements collected at evenly spaced time intervals across the cardiac cycle. The result is a sequence of 3D volumes representing a snapshot of the heart at each time frame.

We argue that temporally adjacent voxels will exhibit the same similarity that exists between spatially adjacent voxels. Consequently, every non-border voxel in this 3D volume sequence will have two extra *temporal* neighbors as depicted in Figure 1. Since region–based segmentation methods use information from neighboring voxels, adding more neighbors to each voxel will enhance the result of the segmentation. In an attempt to determine the amount of correlation between temporally adjacent voxels, we plot the variation over time in the grey level of a voxel on the inner border of the left ventricle (LV) as shown in Figure 2.

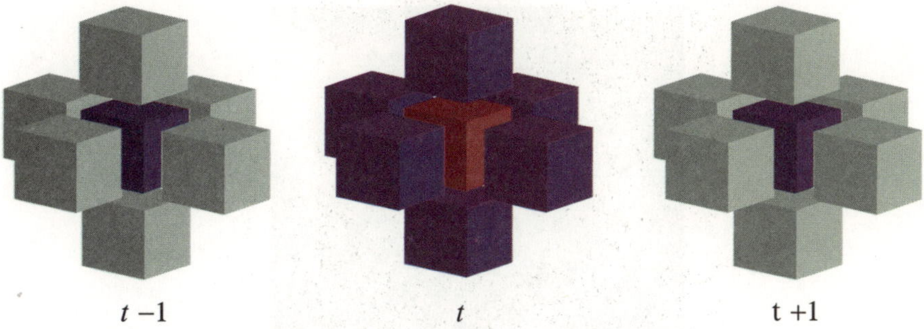

Fig. 1. Each non-boundary voxel has six spatially neighboring voxels in every 3D volume. By making temporal neighbors available, the total number of neighboring voxels increases to eight. The higher number of neighboring voxels increases the amount of information available to the segmentation procedure. In this image, each cube represents a voxel. All the blue voxels are considered immediate neighbors to the red voxel

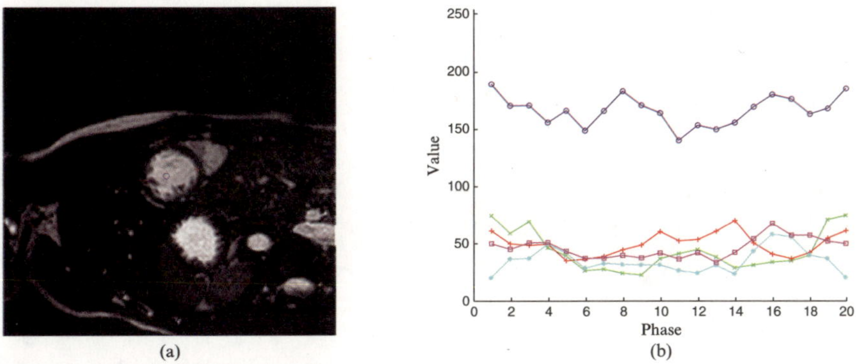

Fig. 2. To verify the correlation between temporally neighboring voxels, the gray level value of the same voxel is tracked over time. The plot on the right shows the change in grey level values of the voxel at the marker location in the image on the left

Temporally neighboring regions are similar when the displacement between the two regions is small enough so values remain correlated. Figure 4 depicts 20 instances of a gated MR slice through a cardiac LV. The movement of a small region on the inner wall of the LV is observed. The movement of several regions on the wall of the LV, identified by the markers shown in Figure 3. The spatial location and displacement of the marker are then measured. A plot of the displacement in the x and y directions is depicted in Figure 4, which shows the similarity between temporally adjacent points.

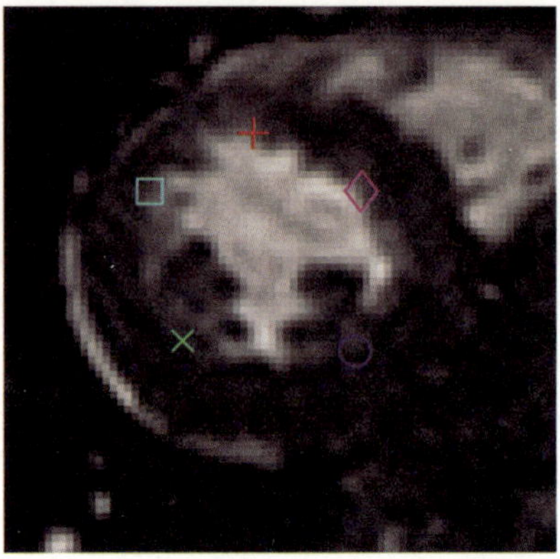

Fig. 3. A slice of a gated MR scan of the left ventricle. A marker is manually placed on the same spot of the LV wall in an effort to track its displacement through the cardiac cycle. Every colored shape represents a region: blue (o) - region 1, green (x) - region 2, red (+) - region 3, cyan (square) - region 4, and magenta (diamond) - region 5

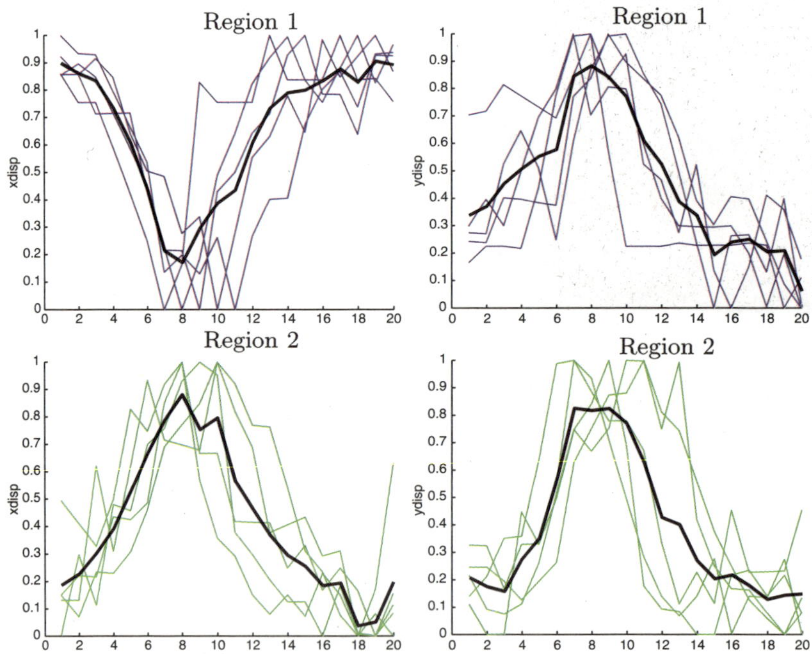

Fig. 4. The x and y displacements of the markers placed on the different parts of the cardiac muscle. Every marker had a similar motion in each data set that was studied. The black curve shows the mean displacement. This similarity in motion can only be used when the time variable is included in the segmentation procedure

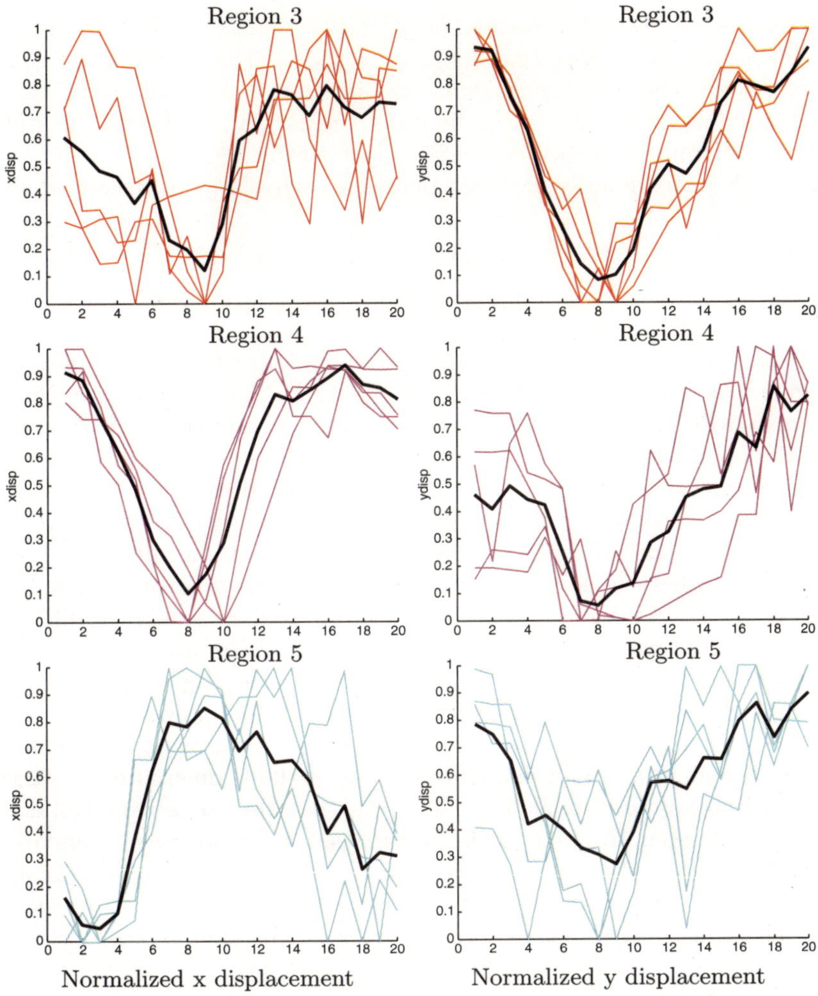

Fig. 4. (Continued)

2 Method

Based on the discussion above, this section describes the method used to incorporate time into 3D shape–based segmentation method to produce a 4D process.

2.1 Combining Space and Time

Let I represent a sequence of 3D volumetric data on a structured cartesian grid. We can denote by $I_t(x, y, z)$ a value of the sequence I at time t and spatial position x, y, z. Similarly, $I_{t-1}(x, y, z)$ would be a value in the volume at time

$(t-1)$ at the same spatial position (x, y, z). If we consider I as a complete 4D data set, we can then write $I_t(x, y, z)$ as $I(x, y, z, t)$. The coordinates (x, y, z, t) pinpoint a specific location in space and in time. In clinical cardiology, the time variable t falls between the limits of the cardiac cycle and the state of the heart at time instance t is referred to as cardiac *phase*.

If voxels in temporally adjacent phases are correlated as discussed above, then we have

$$I(x, y, z, t) \simeq I(x, y, z, t-1)$$
$$I(x, y, z, t) \simeq I(x, y, z, t+1)$$

Figure 5 explains why it is is useful to include the temporal dimension in a segmentation of an object that is in motion. The central image in Figure 5 depicts an MR slice through a left ventricle at a certain time t. It is an image of all the points at position (z, t). It is surrounded by its immediate neighboring slices at locations $(z+1, t)$, $(z, t-1)$, $(z, t+1)$ and $(z-1, t)$. The red arrow points to a blurred and weak boundary that creates a gap in the left ventricle wall. The same part of this wall shows stronger definition in the neighboring slice $(z-1, t)$ below, and in slice at $(z, t+1)$ in the following phase on the right. If only a 2D segmentation algorithm is used, it would be difficult to delineate the LV wall properly since there is no extra information available to guide the segmentation to the proper location of the edge. A 3D segmentation algorithm will have more information about the missing boundary since one of the neighboring slices at $(z-1, t)$, contains a complete edge. However, adding the temporal dimension and performing the segmentation in 4D, provides yet another additional slice, $(z, t+1)$, that also contains a well defined edge. This doubles the amount of information available when compared to the classical 3D segmentation. Now since the boundary in the central slice is weak and not completely missing, a strong edge in a neighboring slice will drive the segmentation to complete this weak gap by defining its location and shape.

A 4D segmentation approach is described in [9]. However, their approach is different from the one discussed in this chapter. They start with a 3D segmentation of one phase of the heart, then progress through the rest of the phases taking into consideration the segmentation of the previous phase. Although this approach uses all 4 dimensions, it is not a pure 4D segmentation since the segmentation actually happens in 3D.

2.2 Adding time to shape–based segmentation methods

The main portion of the research is motivated by the work done in [16]. The authors of [16] use a shape–based model with two and three dimensional data. They used training data to generate shape priors that were incorporated into the segmentation. They also derived a parametric model for an implicit representation of the segmenting curve by applying principal component analysis

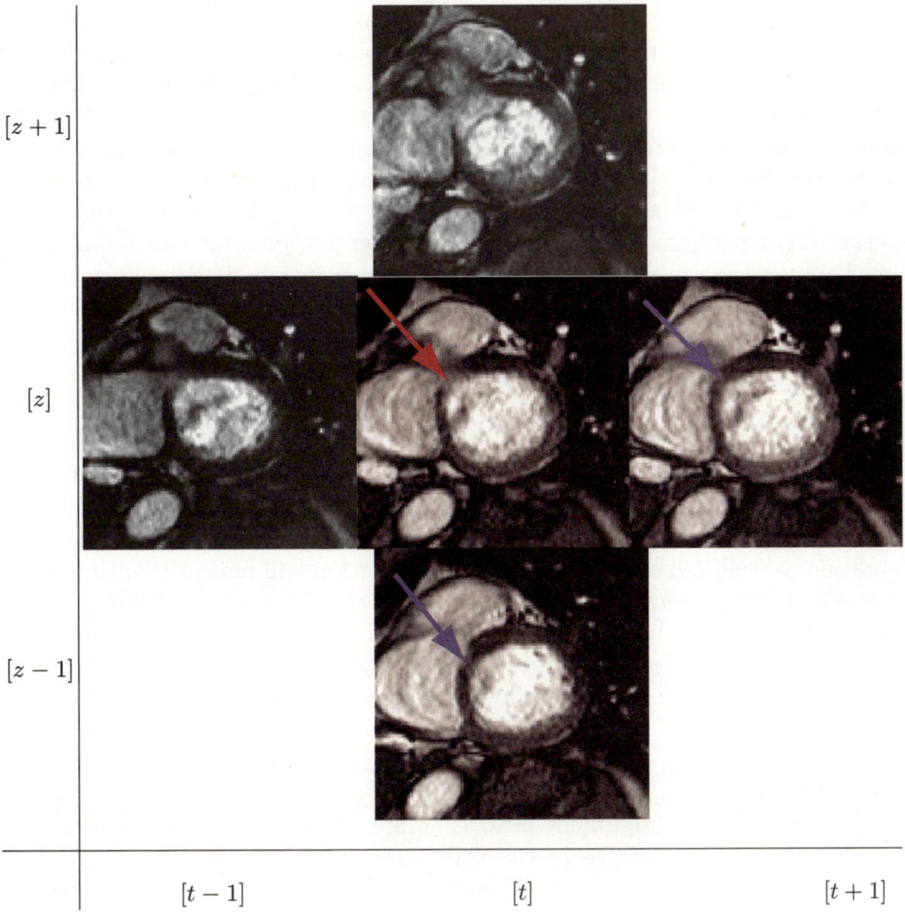

Fig. 5. The image in the center is a slice through the left ventricle that we are trying to segment. As shown by the red arrow, part of the edge of the left ventricle wall is blurred making it difficult to decide how the segmenting curve should go. However, the same part of the wall is much stronger in the neighboring slices, in space $z - 1$ or in time $t + 1$ (blue arrows). By using this information, the segmentation is able to close the weak gap in the central slice. Using information from neighboring slices, we can improve the segmentation results by exploiting similarities between them

(PCA) to a collection of signed distance representations of the training data. Our goal is to extend this procedure and use all available dimensions by treating time the same as a spatial dimension.

The algorithm can be described as follows: Define a set or subspace **L** that includes all the possible shapes, location and orientation of a segmenting contour. The combination of the contour location and orientation is referred to as its pose. The subspace $\mathbf{L}$ is spanned by a set of orthogonal principal components, or eigen–shapes. Any new contour can be generated by moving from the mean shape, or origin of the subspace, using a linear combination or a weighted sum of these "posed" eigen–shapes. The final shape of the contour can be altered by changing the weight of each eigen–shape in the linear combination. The problem can naturally be divided into two parts: Generating the eigen–shapes that span the subspace $\mathbf{L}$ and performing the actual segmentation. The shape of the segmenting contour can be modified by changing the weights used in the linear combination. The final orientation can also be changed by changing the pose of each eigen–shape.

2.3 Generating Training Data

Eigen–shapes are generated using data that has been manually segmented. The result of the manual segmentation is a set of binary images that describe each segmentation. Obviously, not all cardiac MR scans are alike resulting in manual segmentations that are different not only in shape, but in size and orientation as well. Differences in size and orientation will mask the variations in shape and will prevent PCA from capturing them. Therefore, the different segmentations must be be aligned or registered so variations in size and pose will become irrelevant. Pose parameters include translation, rotation and scale. In four dimensions, the pose parameters can be assembled in a vector $\mathbf{p}$ such that $\mathbf{p} = [a, b, c, d, h_x, h_y, h_z, \alpha, \beta, \gamma]^T$, with a, b, c, d corresponding to x-, y-, z-, t-translation, h_x, h_y, h_z corresponding to scale and α, β, γ corresponding to rotation in the xy, xz and yz planes, respectively.

The goal of the alignment is to generate a set of pose parameters used to align the binary images of the training set. If we have a training set $\mathcal{T}$ of n binary images $\mathrm{I}^1, \mathrm{I}^2, ..., \mathrm{I}^n$, then we would have a set of n pose vectors $\mathbf{p}^1, \mathbf{p}^2, \cdots, \mathbf{p}^n$, one for each image respectively. If $\tilde{\mathrm{I}}$ is the transformed image of I using a transformation matrix T based on the pose parameter $\mathbf{p}$, then we have

$$\begin{bmatrix} \tilde{x} \\ \tilde{y} \\ \tilde{z} \\ \tilde{t} \end{bmatrix} = T(\mathbf{p}) \begin{bmatrix} x \\ y \\ z \\ t \end{bmatrix} \qquad \text{and} \qquad \tilde{\mathrm{I}} = T(\mathbf{p})\mathrm{I} \tag{1}$$

The transformation matrix $T(\mathbf{p})$ is the product of three matrices: a translation matrix $M(a, b, c, d)$, a scaling matrix $H(h_x, h_y, h_z)$ and an in-plane rotation matrix $R(\alpha, \beta, \gamma)$. It maps the coordinates $(x, y, z, t) \in \mathbb{R}^4$ into coordinates $(\tilde{x}, \tilde{y}, \tilde{z}, \tilde{t}) \in \mathbb{R}^4$. The transformation is semi–affine and spans all 4 dimensions with only translation applied in the temporal dimension as can be seen in the following:

$$T(\mathbf{p}) = MHR$$

$$M = \begin{bmatrix} 1\,0\,0\,0\,a \\ 0\,1\,0\,0\,b \\ 0\,0\,1\,0\,c \\ 0\,0\,0\,1\,d \\ 0\,0\,0\,0\,1 \end{bmatrix} \qquad H = \begin{bmatrix} h_x & 0 & 0 & 0\,0 \\ 0 & h_y & 0 & 0\,0 \\ 0 & 0 & h_z & 0\,0 \\ 0 & 0 & 0 & 1\,0 \\ 0 & 0 & 0 & 0\,1 \end{bmatrix}$$

$$\begin{aligned} R &= R_z(\alpha)R_y(\beta)R_x(\gamma) \\ &= \underbrace{\begin{bmatrix} C(\alpha) & -S(\alpha) & 0\,0\,0 \\ S(\alpha) & C(\alpha) & 0\,0\,0 \\ 0 & 0 & 1\,0\,0 \\ 0 & 0 & 0\,1\,0 \\ 0 & 0 & 0\,0\,1 \end{bmatrix}}_{R_z(\alpha)} \underbrace{\begin{bmatrix} C(\beta) & 0 & S(\beta) & 0\,0 \\ 0 & 1 & 0 & 0\,0 \\ -S(\beta) & 0 & C(\beta) & 0\,0 \\ 0 & 0 & 0 & 1\,0 \\ 0 & 0 & 0 & 0\,1 \end{bmatrix}}_{R_y(\beta)} \underbrace{\begin{bmatrix} 1 & 0 & 0 & 0\,0 \\ 0 & C(\gamma) & -S(\gamma) & 0\,0 \\ 0 & S(\gamma) & C(\gamma) & 0\,0 \\ 0 & 0 & 0 & 1\,0 \\ 0 & 0 & 0 & 0\,1 \end{bmatrix}}_{R_x(\gamma)} \end{aligned}$$

where $C = \cos(\theta)$ and $S = \sin(\theta)$ [15, 16, 18].

As suggested in [16], a good way to jointly align n binary images is to descend along the following energy functional:

$$E_{align} = \sum_{i=1}^{n} \sum_{\substack{j=1 \\ j \neq i}}^{n} \left\{ \frac{\iint_{\Omega} \left(\tilde{I}^i - \tilde{I}^j\right)^2 \mathrm{d}A}{\iint_{\Omega} \left(\tilde{I}^i + \tilde{I}^j\right)^2 \mathrm{d}A} \right\} \tag{2}$$

where Ω and $\mathrm{d}A$ represent the image domain and unit area respectively. A very good example that intuitively shows the benefits of image alignment is the airplane alignment depicted in Figure 6(a) that is reproduced from [16]. Each airplane shape in Figure 6(a) has a different size, location and orientation. The 2D alignment of all the airplanes to the one on the upper left corner was performed according to (2). The result is shown in Figure 6(b) which also shows how the scale and orientation of all the airplanes match. It is beneficial to note that to avoid getting caught in a local minimum of the energy functional of (2), a multi–resolution approach is followed. Lower resolutions of the training data are obtained by sub–sampling the data at different rates. Alignment starts at the lowest resolution and progresses gradually to higher resolutions until the final native resolution is processed. The initial pose at each resolution stage is the same as the pose when convergence was attained at the previous stage. Extending this example to 4D, if $\mathbf{p}^i_{m,n}$ represents the 4D pose of data set I^i at n^{th} alignment iteration in resolution stage m ($m = 0$ being the alignment stage at the lowest resolution), then

$$\begin{cases} \mathbf{p}^i_{m,0} = \mathbf{p}^i_{m-1,k} \\ \mathbf{p}^i_{0,0} = [0, 0, 0, 0, 1, 1, 1, 0, 0, 0]^T \end{cases}$$

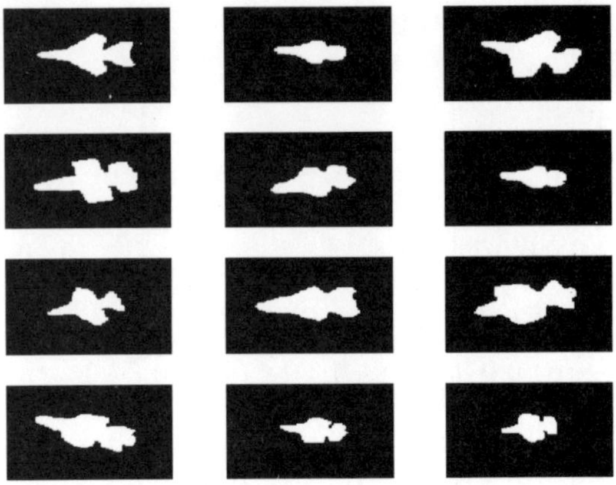

(a) A set of 12 airplane shapes used to show the alignment. Each airplane shape has a size, orientation and position that is different from the rest.

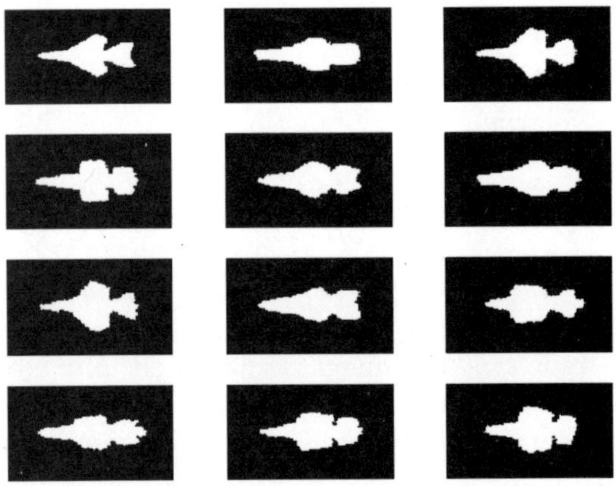

(b) The same shapes all aligned to the one in the upper left corner. Notice how all the airplanes have similar size and orientation.

Fig. 6. Example of image alignment

where k is the number of iterations needed for convergence at any resolution stage. $\mathbf{p}^i_{0,0}$ in the equation above denotes the initial pose of every image I^i at the beginning of the alignment process.

This multi–resolution approach reduces the computational cost of the alignment because it reduces the number of iterations needed to perform alignment at the higher resolution stages due to the fact that at the beginning of each stage, the initial pose is closer to the final result.

The level set approach [10, 12] was used to represent the boundaries of the aligned manual segmentations in the binary images. The idea is to represent the boundary of each binary image by the zero level set of a signed distance function. This function is chosen to be negative inside the shape and positive outside. As a result we have n signed distance functions $\{\Psi_1, \Psi_2, ..., \Psi_n\}$. A mean level set function $\bar{\Phi}$, is then computed as the average of the n distance functions

$$\bar{\Phi} = \frac{1}{n} \sum \Psi_i \tag{3}$$

The shape variabilities were extracted by subtracting the mean signed distance function $\bar{\Phi}$ from each signed distance function $\tilde{\Psi}_i$ to form n separate mean offset functions $\{\tilde{\Psi}_1, \tilde{\Psi}_2, ..., \tilde{\Psi}_n\}$. The mean offset functions then fill the columns $\tilde{\psi}_i$ of a shape variability matrix $\mathcal{S}$. In 4D, the size of $\mathcal{S}$ is $n \times N$, where N = Rows $\times$ Columns $\times$ Slices $\times$ Phases. The columns of each slice of each phase of $\tilde{\Psi}_i$ are stacked lexicographically to form one column of $\mathcal{S}$.

$$\mathcal{S} = [\tilde{\psi}_1 \, \tilde{\psi}_2 \ldots \tilde{\psi}_n]$$

An eigenvalue decomposition is then performed on the covariance matrix $\Sigma = \frac{1}{n}\mathcal{S}\mathcal{S}^T$ such that:

$$\Sigma = \frac{1}{n}\mathcal{S}\mathcal{S}^T = U \Lambda U^T \tag{4}$$

U is an $N \times n$ matrix whose columns are the orthogonal modes of variation in the shape. The images are rebuilt from the columns of U by undoing the lexicographical arrangement mentioned above and rearranging the elements into a rectangular grid producing n eigenshapes $\{\Phi_1, \Phi_2, \ldots, \Phi_n\}$.

These eigenshapes can now be used in the segmentation step. However, as in our case, when each principal component is represented by every voxel, the dimension of the covariance matrix Σ is $N \times N$ is considerably large. With $N \gg n$, the system is under constrained and a number of eigen values will be zero. Computing the eigenvectors of the large matrix Σ is inefficient. A useful trick in [8] shows that performing PCA on a much smaller matrix $\mathcal{W}$ of size $n \times n$ can be used to calculate the eigen values and eigen vectors of the larger matrix Σ. $\mathcal{W}$ is defined as:

$$\mathcal{W} = \frac{1}{n}\mathcal{S}^T\mathcal{S} \tag{5}$$

Following the steps in [8], if $\mathbf{d}$ is an eigen vector of $\mathcal{W}$ with eigen value λ then

$$\begin{aligned}
\Sigma(\mathcal{S}\mathbf{d}) &= \frac{1}{n}\mathcal{S}\mathcal{S}^T(\mathcal{S}\mathbf{d}) \\
&= \mathcal{S}\left(\frac{1}{n}\mathcal{S}^T\mathcal{S}\right)\mathbf{d} \\
&= \mathcal{S}(\mathcal{W}\mathbf{d}) \\
&= \mathcal{S}\lambda\mathbf{d} \\
&= \lambda(\mathcal{S}\mathbf{d})
\end{aligned}$$

It follows that $\hat{\mathbf{d}} = \mathcal{S}\mathbf{d}$ is an eigenvector of the original matrix $\frac{1}{n}\mathcal{S}\mathcal{S}^T$ with the same eigen value λ. If d_i and $\hat{d}_i$ represent the ith element of vectors $\mathbf{d}$ and $\hat{\mathbf{d}}$ respectively and if $\tilde{\psi}_j^i$ represents the ith element of the offset function $\tilde{\psi}_j$, then

$$\hat{d}_i = \sum_{j=1}^{n} d_i \tilde{\psi}_j^i \tag{6}$$

It is important to note here that the vector $\hat{\mathbf{d}}$ in (6) *is* periodic. In 4D, $\tilde{\psi}_j^i$ is a periodic signal since it is the result of the difference of the mean function $\bar{\boldsymbol{\Phi}}$ and the signed distance function Ψ_i, both of which are periodic. Each element $\hat{d}_i$ in (6) becomes a linear combination of elements of periodic functions $\tilde{\psi}_j$. This is an important result for the segmentation phase, since it allows access to values at locations that fall outside the cardiac cycle by wrapping back to the proper place inside the cardiac period to get the required value. This is explained further in Section 3.

2.4 Segmentation

Any new level set function Φ can now be expressed as a linear combination of these orthogonal modes of variations or eigenshapes. For $k \leq n$:

$$\Phi(\mathbf{w}) = \bar{\Phi} + \sum_{i=1}^{k} w_i \Phi_i \tag{7}$$

where $\mathbf{w} = \{w_1, w_2, \ldots, w_k\}$ are the weights for the k eigenshapes. The variances of these weights $\{\sigma_1^2, \sigma_2^2, \ldots, \sigma_k^2\}$ given by the eigenvalues calculated earlier.

The value of k should be large enough to capture the prominent shape variation present in the training set. If it is too large, the shape will start capturing intricate details that are specific to the training data used. In this research, k was chosen empirically. For testing purposes, 5 eigen shapes that correspond to the 5 largest eigen values were chosen. We found that this was a good compromise between capturing important details and the speed of the segmentation. As mentioned before, the weights w_i in the equation above dictate how much influence each eigenshape has and therefore changing the weights will change the shape of the final contour.

Equation (7), however, does not provide any means to change the pose of the segmenting surface. Without any means to change the position and orientation, the segmentation will remain stationary and will only change its shape. To achieve the ability to handle pose changes, a pose parameter ($\mathbf{p}$) is introduced to (7):

$$\Phi(\mathbf{w}, \mathbf{p}) = \bar{\Phi}(\mathbf{p}) + \sum_{i=1}^{k} w_i \Phi_i(\mathbf{p}) \tag{8}$$

To accomplish the segmentation, we then evolve the pose parameters ($\mathbf{p}$) and weights ($\mathbf{w}$) according to a region–based curve evolution model. For the purpose of this research, we used the Chan–Vese [4] piecewise constant model as described in (9). Since the level set function Φ is expressed as a signed distance function, then the segmentation curve or hyper–surface can be defined as

$$\overrightarrow{C} = \{(x, y, z, t) \in \mathbb{R}^4 : \Phi(x, y, z, t) = 0\}$$

The regions inside and outside the segmentation hyper–surface $\overrightarrow{C}$, denoted respectively by R_{in} and R_{out} are given by

$$\begin{aligned} R_{\text{in}} &= \{(x, y, z, t) \in \mathbb{R}^4 : \Phi(x, y, z, t) < 0\} \\ R_{\text{out}} &= \{(x, y, z, t) \in \mathbb{R}^4 : \Phi(x, y, z, t) > 0\} \end{aligned}$$

Other important image quantities can be written as:

$$\begin{aligned} \text{4D volume inside } C: \ V_{\text{in}} &= \iint \mathcal{H}(-\Phi(\mathbf{w}, \mathbf{p}))\, \mathrm{d}V \\ \text{4D volume outside } C: V_{\text{out}} &= \iint \mathcal{H}(\Phi(\mathbf{w}, \mathbf{p}))\, \mathrm{d}V \\ \text{Pixel intensity sum inside } C: \ S_{\text{in}} &= \iint I\mathcal{V}(-\Phi(\mathbf{w}, \mathbf{p}))\, \mathrm{d}V \\ \text{Pixel intensity sum outside } C: S_{\text{out}} &= \iint I\mathcal{V}(\Phi(\mathbf{w}, \mathbf{p}))\, \mathrm{d}V \\ \text{Mean pixel intensity inside } C: \ \mu &= \frac{S_{\text{in}}}{V_{\text{in}}} \\ \text{Mean pixel intensity outside } C: \ \nu &= \frac{S_{\text{out}}}{V_{\text{out}}} \end{aligned}$$

where I is the observed data set and $\mathrm{d}V$ represents a 4D unit hyper–volume. $\mathcal{H}$ is the Heaviside function defined as:

$$\mathcal{H}(\Phi(\mathbf{w}, \mathbf{p})) = \begin{cases} 1 \text{ if } \Phi(\mathbf{w}, \mathbf{p}) \geq 0 \\ 0 \text{ if } \Phi(\mathbf{w}, \mathbf{p}) < 0 \end{cases}$$

The energy functional of the Chan-Vese model can then be written as,

$$E = -\left(\mu^2 H_{\text{in}} + \nu^2 V_{\text{out}}\right) = -\left(\frac{S_{\text{in}}^2}{V_{\text{in}}} + \frac{S_{\text{out}}^2}{H_{\text{out}}}\right) \tag{9}$$

The pose $\mathbf{p}$ and weight $\mathbf{w}$ are then updated by following the steepest descent of the energy functionl in (9) and can be written as,

$$\begin{aligned} \nabla_{\mathbf{w}} E &= -2(\mu \nabla_{\mathbf{w}} S_{\text{in}} + \nu \nabla_{\mathbf{w}} S_{\text{out}}) + (\mu^2 \nabla_{\mathbf{w}} V_{\text{in}} + \nu^2 \nabla_{\mathbf{w}} V_{\text{out}}) \\ \nabla_{\mathbf{p}} E &= -2(\mu \nabla_{\mathbf{p}} S_{\text{in}} + \nu \nabla_{\mathbf{p}} S_{\text{out}}) + (\mu^2 \nabla_{\mathbf{p}} H_{\text{in}} + \nu^2 \nabla_{\mathbf{p}} H_{\text{out}}) \end{aligned}$$

The weights and pose then updated by

$$\begin{aligned} \mathbf{w}_n &= \mathbf{w}_{n-1} - \epsilon_{\mathbf{w}} \nabla_{\mathbf{w}} E \\ \mathbf{p}_n &= \mathbf{p}_{n-1} - \epsilon_{\mathbf{p}} \nabla_{\mathbf{p}} E \end{aligned}$$

where $\epsilon_{\mathbf{p}}$ and $\epsilon_{\mathbf{w}}$ are the update step size and n is the iteration number.

It is important to understand that the signed distance function is computed using all dimensions, spatial and temporal. Unit measurements in all dimensions are normalized to unity. A translation by one unit vector in any dimension will change the position in that dimension by one. For example, a positive translation of one unit in the x-dimension increases the x-position by one. Similarly, a positive translation of one unit in the t-dimension increases the t-position by one. Moving along the spatial dimension keeps the current cardiac phase, but moving along the t-dimension changes to a new phase.

3 Periodic data

Special attention has to be paid when calculations are on the edges of the grid. Spatial first derivatives, for example, are approximated by central differences at locations that are not on the periphery of the grid, and as a forward or a backward difference when the voxel happens to be on the edge. For example, the derivative dx can be computed as follows

$$\mathrm{d}x = \begin{cases} \dfrac{\mathrm{I}(x{+}1,y,z,t) - \mathrm{I}(x{-}1,y,z,t)}{2} & \text{if } 0 < x < \mathrm{X} \quad \text{(central difference)} \\ \mathrm{I}(x{+}1,y,z,t) - \mathrm{I}(x,y,z,t) & \text{if } x = 0 \quad \text{(forward difference)} \\ \mathrm{I}(x,y,z,t) - \mathrm{I}(x{-}1,y,z,t) & \text{if } x = \mathrm{X} \quad \text{(backward difference)} \end{cases} \tag{10}$$

where X denotes the size of the image in the x direction.

However, when calculating derivatives for the variable t, only central differences are used. To compute the correct time index, the modulus operator (mod) is used since the cardiac data is considered periodic. The temporal derivative dt can be calculated as the follows

$$\mathrm{d}t = \frac{\mathrm{I}(x,y,z,(t{+}1)\,\mathtt{mod}(T)) - \mathrm{I}(x,y,z,(t{-}1{+}T)\,\mathtt{mod}(T))}{2} \tag{11}$$

where T is the length of the cardiac cycle.

4 Results

The performance of the 4D segmentation was evaluated by conducting several consistency and accuracy tests. The results of the new 4D segmentation were compared to the results from a classical 3D segmentation as well as results from a manual segmentation performed by an expert. Consistency tests measure the similarity in the results of the same segmentation method on the same data set but when different initial conditions are used. Accuracy tests measure the similarity between an automatic segmentation and a manual segmentation of the same data set. To isolate the contribution of the time variable in the final result, the automatic segmentations were performed using the same initial conditions. In other words, the same shape priors and energy functionals that were used in the 4D segmentation were also used in 3D and the only difference was the inclusion of the time variable. The consistency and accuracy of all 3 segmentation methods were compared in pairs.

A total of 15 data sets were used to assess the performance of the 4D automatic segmentation method. Each data set consisted of 20 time frames equally spread over the cardiac cycle. Each time frame consisted of 18, 256×256 slices spanning from the apex to the base of the heart. Using the overlap method of (12), 5400 slices were processed.

4.1 Consistency Tests

Consistency tests on the manual segmentations were performed by comparing the segmentations of five data sets performed by the same user at different times. There was a 10 month time difference between the two segmentations in an attempt to minimize the effect of memory and to have the two segmentations as independent of each other as possible. The second segmentation was carried out without any regard to the results from the previous one. Figure 7 depicts an example of the manual segmentation results of a sequence of slices close to the base of the heart where there is an abundance of incomplete region boundaries. These images portray the difference between the two segmentations, even when they were performed by the same user. The majority of the differences occur in areas where shape and location of the curve have to be approximated. Missing boundaries make it necessary to approximate the contour, resulting in different approximations at different times.

A popular method to compare two different studies is called the Bland and Altman method [2]. It measures the similarity between two methods and can be used to measure consistency. The method states that for two clinical methods to be consistent, the average difference between the results of two measurements should be zero and that 95% of the difference should be less than two standard deviations. We will refer to this as the Bland–Altman consistency test. Figure 8(a) shows a scatter plot of the areas of the slices resulting from the two manual segmentations and Figure 8(b) shows the plot of the Bland–Altman consistency test. Most of the points in Figure 8(b) are within two standard deviations which indicates that the two manual tests are consistent despite the visual difference in Figure 7. Consistency of the automatic segmentation was measured using cross validation, a method that is also adequate to check of how well the algorithm generalizes to new data. A total of 15 data sets were used. In each data set, the endocardium or the inner boundary of the left ventricle (LV) was segmented using the 3D and 4D shape–based methods mentioned above. The other 14 data sets were used as training data for the generation of the shape priors. For every data set, a total of six automatic segmentations were performed, a set of three 3D segmentations and another set of three 4D segmentations. Each segmentation had a different starting position of the initial mean shape $\bar{\Phi}$ of (3). The different centers

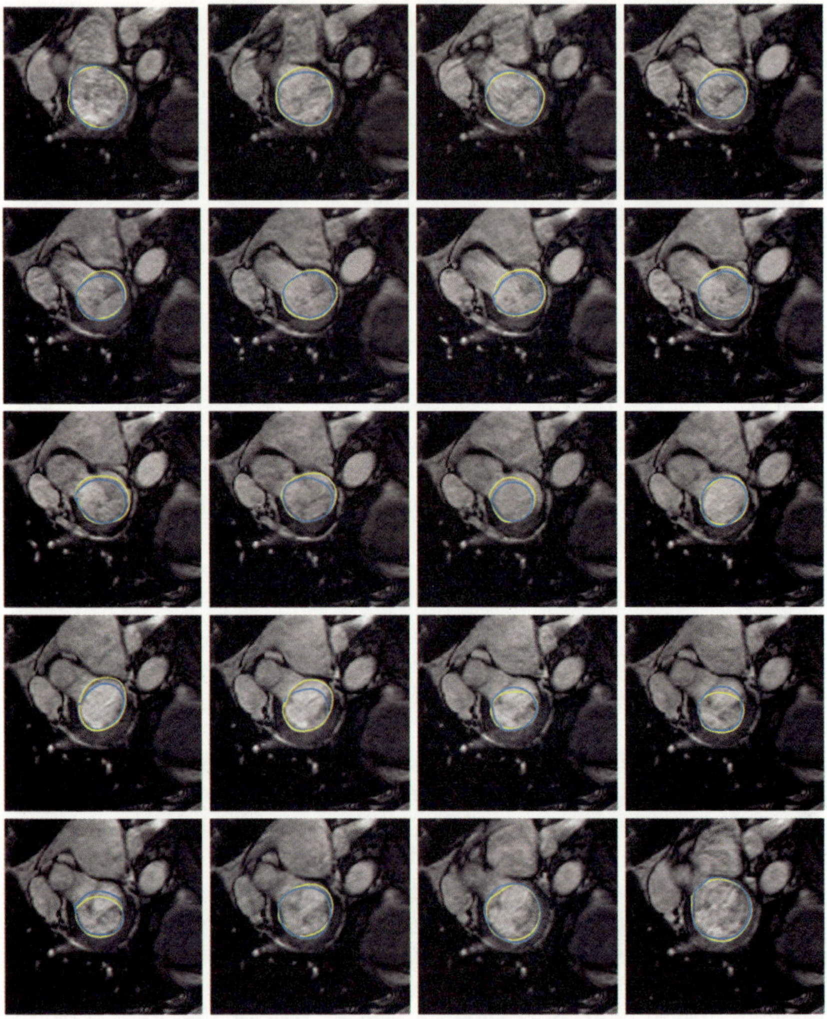

Fig. 7. A sample result of two manual segmentations of a sequence of MR slices near the base of the heart. The segmentations were performed 10 months apart to try to minimize the dependency of one segmentation on the other. The yellow curve is the initial segmentation. The second segmentation done without regard of the previous one and is shown in blue. This figure is a good example of the difference between manual segmentations especially in places with missing boundaries

of the mean shape are shown in Figure 9. The results of the segmentations were then compared for similarity. Figure 10 depicts the consistency results of the 3D segmentation. The overall result shows that the segmentations were, in general, consistent, despite the fact that one segmentation failed to accurately segment the left ventricle. For this particular case, the initial position of the

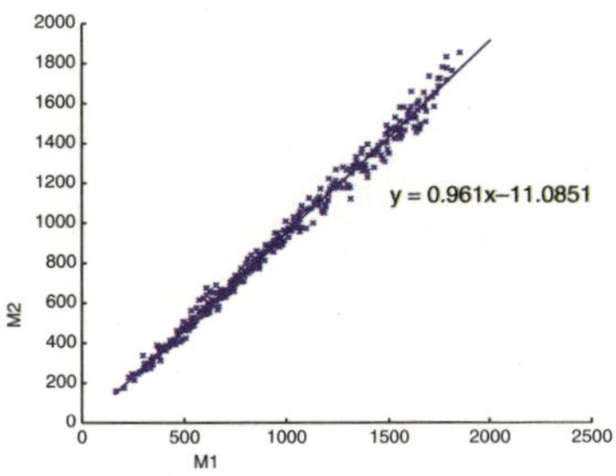

(a) A plot to show consistency of two manual segmentations performed 10 months apart of the same data sets by the same user. Many of the differences in values occur at places at the apex or base of the heart where the boundaries are undefined and have to be approximated.

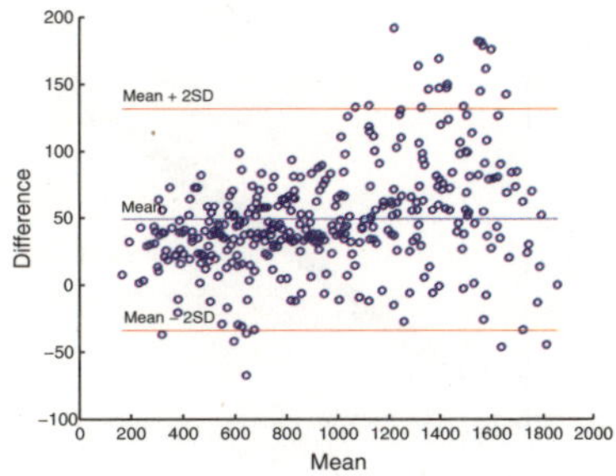

(b) A plot of the Bland–Altman consistency test for the manual segmentation. It shows that the two manual segmentations are consistent since the mean difference of the left ventricle areas is zero and that 95% of the difference are less than two standard deviations.

Fig. 8. Scatter and Bland–Altman plots that depict the consistency in two manual segmentations

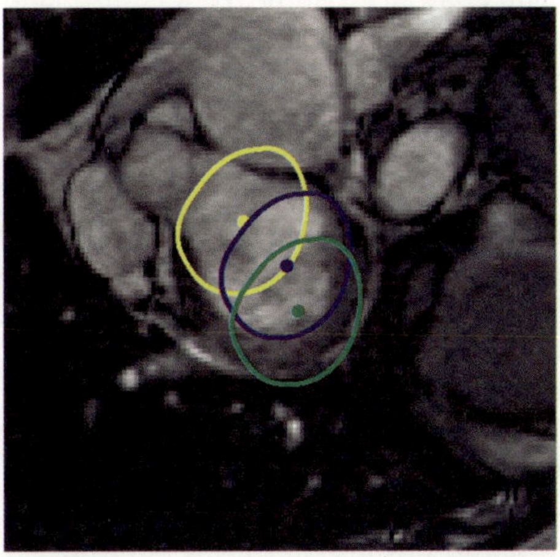

Fig. 9. The different locations of the initial contour used to test for consistency of automatic 3D and 4D segmentations

segmentation included only part of the blood pool inside the left ventricle. The rest of the segmentation included the cardiac muscle as well as regions with abundant blood activity such as the arteries. The Chan–Vese method used to evolve the surface tries to match the means inside the contour to the data. By doing this, if the initial mean of the voxel values inside the contour was too different from the mean of the the values inside the left ventricle, the method will try to put the left ventricle on the outside of the contour.

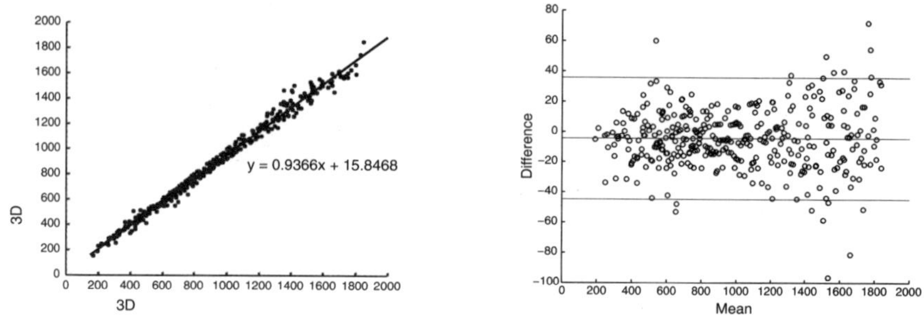

(a) Scatter plot to show the consistency of 3D automatic segmentation

(b) Bland–Altman plot for 3D segmentation consistency test

Fig. 10. Consistency results of the 3D segmentation

Figure 16 shows the position of the initial contour (green), and the results of a 3D (red), 4D(blue) and manual (yellow) segmentations. Note how a big portion of the initial contour falls outside the left ventricle and that other structures surrounding the left ventricle had significant effect on the values inside and outside the contour. The heart was also close to an end systolic state, a state where the left ventricle has a minimum volume with the minimum amount of blood inside the left ventricle. This situation forced the contour to try to enclave the area outside the left ventricle that includes the regions where the arteries connect to the ventricles, regions with high blood presence and activity. The result of the failure was not used in Figure 10.

Figure 11 shows the consistency results of the 4D automatic segmentation. The graph shows that the consistency of the segmentation was higher than in the 3D case simply because the method takes advantage of the coherency of the left ventricle across space and time giving it the extra information needed to locate the ambiguous boundaries. It is important to note that there were no failures in any of the results of the 4D method because the 4D hyper–surface segments space and time *simultaneously.* The hyper–surface was able to take advantage of the information of the left ventricle when it was in a state other than the systolic state, an advantage that a pure 3D segmentation does not have. Another method to measure performance is the overlap method, which is described in (12) and depicted in Figure 12.

$$O = \frac{A_1 \cap A_2}{\frac{1}{2}(A_1 + A_2)} \tag{12}$$

Figure 13 depicts a bar graph that shows the different levels of consistency of the 3 different segmentation methods. The results are summarized by the 95% confidence interval on mean of overlap ratios. The consistency of the 4D segmentation is 0.95 ± 0.02 when using different initial locations of the segmentation curve. The 3D segmentation follows at 0.89 ± 0.03. The manual segmentation consistency measure was only 0.78 ± 0.03.

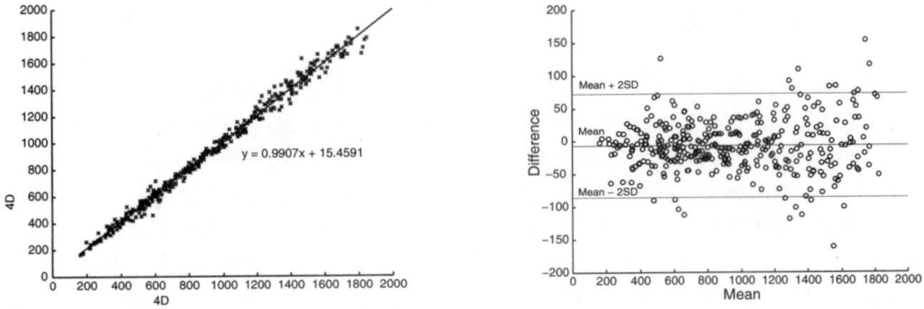

(a) Scatter plot to show the consistency of 4D automatic segmentation

(b) Bland–Altman plot for 4D segmentation consistency test

Fig. 11. Consistency results of the 4D segmentation

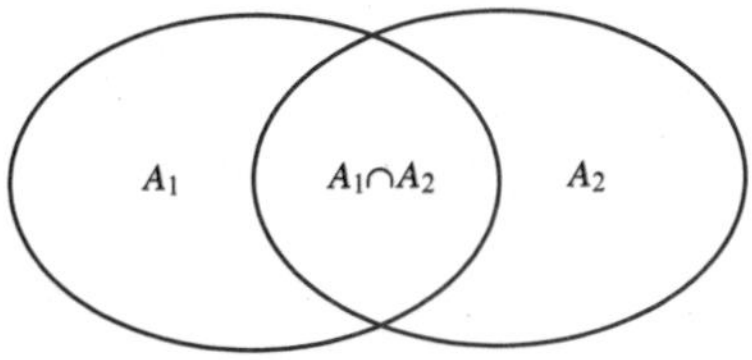

Fig. 12. Overlap method used to calculate similarity between two segmentations

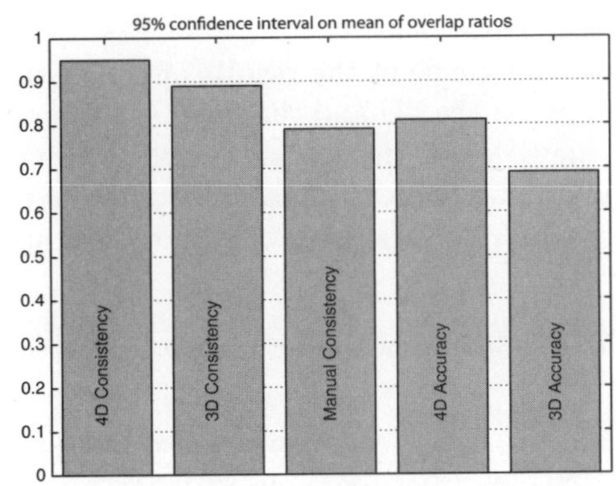

Fig. 13. Bar graph showing the different consistency test measures. The 4D segmentation had the highest consistency of 0.95 ± 0.02 (95% confidence interval on the mean) followed by the classical 3D segmentation 0.89 ± 0.03. The manual segmentation was the least consistent 0.78 ± 0.04. Accuracy measures show that the 4D segmentation is more comparable to an expert manual segmentation than the 3D method. The overlap ratios between the manual and the 4D result is 0.81 ± 0.02, a result that is comparable between the overlap between the two manual segmentations. The manual result and the 3D segmentation had a 0.69 ± 0.02 mean overlap ratio

4.2 Accuracy Tests

To measure accuracy, the results of the automatic segmentations were compared to the results from the manual segmentation. Two values were used for comparison: ejection fraction and the short axis slice area of the slice that falls in the middle between the apex and the base. The ejection fraction (EF) is the fraction of the volume of the blood pumped out of the heart with every beat, or the stroke volume (SV) to the volume of the blood after filling, which is also called the end diastolic volume of (EDV) [1]. It can be calculated by the following:

$$\mathrm{EF} = \frac{V_d - V_s}{V_d} \tag{13}$$

where V_d and V_s are the end diastolic and end systolic volumes of the inner walls of the LV. The area of the endocardium was calculated for every slice since the manual segmentation is done on each slice in 2D. It is measured in mm^2.

To assess accuracy of the different segmentation procedures, the endocardium of 5 data sets out of the total 15 were traced manually. The same data sets were also segmented using 3D and 4D automatic segmentations. Shape priors for each of the 5 automatic segmentations were produced using the 14 leftover data sets.

Table 1 shows the signed average and the root mean square error of the ejection fraction and the endocardium area of the results from the manual and 3D automatic segmentations.

Table 2 shows the difference in the results of the manual and 4D segmentations. The error between the 4D and the manual segmentation is lower than the error between the 3D and the manual segmentation for the same data sets. Figures 14 and 15 show linear analysis between automatic and manual segmentations. Notice how the scatter plot is more widespread in the 3D case

Table 1. Accuracy of the 3D Segmentation $\pm$ 1SD (σ) measured versus a manual segmentation

	Average Signed Error	RMS Error
Ejection Fraction $\pm\sigma_{\mathrm{EF}}$	1.673 ± 0.837	4.298
Endocardium Area $\pm\sigma_{\mathrm{EA}}$	4.319 ± 4.98	5.732

Table 2. Accuracy of the 4D Segmentation $\pm$ 1SD (σ) measured versus a manual segmentation

	Average Signed Error	RMS Error
Ejection Fraction $\pm\sigma_{\mathrm{EF}}$	1.561 ± 0.712	2.974
Endocardium Area $\pm\sigma_{\mathrm{EA}}$	3.427 ± 3.81	4.146

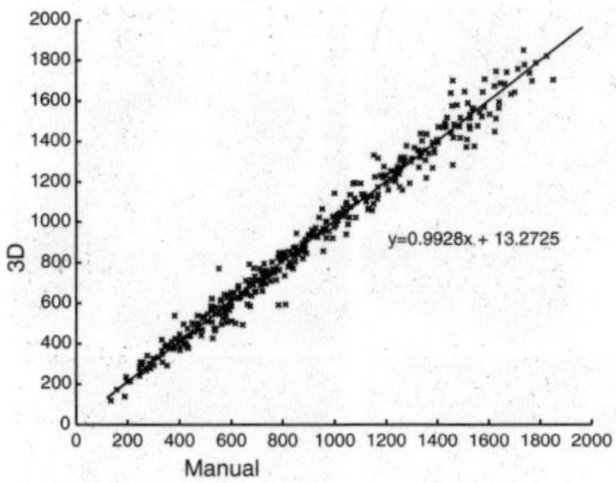

(a) Scatter plot to show the accuracy of 3D automatic segmentation

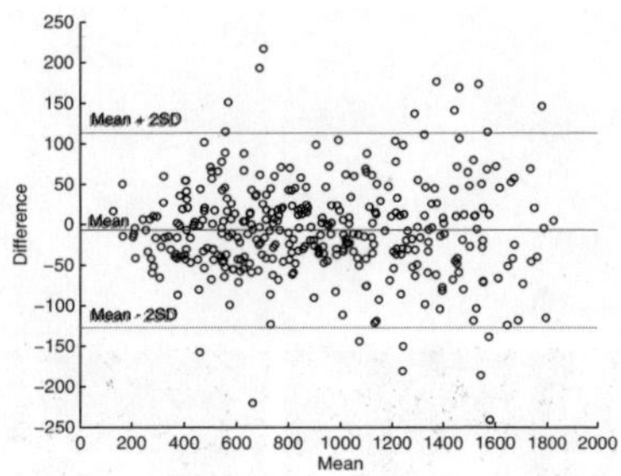

(b) Bland–Altman

Fig. 14. Accuracy results of the 3D segmentation

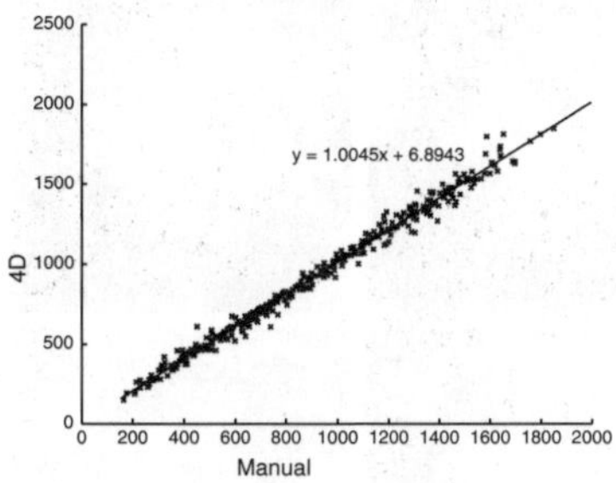

(a) Scatter plot to show the accuracy of accuracy of 4D automatic segmentation

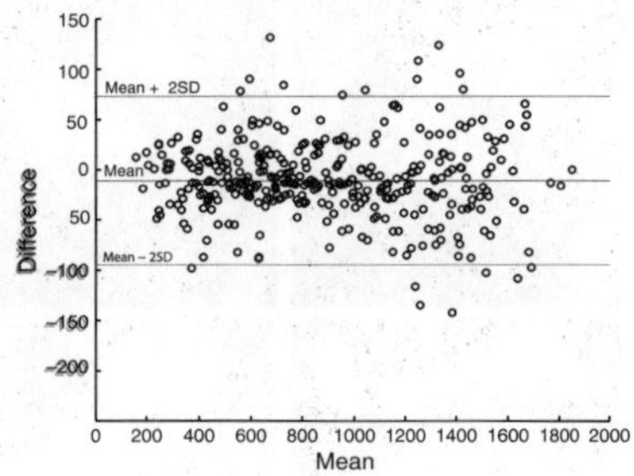

(b) Bland–Altman

Fig. 15. Accuracy results of the 4D segmentation

compared to the 4D. The Bland–Altman plot in Figure 15 is also not as widespread as the plot in Figure 15. This is attributed to the fact that the 4D segmentation has a higher resemblance to the manual segmentation especially at the apex and base of the heart where there are missing boundaries and confusing structures nearby and also explains why the 3D segmentation had a lower correlation value r = 0.96 with the manual segmentations, compared to the 4D correlation value of 0.98. It is also worth noting that watching an expert perform a manual segmentation, the user actually flips through the neighboring slices, in space and time, to help decide the location of the contour in places where it is ambiguous to do so.

Using the overlap method, the accuracy of the automatic segmentations were compared against the manual segmentation. Figure 13 also conveys that the 4D automatic segmentation is close to a manual segmentation performed by an expert. This means that the results of the 4D segmentation method can be used as an expert interpretation.

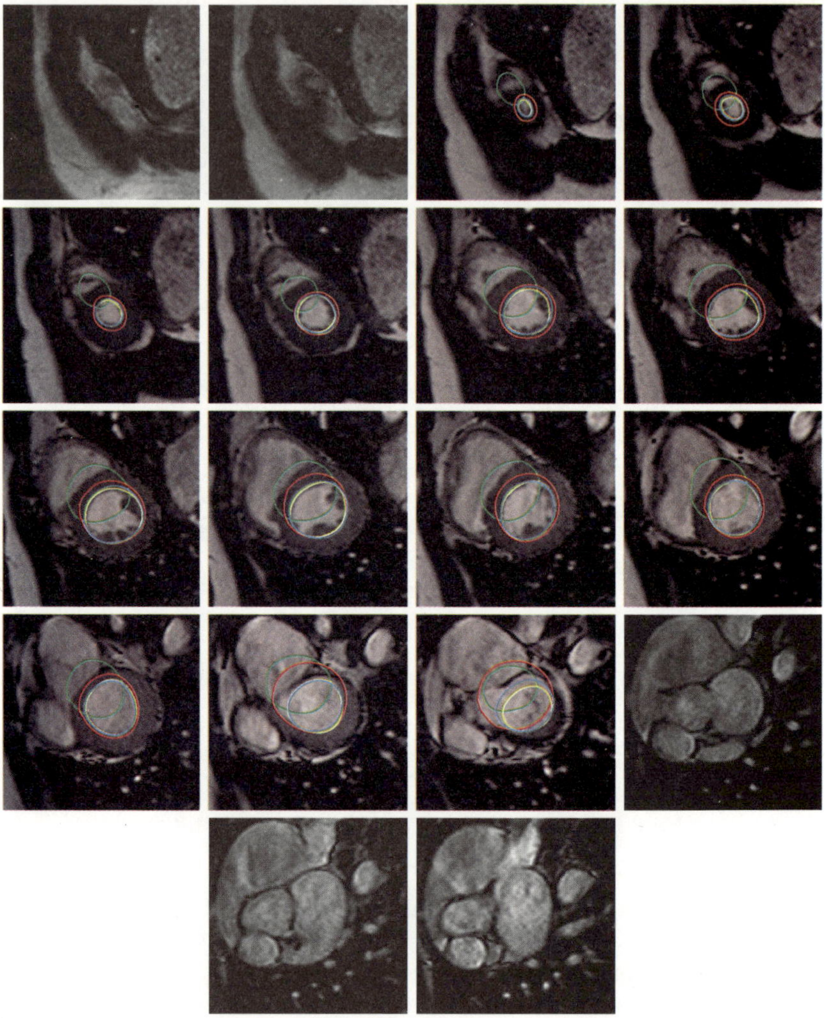

Fig. 16. The red curve shows the erroneous result of a classical 3D segmentation. The green curve shows the initial position of the segmentation which is less than optimal since it overlapped both ventricles. The location of the green curve overlapped both ventricles. This initial condition is difficult to overcome since the statistics of the pixels inside both ventricles are similar. The yellow curve depicts the result of a manual segmentation and considered as a gold standard. The red curve is larger and is more biased to include parts of the right ventricle. The blue curve is the result of a 4D segmentation with the same energy functional as the 3D. However, because the 4D has more information from temporally neighboring phases, the result was more accurate and in par with the manual segmentation

5 Conclusion

In this paper we determined the advantage gained from maximizing the amount of information available to a shape–based segmentation method. By including the time variable in the segmentation procedure, we were able to take advantage of the coherency of the data not only spatially across slices, but also temporally across time frames as well. Performing the segmentation using a 4D hyper–surface produced results that are closer to the results produced by a trained user. The benefit of including the time variable was isolated by performing a comparative study with a 3D segmentation procedure. The only difference between the two procedures was that the 3D was a pure spatial segmentation method, while the 4D included the temporal domain as well. Consistency results proved that the 4D segmentation method is more consistent. In fact, there was one instance where the 3D segmentation failed to define the boundaries of the endocardium. The data resembled a systolic state of the heart where the left ventricle has the minimum volume. The segmenting surface was confused and located a structure outside the left ventricle. This did not happen when the 4D segmentation was used since, in the 4D case, the entire data is available, from systolic to diastolic. Knowing about the other states of the endocardium, the 4D hyper–surface was able to avoid the confusion and produce a correct segmentation.

Accuracy tests also showed that segmentation in 4D performed better than in 3D. The correlation between the 4D results and the results of the manual segmentation was higher than the 3D case. This is attributed to the fact that the segmentation in 4D has a higher resemblance to the manual segmentation than the one in 3D. By utilizing information from all available dimensions, a 4D segmentation mimics a manual segmentation performed by an expert where the user actually looks at neighboring slices in space and also time to determine the location of the contour in places where the boundary is unclear.

References

1. *Merriam-Webster's Medical Dictionary.* Merriam-Webster Inc., 2002
2. J. Bland and D. Altman. Statistical methods for assessing agreement between two methods of clinical measurements. *Lancet*, pages 307–310, 1986
3. A. Chakraborty, L.H. Staib, and J.S. Duncan. An integrated approach to boundary finding in medical images. In *IEEE Workshop on Biomedical Image Analysis*, pages 13–22, 1994
4. T.F. Chan and L.A. Vese. Active contours without edges. *IEEE Transactions on Medical Imaging*, 10(2):266–277, 2001
5. T.F. Cootes, A. Hill, C. Taylor, and J. Haslam. Use of active shape models for locating structure in medical images. *Image and Vision Computing*, 12(6):355–366, 1994
6. A. Gupta, L. von Kurowski, A. Singh, D. Geiger, C.-C. Liang, M.-Y. Chiu, L. Adler, M. Haacke, and D. Wilson. Cardiac MR image segmentation using deformable models. In *Computers in Cardiology*, pages 747–750, 1993

7. T. Kapur, W.E.L. Grimson, R. Kikinis, and W.M. Wells. Enhanced spatial priors for segmentation of magnetic resonance imagery. *Medical Image Computing and Computer-Assisted Intervention (MICCAI)*, pages 457–468, 1998
8. M. Leventon. Statistical models in medical image analysis. PhD thesis, MIT, Cambridge, MA, 2000
9. M. Lorenzo-Valdes, G.I. Sanchez-Ortiz, A. Elkington, R. Mohiaddin, and D. Rueckert. Segmentation of 4d cardiac mr images using a probabilistic atlas and the em algorithm. *Medical Image Analysis*, 8(3):255–265, 2004
10. S. Osher and J. Sethian. Fronts propagating with curvature dependent speed: Algorithms based on Hamilton–Jacobi formulations. *Journal of Computational Physics*, 79:12–49, 1988
11. L.H. Saib. Prior shape models for boundary finding. In *International Symposium on Biomedical Imaging*, pages 30–33, 2002
12. J. Sethian. *Level Set Methods and Fast Marching Methods*, 2nd edn. Cambridge University Press, 1999
13. L.H. Staib and J.S. Duncan. Boundary finding with parametrically deformable models. *IEEE Transactions on Pattern Analysis and Machine Intelligence*, 14(11):1061–1075, 1992
14. G. Szekely, A. Kelemen, C. Brechbuhler, and G. Gerig. Segmentation of 3D Objects from MRI volume data using constrained elastic deformations of flexible fourier surface models. In *CVRMed95*, pages 495–505, 1995
15. A. Tasi, A. Yezzi, W.W. III, C. Tempany, D. Tucker, A. Fan, W.E. Grimson, and A. Wilsky. Model-based curve evolution technique for image segmentation. In *CVPR*, volume 1, pages I–463–I–468, 2001
16. A. Tasi, A. Yezzi, W. Wells, C. Tempany, D. Tucker, A. Fan, W.E. Grimson, and A. Willsky. A shape-based approach to the segmentation of medical imagery using level sets. *IEEE Transactions on Medical Imaging*, 22(2):137–154, 2003
17. A. Tsai, A. Yezzi, and A. Willsky. Curve evolution implemntation of the mumford–shah functional for imafe segmentation, denoising, interpolations and magnification. *IEEE Transactions on Image Processing*, 10(8):1169–1186, August 2001
18. A. Tsai, A. Yezzi, and A.S. Willsky. A curve evolution approach to smoothing and segmentation using the mumford-shah functional. In *IEEE Conference on Computer Vision and Pattern Recognition*, volume 1, pages 119–124, 2000
19. Y. Wang and L.H. Staib. Boundary finding with prior shape and smoothness. *IEEE Transactions on Pattern Analysis and Machine Intelligence*, 22(2):738–743, 2000
20. A. Yezzi, L. Zllei, and T. Kapur. A variational framework for joint segmentation and registration. In *Mathematical Methods in Biomedical Image Analysis*, pages 44–51, 2001

Measuring Similarity Between Trajectories of Mobile Objects

Sigal Elnekave[1], Mark Last[2] and Oded Maimon[3]

[1] Department of Information Systems Engineering
Ben Gurion University of the Negev, Israel
elnekave@bgu.ac.il

[2] Department of Information Systems Engineering
Ben Gurion University of the Negev, Israel
mlast@bgu.ac.il

[3] Department of Industrial Engineering
Tel Aviv University, Israel
maimon@eng.tau.ac.il

Summary. With technological progress we encounter more available data on the locations of moving objects and therefore the need for mining moving objects data is constantly growing. Mining spatio-temporal data can direct products and services to the right customers at the right time; it can also be used for resources optimization or for understanding mobile patterns. In this chapter, we cluster trajectories in order to find movement patterns of mobile objects. We use a compact representation of a mobile trajectory, which is based on a list of minimal bounding boxes (MBBs). We introduce a new similarity measure between mobile trajectories and compare it empirically to an existing similarity measure by clustering spatio-temporal data and evaluating the quality of resulting clusters and the algorithm run times.

1 Introduction

With technological progress, more data is available on the location of moving objects at different times, either via GPS technologies, mobile computer logs, or wireless communication devices. This creates an appropriate basis for developing efficient new methods for mining moving objects.

Spatio-temporal data can be used for many different purposes. The discovery of patterns in spatio-temporal data, for example, can greatly influence such fields as animal migration analysis, weather forecasting, and mobile marketing. Clustering spatio-temporal data can also help in social networks discovery, which is used in tasks like shared data allocation, targeted advertising, and personalization of content and services.

Bennewitz et al. [4] learn motion patterns of persons in order to estimate and predict their positions and used them for improving the navigation

S. Elnekave et al.: *Measuring Similarity Between Trajectories of Mobile Objects*, Studies in Computational Intelligence (SCI) **91**, 101–128 (2008)
www.springerlink.com

behavior of a mobile robot. Krumm and Horvitz [16] describe a method called Predestination that uses the history of a driver's destinations, along with data about driving behaviors, in order to predict where a driver is going in the progress of his/her trip. This can reduce the cognitive load on the driver by eliminating information about places that he or she is unlikely to visit. Destination prediction can also be used to detect if a user is deviating from the route to an expected location.

Niculescu and Nath [24] present the Trajectory Based Forwarding (TBF) technique, a novel method to forward packets in a dense ad hoc network that makes it possible to route a packet along a predefined curve. The trajectory is set by the source (in a parametric function form), but the forwarding decision is local and greedy. TBF requires that nodes know their position relative to a coordinate system. TBF is suggested as a layer in position centric ad hoc networks as a support for basic services: routing (unicast, multicast, multipath), broadcasting, and discovery.

Peng and Chen [27] search for user moving patterns in a mobile computing environment to develop data allocation schemes that can improve the overall performance of a mobile system.

In Zhang and Gruenwald [39] a novel location management scheme is provided, using spatial and temporal context-aware profiles of mobile users. The unique feature is to subset mobility profiles under a given spatial and temporal mobility context. The reduction of the candidate list length of mobility profile and the improvement of its accuracy should reduce paging cost and latency. They also propose a geography-based method to generate initial profiles and an incremental and streamed location update method for online profile updating while reducing storage requirement. Baggio et al. [2], present a location service that supports tracking mobile objects, partly by dynamically adapting to the mobile behavior of each object separately. Current location services have limited scalability due to poor exploitation of locality and ineffective caching. Caching, in the presence of mobility, requires the identification of boundaries of the region within which a mobile object usually remains. Caching a reference to such a region rather than to the object itself ensures that the cached entry remains stable.

Odoi et al. [25] investigate the geographical and temporal distribution of human giardiasis in Ontario in order to identify possible high risk areas and seasons. Two spatial scales of analyses and two disease measures are used to identify geographical patterns of giardiasis in Ontario. Global Moran's I and Moran Local Indicators of Spatial Associations were used to test for evidence of global and local spatial clustering, respectively. The study identified spatial and temporal patterns in giardiasis distribution. This information is important in guiding decisions on disease control strategies. The study also showed that there is a benefit in performing spatial analyses at more than one spatial scale to assess geographical patterns in disease distribution and that smoothing of disease rates for mapping in small areas enhances visualization of spatial patterns.

Previous work on mining spatio-temporal data includes querying data using special indexes for efficient performance, recognizing trajectory patterns, and clustering trajectories of closely moving objects as 'moving micro-clusters'. More extensive research has been done on spatial data and temporal data separately, including periodicity that has been studied only with time series databases. But the field of spatio-temporal mining is relatively young, and requires much more research. In this chapter we incrementally build a trajectory by pre-processing an incoming spatio-temporal data stream. We first generate a synthetic spatio-temporal dataset to be used as a benchmark for testing and evaluating the proposed algorithm. Due to the sensitivity of this kind of data, it was not possible to obtain a real-world spatio-temporal dataset for research purposes. Next, we use a compact representation of a spatio-temporal trajectory and define an algorithm for incrementally building it. Then we define a new, data-amounts-based similarity measure between trajectories according to proximity of trajectories in time and space. This measure allows the discovery of groups that have similar spatio-temporal behavior. Finally we evaluate the proposed similarity measure by conducting experiments on a synthetic data stream. We compare trajectory clusters built using our suggested similarity measure to clusters based on the minimal distances similarity measure.

2 Related Work

2.1 Spatio-temporal data collection

Several techniques for tracking moving objects are described in literature. Thus, in [35] the computer science researchers have created TurtleNet, a network of postcard-sized waterproof computers that have been attached to the shells of turtles with a combination of orthodontic cement and duct tape. The computers are lightweight and do not weigh down the turtles so that the gadgets do not disturb their mating habits. The devices are designed to periodically note the location and body temperature of the turtles, and when they come within a tenth of a mile of each other they swap information, which helps extend the battery life of the computers. The units also feature solar panels to recharge the batteries. The relay of information between turtles ends when they pass a single base station, where the data is accumulated before it is transmitted back to the UMass-Amherst campus about 15 miles away.

The present Lower Manhattan Security Initiative [23] will include not only license plate readers but also 3,000 public and private security cameras below Canal Street, as well as a center staffed by the police and private security officers, and movable roadblocks. The license plate readers will check the plates numbers and send out alerts if suspicious vehicles are detected. The city is already seeking the state approval to charge drivers a fee to enter Manhattan below 86th Street, which will require the use of license plate readers. If the plan is approved, the police will most likely collect information from those readers too. Unlike the 250 or so cameras the police have already placed in high-crime areas throughout the city, which capture moving images that have to be downloaded, the security initiative cameras would transmit live information instantly. The Police Department is still considering whether to use face-recognition technology or biohazard detectors in its Lower Manhattan network.

Cohen and Medioni [8] address the problem of detecting and tracking moving objects in a video stream obtained from a moving airborne platform. The framework proposed is based on a graph representation of the moving regions extracted from a video acquired by a moving platform. The integration of detection and tracking in this graph representation allows to dynamically infer a template of all moving objects in order to derive a robust tracking in situations such as stop and go motion and partial occlusion. Finally, the quantification of the results provides a confidence measure characterizing the reliability of each extracted trajectory.

2.2 Representing spatio-temporal data

Wang et al. [37] and Mamoulis et al. [21] use a set of time series of locations, one for each user, where a time series contains triplets (t, x, y). For simplicity, they assume that all user locations are known at every time point and that the time interval between every t and $t + 1$ is fixed, as opposed to Braz et al. [5] who assumed that observations are taken at irregular rates for each object and that there is no temporal alignment between the observations of different objects. In Nehme and Rundensteiner [22] moving objects are assumed to move in a piecewise linear manner on a road network. Their movements are constrained by roads, which are connected by network nodes. The updates of moving objects location arrive in the form of *o.ID, o.Location, o.t, o.Speed, o.destination(position of the connection node), o.Attributes.* In Ma et al. [19] and Bakalov et al. [3], the user moving history is an ordered (c, t) list where c is the cell ID and t is the time when the object reaches the cell c.

Representing spatio-temporal data in a more concise manner can be done by converting it into a trajectory form. In Hwang et al. [14] and in Pelekis et al. [26], a trajectory is a function that maps time to locations. To represent object movement, a trajectory is decomposed into a set of linear functions, one for each disjoint time interval. The derivative of each linear function yields the direction and the speed in the associated time interval. A trajectory is a disjunction of all its linear pieces. For example, a trajectory of the user moving on a 2-D space may consist of the following two linear pieces:

$$[(x = t - 3) \cap (y = t + 3) \cap (0 < t < 2)] \bigcup [(x = 6) \cap (y = -t) \cap (3 < t < 5)].$$

In Braz et al. [5], Lee et al. [17], Pfoser [28], Pfoser et al. [29], Rasetic et al. [31], a linear interpolation is used. The sampled positions then become the endpoints of line segments of polylines and the movement of an object is represented by an entire polyline in 3D space. A trajectory T is a sequence $< (x_1, y_1, t_1), (x_2, y_2, t_2), ..(x_k, y_k, t_k) >$. Objects are assumed to move straight between the observed points with a constant speed. The linear interpolation seems to yield a good tradeoff between flexibility and simplicity.

In Li et al. [18], the objects are also assumed to move in a piecewise linear manner. Namely, an object moves along a straight line with some constant speed till it changes its direction and/or speed. If an object deviates significantly from the expected position, it is responsible for reporting the new velocity. Each moving object o (in the 2-D space) is represented by a 5-tuple $(x_o, y_o, v_{x_o}, v_{y_o}, t_o)$. It is also called the profile of moving object o, because it uniquely determines the track of o.

D'Auria, et al. [9] use a set of triples *(id; loc; t)*. Starting from the set of triples for a given object id it is therefore possible, in principle, to approximate a function $fid : time \rightarrow space$, which assigns a location to an object id for each moment in a given time interval. Such a function is called a trajectory.

In Niculescu and Nath [24] the trajectory is expressed in the parametric form $X(t); Y(t)$. For example, to move along a line with slope α passing through the source with coordinates (x_1, y_1), the trajectory would be described by $X(t) = x_1 + tcos(\alpha)$; $Y(t) = y_1 + tsin(\alpha)$. α, x_1, y_1 are constants, and the parameter t actually describes the Euclidean distance traveled along the line.

In Porkaew et al. [30] motion information of objects is represented in the database using motion parameters and location functions which compute the spatial position of an object at any time. A location function corresponds to a type of motion. Motion parameters specify the instantiation of the location function, e.g., a starting location and speed of motion in the linear translation. As an example, consider an object that translates linearly with constant velocity in 2-dimensional space. The motion parameters in this case correspond to the object's starting location (x_s, y_s) at some initial time t_s and its velocity $(v_x; v_y)$ along the spatial dimensions. Using these parameters, the location of the object at any future time $t > t_s$ can be determined. The authors are primarily concerned with object trajectories that correspond to linear motion with a constant speed. As time progresses, the motion of an object may deviate from its representation in the database. When the deviation exceeds a threshold, an object updates its motion parameters and/or the location function stored in the database to reflect its most current motion information.

In Peng and Chen [27] a movement log contains pairs of $(oldVLR$, $newVLR)$, where VLR is a visitor location register representing the time of the visitor's presence within a location. In the beginning of a new path, the old VLR is null. For each mobile user, a moving sequence can be obtained from the movement log. Each node in the network topology of a mobile computing system can be viewed as a VLR and each link is viewed as a connection between $VLRs$.

A set of maximal moving sequences is represented as a string (e.g., ABCDHG), where each letter represents a node.

2.3 Spatio-temporal data summarization

In Baggio et al. [2], a location service for tracking and locating mobile objects is based on a hierarchical organization of the network into domains, similar to the organization of domains in the Domain Name System (DNS). Domains divide the underlying network into geographical, administrative, or network-topological areas. The network is subdivided into domains that are aggregated into larger, non overlapping domains. Each domain is represented by a node in the tree. The root node represents the entire network. For example, a lowest-level domain (called a leaf domain) may represent the network of a city, whereas the next higher-level domain represents the country or state in which this city is located. The highest-level domain represents the entire wide-area network, such as the Internet.

Goh and Taniar [13] perform a minor data analysis and summary at the mobile node level before the raw data is processed by the data mining machine (at the server). By the time the mobile object data arrives to the data mining machine, it will be in the form of summary transactions, which reduce the amount of further processing required to perform data mining. Performance evaluation shows that this model is significantly more efficient than the traditional model to perform mobile data mining. Yet, in this model the summarization is per mobile node, implying that only times are summarized, but there is no summarization of locations for multiple nodes, since each node represents a single location.

Anagnostopoulos et al. [1] use segmentation that leads to simplification of the original objects into smaller, less complex primitives that are better suited for storage and retrieval purposes. Different distance-based segmentation algorithms are presented that operate at various scales of computational granularities. The distance-based segmentation criterion attempts to create minimal bounding rectangles (MBRs) that bound close data points into rectangular intervals in such a way that the original pairwise distances between all trajectories are preserved as much as possible. A variance-based hybrid variation is presented that can provide a compromise between running time and approximation quality, with running times of $O(m^2nlogn)$ in the worst case.

Rasetic et al. [31] split spatio-temporal trajectories in order to improve the performance of queries using MBR-based access structures to index these trajectories. They argue that splitting trajectories with the goal of minimizing the volume of the resulting MBRs alone is not the best strategy, and suggest taking into account average query sizes. They derive a linear time splitting algorithm that consistently outperform other previously proposed policies (5-6 times less disk I/Os).

Faloutsos et al. [11] split trajectories incrementally whenever the latest point increases the marginal cost in the current sub-trail.

Braz et al. [5] discuss how data warehousing technology can be used to store aggregate information about trajectories and perform OLAP operations over them. They define a data cube with spatial and temporal dimensions, discretized according to a regular grid. For simplicity, they use a regular subdivision of space and time, but they mention that they could extend the model with cells consisting of more complex areas considered at different time intervals.

Ma et al. [19] discretize the time attribute of the moving histories using clustering. For each cell c in the cell set, all the elements in the moving history database D are collected, and the CURD algorithm (Clustering Using Reference and Density) is used to cluster the element set of each cell, where Euclidean distance at time t is used as the similarity function between two elements in a cell. Clustering results are triples of cell, start time, and end time, implying that the mobile user often enters cell c during the period $[T_s, T_e]$. The elements in the dataset are replaced with the clusters that they belong to, and the data is represented as moving sequences.

Bakalov et al. [3] utilize the Piecewise Aggregate Approximation (PAA) technique and then transform PAA into a string. The PAA accepts as input a time-series of length n and produces as output an approximation of reduced size m. The algorithm divides the input sequence into m equi-sized "frames" and replaces the values contained in each frame using the average of these values. The advantage of PAA is that the length of the reduced time-series m can be chosen at will, thus the accuracy of the resulting approximation can be tuned easily. As a second step, the PAA approximation can be discretized using a symbolic representation based on a uniform grid and assigning a unique symbol to every partition of the grid.

Cao et al. [6] propose to trade off accuracy for efficiency in order to solve data size problems, using line simplification. The goal is to approximate a polygonal curve by another polygonal curve which is "not very far" from the original, but has fewer points. Since there is no a single distance that is sound for the four spatio-temporal query types studied, a combination of distances is used for simplifying a line. The authors also handle the aging of trajectories by producing increasingly compact approximations of trajectories over time. Experimental results show that the optimal algorithm is several times faster than the DP (Douglas-Peuker) heuristic, which is the best known and studied heuristic-based algorithm for finding an optimal line simplification.

2.4 Querying spatio-temporal data

Porkaew et al. [30] explore techniques for indexing and query processing over spatio-temporal databases that store mobile objects. The primary difficulty in developing query processing techniques arises from the fact that the location of an object changes continuously as a function of time without an explicit update to the database. They explore algorithms for selection queries. Selection queries may be classified as either range or nearest neighbor queries. In range queries, objects that fall within a given spatio-temporal range are retrieved. In the nearest neighbor queries, we search for objects that are relatively closer to the query point. The notion of proximity differs between time and spatial dimensions. In the temporal sense, proximity means co-occurrence within a certain time period. Spatially, it refers to geographical/geometrical closeness. Thus, it is highly desirable to separate the selection predicate into spatial and temporal components, each of which can be specified either as a k-Nearest Neighbor or as a Range predicate.

Pfoser et al. [29] present a set of spatio-temporal queries. A typical search on sets of objects trajectories includes a selection with respect to a given range, a search inherited from spatial and temporal databases. An important query type in temporal databases is the time-slice query, to determine the positions of (all) moving objects at a given time point in the past. In addition, novel queries become important due to the specific nature of spatio-temporal data. Trajectory-based queries are classified in "topological" queries, which involve the whole information of the movement of an object in order to determine whether an object enters, crosses, or bypasses a given area, and "navigational" queries, which involve derived information, such as speed and heading. The two types of spatio-temporal queries are coordinate-based queries, such as point, range, and nearest-neighbor queries in the resulting 3D space and trajectory-based queries, involving the topology of trajectories (topological queries) and derived information, such as speed and heading of objects (navigational queries).

Bakalov et al. [3] consider how to efficiently evaluate trajectory joins, i.e., how to identify all pairs of similar trajectories between two datasets. They represent an object trajectory as a sequence of symbols (i.e., a string). Based on special lower-bounding distances between two strings, they propose a pruning heuristic for reducing the number of trajectory pairs that need to be examined.

2.5 Indexing trajectories of moving objects

Pfoser [28] reviews existing methods for indexing trajectories. He groups these approaches according to three movement scenarios: constrained movement, unconstrained movement, and movement in networks. Unconstrained movement brought the definition of new access methods. In order to obtain information of the entire movement path (not only about the known positions, but also between them), we need to interpolate. The simplest approach is the linear interpolation. The sampled positions then become the endpoints of the segments of a polyline and the object's movement is represented by an entire polyline in the 3D space. Trajectories can be indexed using spatial access methods, but they are decomposed into their constituent line segments, which are then indexed. R-tree approximates the data objects by Minimum Bounding Boxes (MBB), which are 3D intervals.

The algorithms for selection queries suggested in Porkaew et al. [30] are developed under two different representation and indexing techniques: native space and parametric space indexing (NSI/PSI). While NSI indexes motion in the original space in which it occurs, the PSI strategy indexes it in the space defined by the motion parameters (start location and velocity). NSI preserves the locality of objects but also indexes "dead space" in the non-leaf levels of the index tree. PSI uses the motion parameters to represent an object but suffers from the fact that objects that are close in parametric space might be somewhat divergent in real space and that objects that are far away in parametric space may be at some time very close to each other. Additionally, more complex types of motion (e.g., with acceleration) cannot be easily represented. NSI outperforms PSI according to the experiments exhibited in that paper.

Pfoser et al. [29] describe the R-tree, a height-balanced tree with the index records in its leaf nodes containing pointers to actual data objects. Leaf node entries are of the form (id, MBB), where id is an identifier that points to an actual object and MBB (Minimum Bounding Box) is an n-dimensional interval. Non-leaf node entries are of the form (ptr, MBB), where ptr is the pointer to a child node and MBB is the covering n-dimensional interval. A node in the tree corresponds to a disk page. Every node contains between m and M entries. The insertion of a new entry into the R-tree is done by traversing a single path from the root to the leaf level. In case an insertion causes splitting of a node, its entries are reassigned to the old node and a newly created one. In case the deletion causes an underflow in a node, i.e., node occupancy falls below m, the node is deleted and its entries are re-inserted. When searching an R-tree, we check whether a given node entry overlaps the search window. If it does, we visit the child node and thus recursively traverse the tree. Since overlapping MBBs are permitted, at each level of the index there may be several entries that overlap the search window. Efficient processing of trajectory-based and combined queries requires indices and access methods for spatio-temporal data; therefore the writers propose the STR-tree and the TB-tree for indexing the trajectories of moving point objects. The STR-tree is an extension of the R-tree to support efficient query processing of trajectories of moving points. The two access methods differ in their insertion/split strategy. The insertion strategy of the R-tree is based on the least enlargement criterion. On the other hand, insertion in the STR-tree not only considers spatial closeness, but also partial trajectory preservation. The authors try to keep line segments belonging to the same trajectory together. As a consequence, when inserting a new line segment, the goal should be to insert it as close as possible to its predecessor in the trajectory. The TB-tree takes a more radical step. An underlying assumption when using an R-tree is that all inserted geometries are independent. In the context of [29] this translates to all line segments being independent. However, line segments are parts of trajectories and this knowledge is only implicitly maintained in the R-tree and the STR-tree structures. The TB-tree provides an access method that strictly preserves trajectories so that a leaf node only contains segments belonging to the same trajectory and an index is interpreted as a trajectory bundle. The TB-tree access method proves to be well suited for trajectory-based queries, and it also has a good spatial search performance.

Bakalov et al. [3] present an algorithm and an index structure for efficiently evaluating trajectory join queries. They propose a technique that uses symbolic trajectory representations to build a very small index structure that can help evaluate approximate answers to the join queries. Then, by using a post filtering step and loading only a small fraction of the actual trajectory data, the correct query results can be produced. Their techniques utilize specialized lower bounding distance functions on the symbolic representations to prevent false dismissals.

2.6 Clustering moving objects and trajectories

Li et al. [18] study the problem of clustering moving objects. Clustering analysis had been studied successfully before on static data [15], but without the support of spatio-temporal information. There are solutions that use moving micro-clusters (MMC) for handling very large datasets of mobile objects. A micro-cluster denotes a group of objects that are not only close to each other at current time, but are also likely to move together for a while. In principle, those moving micro-clusters reflect some closely moving objects, naturally leading to high quality clustering results. The authors of [18] propose incremental algorithms to keep the moving micro clusters geometrically small by identifying split events when the bounding rectangles reach some boundary and by using knowledge about collisions between the MMCs (splitting or merging MMCs when those events occur). In experiments conducted on synthetic data with the K-Means as the generic algorithm used in micro-clustering, MMCs showed improvement in running times compared to NC (normal clustering), though with a slight deterioration in performance.

The problem of trajectory clustering is also approached at D'Auria, et al. [9]. They propose clustering trajectory data using density-based clustering, based on distance between trajectories. Their OPTICS system uses reachability distance between points and presents a reachability plot showing objects ordered by visit times *(X)* against their reachability measure *(Y)*, allowing the users to see the separation to clusters to decide on a separation threshold. The authors of [9] propose to cluster patterns by the temporal focusing approach to improve the quality of trajectories. Some changes to OPTICS are suggested, by focusing on the most interesting time intervals instead of examining all intervals, where the interesting intervals are those with the optimal quality of the obtained clusters. A comparison between K-Means, three versions of hierarchical agglomerative clustering, and the trajectory version of OPTICS shows that OPTICS improves purity, with a decrease in completeness.

In Nehme and Rundensteiner [22], the SCUBA algorithm is proposed for efficient cluster-based processing of large numbers of spatio-temporal queries on moving objects. The authors describe an incremental cluster formation technique that efficiently forms clusters at run-time. Their approach utilizes two key thresholds, distance and speed. SCUBA combines motion clustering with shared execution for query execution optimization. Given a set of moving objects and queries, SCUBA groups them into moving clusters based on common spatio-temporal attributes. To optimize the join execution, SCUBA performs a two-step join execution process by first pre-filtering a set of moving clusters that can produce good results in the join-between moving clusters stage and then proceeding with the individual join-within execution on those selected moving clusters. Experiments show that the performance of SCUBA is better than the traditional grid-based approach where moving entities are processed individually.

2.7 Spatio-temporal group patterns mining

The problem of mining group patterns by deriving grouping information of mobile device users based on the spatio-temporal distances between them is studied by Wang et al. [37]. They define a valid segment as a set of consecutive time points $[t, t+k]$ that group users which are not farther than a maximal threshold distance from each other during the time of the interval and some of the users are farther than that at the time *t-1*. Two algorithms are presented for this task: AGP (Apriori-like algorithm for mining valid group patterns) and VG-growth (valid group graph data structures, built from a set of valid 2-groups). Evaluation on a synthetic dataset shows that VG growth outperforms AGP, especially when the minimum weight (used for validating a pattern) becomes smaller. Both algorithms spend most time on searching for valid groups of two elements, which is the main cost of mining here.

The problem of deriving group information on mobile devices based on the trajectory model is approached in Hwang et al. [14]. The trajectory model was chosen for saving the storage space and coping with untracked location data. A trajectory T is a set of linear functions, each of which maps from a disjoint time interval to an n-dimensional space. The problem is to find all valid mobile groups given minimal weight and duration thresholds, maximal distance threshold, and trajectory-based dataset. A valid mobile group exists when the weight of a mobile group exceeds a minimum weight threshold. In order to find all valid mobile groups under such a model the authors present two algorithms. Performance evaluation comparing the two algorithms shows that TVG growth (based on VG-growth) outperforms TAGP (based on AGP) in terms of running times.

The problem of mining trajectory patterns from a set of imprecise trajectories is studied by Yang and Hu [38]. They define the concept of a pattern group to represent trajectory patterns in a concise manner. Since the Apriori property no longer holds on the trajectory patterns, a new mean-max property is identified for trajectory patterns, and it is the basis for the developed TrajPattern algorithm that mines trajectory patterns by first finding short patterns and then extending them in a systematic manner.

Lee et al. [17] present a method for mining temporal pattern of moving objects with temporal and spatial components, i.e. finding significant patterns from users' location information that also change over time. They define moving pattern as the frequent regularity of location change over time. Their work consists of four stages. First, the database is arranged into object identifier and valid time. Second, using spatial operation, moving objects' location information is transformed into area codes in order to discover significant information. Each area includes in its scope several locations. Third, time constraint is imposed to extract effective moving sequence. Finally, the frequent moving patterns are extracted from the generated moving sequences. Results show that as the length of the moving sequence grows, especially when the length exceeds 10, the execution time increases, thus indicating the need for developing efficient algorithms and improved storage capacity.

In Ma et al. [19] the mining of mobile user maximal moving sequential patterns is done using PrefixTree algorithm, by first scanning the database once to find the set of frequent items, and ordering them according to their values of time attribute and then, computing candidate consecutive items in terms of the after relation. Afterwards scanning the database again to generate frequent length-2 moving sequences, where the candidate frequent length-2 moving sequences are generated based on candidate consecutive items. Based on the frequent length-2 moving sequences, candidate consecutive items are reduced again based on Pseudo-Apriori property. Then, scanning the database to construct prefix trees based on candidate consecutive items and Pseudo-Apriori property, which is the last time to scan the database. Next, generating moving sequential patterns based on the prefix trees and finally, generating maximal moving sequential patterns. The performance study shows that the PrefixTree is more efficient and scalable than the Revised LM (for identifying large moving sequences).

Bennewitz et al. [4] present a method for learning and utilizing the motion patterns of persons. Their approach applies the expectation maximization (EM) algorithm to cluster trajectories recorded with laser-range sensors into a collection of motion patterns. Furthermore, they introduce a method for automatically deriving HMMs (hidden Markov models) from these typical motion patterns of persons. To update the resulting HMMs based on laser-range data and vision information, they apply JPDAFs (joint probabilistic data association filters). Practical experiments carried out with a mobile robot in different environments demonstrate that their method is able to learn the typical motion patterns of persons and to reliably use them for maintaining a probabilistic belief about the current positions of persons. Furthermore, they presented experiments illustrating that the behavior of a mobile robot can be improved by predicting the motions of persons based on learned motion patterns.

2.8 Incremental maintenance of mobile patterns

In Peng and Chen [27] a new data mining algorithm is presented which involves incremental mining for user moving patterns in a mobile computing environment and utilizes the mining results to develop data allocation schemes and, subsequently, to improve the overall performance of a mobile system. Algorithms are proposed to capture the frequent user moving patterns from a set of log data in a mobile environment (MM for determining maximal moving sequences and LM to identify large moving sequences). The algorithms are incremental and are able to discover new moving patterns without compromising the quality of obtained results.

Cheng et al. [7] investigate the issues of incremental mining of sequential patterns in large databases and address the inefficiency problem of mining the appended database from scratch. Several novel ideas are introduced in their proposed algorithm IncSpan: (1) maintaining a set of "almost frequent" sequences as the candidates in the updated database, and (2) two optimization techniques, reverse pattern matching and shared projection, are designed to improve the performance. Reverse pattern matching is used for matching a sequential pattern in a sequence, and prunes some search space. Shared projection is designed to reduce the number of database projections for some sequences which share a common prefix. Experiments show that IncSpan outperforms the non-incremental method (using PrefixSpan) and an incremental mining algorithm ISM (Incremental and interactive sequence mining).

Ma et al. [20] study a fast incremental updating technique for maintenance of maximal moving patterns. They propose the "PrefixTree+" algorithm that uses the previous mining results to improve the mining efficiency. Its novelty is materializing prefix trees and using the lemma that any moving sequence that is potentially frequent with respect to a database must occur as a frequent moving sequence in at least one of its partitions. In this technique, maximal moving sequential patterns are stored in prefix trees and new moving sequences can be combined with the existing patterns. A performance study has shown that this algorithm is more efficient and scalable than "LM" and "PrefixTree".

2.9 Predicting spatio-temporal data

While traditional methods for prediction in spatio-temporal databases assume that objects move according to linear functions, in practice, individual objects may follow non-linear motion patterns. Tao et al. [34] introduce a general framework for monitoring and indexing moving objects, where first, each object computes individually the function that accurately captures its movement and second, a server indexes the object locations at a coarse level and processes predictive queries that forecast the objects that will qualify a spatial condition at some future time based on the current knowledge, using a filter-refinement mechanism. A novel recursive motion function is suggested that supports a broad class of non-linear motion patterns. The function does not presume any a-priori movement but can postulate the particular motion of each object by examining its locations at recent timestamps. An indexing scheme is suggested that facilitates the processing of predictive queries without false misses.

Krumm and Horvitz [16] describe a method called Predestination that uses a history of a driver's destinations, along with data about driving behaviors, to predict where a driver is going as a trip progresses. Driving behaviors include types of destinations, driving efficiency, and trip times. Four different probabilistic cues are considered and combined in a mathematically principled way to create a probability grid of likely destinations. The authors introduce an open-world model of destinations that helps the algorithm to work well in spite of a paucity of training data at the beginning of the training period by considering the likelihood of users visiting previously unobserved locations based on trends in the data and on the background properties of locations. The best performance on 3,667 different driving trips gave an error of two kilometers at the trip's halfway point.

2.10 Spatio-temporal similarity measures

Anagnostopoulos et al. [1] define the distance between two trajectory segmentations at time t as the distance between the rectangles at time t. Formally:

$$d\left(s\left(T_i\right), s\left(T_j\right), t\right) = \min_{x_i \in P(s(Ti),t), x_i \in P(s(Ti),t)} d\left(x_i, x_j\right) \tag{1}$$

Finally, the distance between two segmentations is the sum of the distances between them at every time instant:

$$d\left(s\left(T_i\right), s\left(T_j\right)\right) = \sum_{t=0}^{m-1} d\left(s\left(T_i\right), s\left(T_j\right), t\right) \tag{2}$$

The distance between the trajectory MBRs is a lower bound of the original distance between the raw data, which is an essential property for guaranteeing correctness of results for most mining tasks.

In D'Auria et al. [9] the similarity of trajectories along time is computed by analyzing the way the distance between the trajectories varies. More precisely, for each time instant they compare the positions of moving objects at that moment, thus aggregating the set of distance values. The distance between trajectories is computed as the average distance between moving objects.

Vlachos et al. [36], suggest the use of non metric similarity functions based on the Longest Common Subsequence (LCSS) that give more weight to similar parts of the sequence and are robust to noise that usually appears in the case of trajectories in two or three dimensional space. This measure allows stretching sequences over time (different speeds) and translating the sequence in space (similar motion in different space regions). Comparing these new methods to

Euclidean and Dynamic Time Warping distance functions (DTW) has shown the superiority of the suggested approach, especially under strong presence of noise.

In Li et al. [18] the similarity of objects within a moving micro cluster is measured by distance on profiles of objects. Basically, similar objects are expected to have similar initial locations and velocities. Specifically,

$$dist(o1, o2) = (x_{o_1} - x_{o_2})^2 + (y_{o_1} - y_{o_2})^2 + (\alpha \cdot (v_{x_{o_1}} - v_{x_{o_2}}))^2 + (\alpha \cdot (v_{y_{o_1}} - v_{y_{o_2}}))^2$$

where $\alpha(\alpha > 1)$ is a weight associated with the velocity attributes since it plays a more important role than the initial locations in determining the spatial distances between o_1 and o_2 in the future.

Pelekis et al. [26] define a set of trajectory distance operators based on the primitive (space and time) as well as the derived parameters of trajectories (speed and direction). The Locality In-between Polylines (LIP) distance operator is proposed. Intuitively, two moving objects are considered spatially similar when they move close at the same location irrespective of time and direction. The idea is to calculate the area of the shape formed by two 2D polylines. Since the LIP distance operator does not include the notion of time, the spatiotemporal LIP (STLIP) distance operator is proposed, according to which two moving objects are considered similar in both space and time when they move close at the same time and location. STLIP is calculated by multiplying LIP by a factor having values greater than 1, that is calculated according to the maximum duration of the temporal element (within a time window) and the durations of the two meeting trajectories. The authors also suggest two variations of this operator that enhance their distance functions by taking into consideration the rate of change (speed, acceleration) and directional characteristics of the trajectories (turn).

2.11 Spatio-temporal data generation

The problem of generating synthetic spatio-temporal data in order to create available datasets for research purposes is approached by Giannotti et al. [12], where the CENTRE system is provided (CEllular Network Trajectories Reconstruction Environment). Their aim is to generate benchmark datasets for cellular devices positioning data, which is not publicly available for scientific research due to privacy concerns. Spatio-temporal data is represented as a set of records of the form *objectid*, *antennaid*, *time*, *distance from zeropoint*. The system aims at simulating semantic-based movement behaviors using a set of user-specified parameters and allows adding user preferences which may influence random distributions or domain semantics like cartography or geographic constraints. CENTRE is composed of 3 modules. The first, Synthetic Trajectories Generation, is based on the Generator for Spatio-Temporal Data (GSTD) and generates objects simulating people movements. The second,

Logs Generation, simulates the cell phone detection by the network and produces the position log. The third, Approximated Trajectories Reconstruction, reconstructs trajectories from the logs, considering the approximation of the data (smaller and denser antennas typically produce a better approximation of original trajectories).

2.12 Summary of related work

In the previous sections we have reviewed different research problems in the area of mining behavior of multiple moving objects. Existing techniques for summarizing spatio-temporal data points like segmentation, piecewise linear approximation, hierarchical organization of the network into domains, clustering the element set of each cell, warehousing, line simplification, etc. may not scale well for mining massive spatio-temporal data streams, especially when incremental learning is needed. In the case of spatio-temporal clustering, measuring similarity between mobile trajectories is the most computationally expensive operation. In this chapter, we describe an incremental algorithm for efficiently representing a mobile object trajectory as a set of MBBs. This preprocessing stage is aimed at reducing the running times of the spatio-temporal data mining algorithms, used by real-time applications. We proceed with presenting an incremental trajectory clustering algorithm, where we define a new "data-amount-based" similarity measure which is shown to outperform the "minimal distances" similarity measure [1] in terms of clustering validity indices.

3 Specific methods

3.1 An Algorithm for MBB-Based Trajectory Representation

A spatio-temporal trajectory is a series of data-points traversed by a given moving object during a specific period of time (e.g., one day). Since we assume that a moving object behaves according to some periodic spatio-temporal pattern, we have to determine the duration of each spatio-temporal sequence (trajectory). Thus, in the experimental part of this paper, we assume that a moving object repeats its trajectories on a daily basis, meaning that each trajectory describes an object movement during one day. In a general case, each object should be examined for its periodic behavior in order to determine the duration of its periodicity period. The training data window is the period which is used to learn the object's periodic behavior based on its recorded trajectories (e.g., daily trajectories recorded during one month).

As a part of the preprocessing technique introduced by us in [10] we represent a trajectory as a list of minimal bounding boxes. A minimal bounding box (MBB) represents an interval bounded by limits of time and location.

By using this structure we can summarize close data-points into one MBB, such that instead of recording the original data-points, we only need to record the following six elements:

$$
\begin{aligned}
i.x_{min} &= \min(\forall m \in i, m.x_{min}) \\
i.x_{max} &= \max(\forall m \in i, m.x_{max}) \\
i.y_{min} &= \min(\forall m \in i, m.y_{min}) \\
i.y_{max} &= \max(\forall m \in i, m.y_{max}) \\
i.t_{min} &= \min(\forall m \in i, m.t_{min}) \\
i.t_{max} &= \max(\forall m \in i, m.t_{max})
\end{aligned} \tag{3}
$$

Where i represents a MBB, m represents a member in a box, x and y are spatial coordinates, and t is time.

We also add other properties to the standard MBB-based representation that improve our ability to perform operations on the summarized data, like measuring similarity between trajectories:

$$i.p = aggregation(\forall m \in i, m.state) \tag{4}$$

where p stands for the value of a property variable in a minimal bounding box i, m represents a member in a box, and *state* is the data-point's property that is being aggregated. In our algorithm, p represents the number of data points (data♯) that are summarized by a given MBB:

$$i.data\sharp = count(\forall m \in i, 1) \tag{5}$$

A daily trajectory of an object is identified by an object ID O and a date D and it can be stored as a list of n MBBs: O_1, D_1, $[t_1, t_2, x_1, x_2, y_1, y_2, N_1]$, $[t_3, t_4, x_3, x_4, y_3, y_4, N_2]$.., $[t_{n-1}, t_n, x_{n-1}, x_n, y_{n-1}, y_n, N_n]$ where t represents time, x and y represent coordinates (1 for minimal and 2 for maximal), and N represents the amount of data points belonging to each MBB. Figure 1 demonstrates an object's trajectory and its MBB-based representation for a given period.

Incoming data-points update the MBBs in the order of their arrival times. Therefore, the minimal time bound of the first MBB is the time of the earliest data-point in the dataset and the maximal time bound of this MBB is

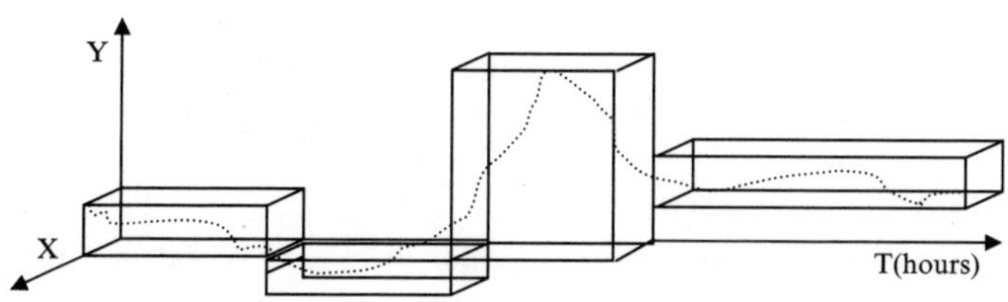

Fig. 1. Object's trajectory

extended until the time or the space distance between the new inserted data points and the bounds of this MBB reaches some pre-defined segmentation thresholds. When one of these thresholds is exceeded, a new minimal bounding box is created with the time of the subsequent data-point as its minimal time bound. The larger the threshold is, the more summarized the trajectories are, meaning that we increase the efficiency of the next mining stages (shorter running times for less transactions) but also decrease their precision.

We set the segmentation threshold in each dimension to be directly proportional to the variance of the values in that dimension normalized by the data range in that dimension. We do so in order to maintain the variance inside the MBBs, and also to maintain the relative MBB size in proportion to the data range. The segmentation thresholds are calculated as follows:

$$bound = \frac{var(D)}{(\max(D) - \min(D)) \cdot b} \tag{6}$$

where D is a set of the data points values in the corresponding dimension, var retrieves the variance of these values, max returns the maximal value in that dimension, min retrieves the minimal value, and b is a user-defined parameter that we try to optimize in the experiments section.

In [10], we have presented an enhanced algorithm for representing an object trajectory as a set of MBBs from a spatio-temporal dataset D covering object movement during a pre-defined period (e.g., 24 hours). This algorithm is described below:

Input: a spatio-temporal dataset (D), a threshold of x and y distances and of time duration of a MBB.
Output: new objects' trajectory (T)
Building an object's trajectory:

1. $T.addMBB(item)$ – First item updates first MBB
2. $ForeachiteminD$ – Except for first item
3. $while(|item.X - T.lastMBB.maxX| < XdistThreshold$ $and\,|item.Y - T.lastMBB.maxY| < YdistThreshold$ $and\,|item.T - T.lastMBB.maxT| < durationThreshold)$
4. $T.lastMBB.addPoint(item)$ – Insert into current MBB
5. $T.addMBB(item)$ – Create MBB when out of thresholds

The algorithm processes each data point in the data stream and inserts a data point into an existing MBB as long as its bounds are within the threshold defined as algorithm parameters; otherwise it creates a new MBB. The *"lastMBB"* function returns the MBB with the maximal (latest) time bounds in the trajectory, the *"addMBB"* function initializes a new MBB in the trajectory with bounds and properties updated by the first incoming data-point (on the first arrival, the minimum and the maximum are equal to the data-point values), and the *"addPoint"* function updates MBB properties (bounds and data amount).

3.2 Defining a new similarity measure

We define similarity between two trajectories as the sum of the similarities between the trajectories' MBBs, divided by the amounts of MBBs in each of the compared trajectories, where the two compared trajectories are described as shown in Figure 2.

In this chapter, we empirically compare two similarity measures between two MBBs. The first similarity measure is called "minimal distances" and it was suggested in [1] and described in section 2.10 above.

If we treat each MBB as a segment we can use the following formula, where t_m is the time when the two MBBs start to overlap and t_n is the time when its overlapping ends:

$$sim\left(MBB\left(T_i\right), MBB\left(T_j\right)\right) = minD\left(MBB\left(T_i\right), MBB\left(T_j\right)\right) \cdot \left|t_m - t_n\right| \quad (7)$$

The minimal distance and the times t_m and t_n are described in Figure 3. We define the minimal distance similarity (minD) between two MBBs as:

$$\begin{aligned} &minD\left(MBB\left(T_i\right), MBB\left(T_j\right)\right) = \\ &\quad XminD\left(MBB\left(T_i\right), MBB\left(T_j\right)\right) + YminD\left(MBB\left(T_i\right), MBB\left(T_j\right)\right) \end{aligned} \quad (8)$$

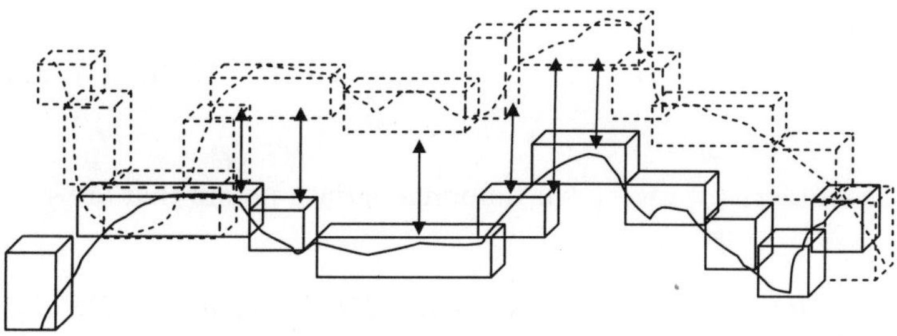

Fig. 2. Similarity between two trajectories

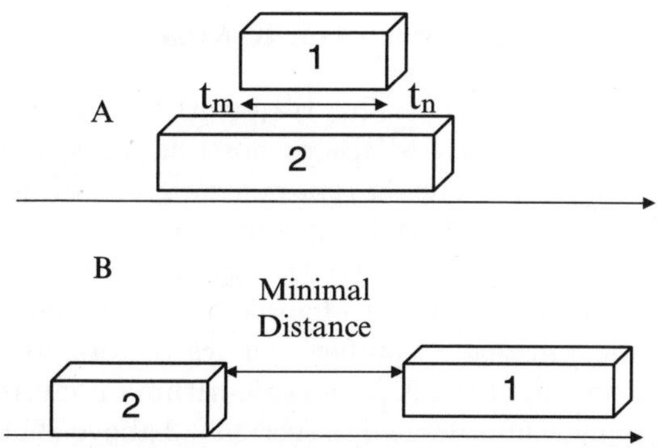

Fig. 3. A. Times of overlapping between two MBBs; B. minimal distance between two MBBs

where the minimal distance similarity measure between two MBBs in X and Y dimensions is calculated as follows:

$$\begin{aligned} XminD\,(MBB\,(T_i), MBB\,(T_j)) = 1 - \mathit{normalized}(\max(0, \\ \max(MBB\,(T_i.x_{min}), MBB\,(T_j.x_{min})) - \min(MBB\,(T_i.x_{max}), \\ MBB\,(T_j.x_{max})))) \end{aligned} \tag{9}$$

$$\begin{aligned} YminD\,(MBB\,(T_i)\,, MBB\,(T_j)) = 1 - \mathit{normalized}(\max(0, \\ \max(MBB\,(T_i.y_{min})\,, MBB\,(T_j.y_{min})) - \min(MBB\,(T_i.y_{max}), \\ MBB\,(T_j.y_{max})))) \end{aligned} \tag{10}$$

Using the enhanced representation of trajectories (as described in the previous section) we propose to improve the similarity measure between trajectories as follows. We multiply the *minimal-distances* measure (2) by the similarity between the amounts of data points of the two compared MBBs (data♮D). Since each MBB summarizes some data points, the more data points are included in both of the compared MBBs, the stronger support we have for their similarity. Our *"data-amount-based"* similarity is calculated as:

$$\begin{aligned} sim\,(MBB\,(T_i)\,, MBB\,(T_j)) = \\ minD\,(MBB\,(T_i)\,, MBB\,(T_j)) \cdot |t_m - t_n| \cdot data\natural D\,(MBB\,(T_i)\,, MBB\,(T_j)) \end{aligned} \tag{11}$$

Where the similarity between the amounts of data points that are summarized within two MBBS is:

$$data\natural D\,(MBB\,(T_i)\,, MBB\,(T_j)) = min(MBB(T_i) \cdot data\natural, MBB(T_j) \cdot data\natural) \tag{12}$$

3.3 Clustering trajectories with the K-Means Algorithm

A trajectories cluster contains similar trajectories. Trajectories in the same cluster contain as much similar MBBs as possible (close in space and time). The centroid of a trajectories cluster represents a group of similar trajectories, meaning that this cluster's centroid can represent a movement pattern.

Since in order to run generic clustering algorithms on the summarized trajectories data, the algorithm needs to handle an input that consists of bound intervals (trajectories) instead of numeric values vectors, we developed in [10] a spatio-temporal version of the K-Means algorithm for clustering trajectories using the similarity measures defined in section 3.2 above. This version handles interval-bounded data represented by a variable amount of attributes. It uses the data-amount-based similarity measure and a new centroid structure and updating method [10].

Each summarized trajectory is represented by an id and a set of MBBs (as described below). This algorithm receives as input a set of object trajectories of the form: T_1, $[t_1, t_2, x_1, x_2, y_1, y_2, N_1]$, $[t_3, t_4, x_3, x_4, y_3, y_4, N_2]$.., $[t_{n-1}, t_n, x_{n-1}, x_n, y_{n-1}, y_n, N_n]$ T_2,$[t_1, t_2, x_1, x_2, y_1, y_2, N_1]$, $[t_3, t_4, x_3, x_4, y_3, y_4, N_2]$.., $[t_{n-1}, t_n, x_{n-1}, x_n, y_{n-1}, y_n, N_n]$.., where a summarized trajectory T_i represents the object movement during a period (e.g., day i). The algorithm outputs trajectory clusters of the form: [c_1,(T_1, T_2, T_9), trajectory centroid of c_1 containing three object trajectories $T_1, T_2, and\ T_9$], [c_2, (T_3, T_7, T_8), trajectory centroid of c_2], etc.).

3.4 Using incremental approach for clustering

We adapt the incremental approach in order to benefit from the difference between the clustering for the first training data window (e.g. trajectories during the first month of data collection), where no previous data is available, and clustering for the subsequent windows, where using previous clustering centroids can help performing a more efficient incremental clustering process, since less updates are needed assuming that the movement behavior of the same object stays relatively stable. With the non-incremental approach, the clustering algorithm is applied repeatedly to each new data window (e.g., on a monthly basis) without utilizing clustering results of earlier data windows (the centroids are re-calculated for each new window).

An algorithm for incrementally discovering periodic movement patterns during a given data window includes the initialization of cluster centroids according to the previous cluster results of the same objects, where early clustering results exist. In our previous work [10] we performed experiments that showed that the incremental approach decreases clustering run times and improves cluster's validity.

4 Evaluation Methods

Our experiments were run on a PC Intel Pentium at 1.86GHz with 1GB RAM and 60GB hard disk. Since clustering is an unsupervised machine learning technique, there is no set of correct ("ground truth") answers that can be compared to the obtained results. We can, though, generate data sets by some logic that will help building the correct partition into groups, and then evaluate the validity of clustering results using the Rand index that measures clustering accuracy with respect to the ground truth. The Rand index performs a comparison of all pairs of objects in the data set after clustering. An "Agreement" (A) is a pair of objects that are both in the same or in different clusters under the actual and the induced clusterings and "disagreement" (D) is the opposite case. The Rand index is computed as: [32]

$$R_I = \frac{A}{A+D} \tag{13}$$

The Dunn index [32] measures the overall worst-case compactness and separation of a clustering, with higher values being better. The Dunn index can be used disregarding the ground truth clustering.

$$D_I = \frac{D_{min}}{D_{max}} \tag{14}$$

where D_{min} is the minimum distance between any two objects in different clusters and D_{max} is the maximum distance between any two items in the same cluster. The numerator captures the worst-case amount of separation between clusters, while the denominator captures the worst-case compactness of the clusters.

We evaluate clustering efficiency by measuring run times of the clustering algorithms as a function of different parameter values.

5 Evaluation Experiments

5.1 Generating spatio-temporal data

In the absence of real spatio-temporal datasets, mainly due to privacy issues, we have generated synthetic data for our experiments. We imitated the idea behind the CENTRE system [12] that simulates the behavior of cellular users. We built our data using movement formulas of the form:

$$\begin{aligned} t_1 &= t_0 \cdot rate + noise \\ x_1 &= x_0 + t_1 \cdot v_x + noise \\ y_1 &= y_0 + t_1 \cdot v_y + noise \end{aligned} \tag{15}$$

Where x_0, y_0 are the previous coordinates, x_1, y_1 are the current coordinates, t_0 is the time when the previous data point was sampled, and t_1 is the time when the current point is sampled. v_x, v_y are the velocities of the movement on X and Y axes (that change along the movement), and rate is the time between samplings. The data is asynchronous. Noise is a number that is randomly chosen from a range that is defined as 15 percents of the range of data in the corresponding dimension.

This synthetic data was used for the empirical evaluation of the proposed algorithm for incremental representation of trajectories with the new *"data-amount-based"* similarity measure. We ran 20 simulations, the first 10 runs simulated daily movements (trajectories) of 10 mobile objects during 25 days, and the other 10 runs simulated daily movements (trajectories) of 10 mobile objects during 45 days. The trajectories of each object belonged to five different movement patterns. Trajectories belonging to the same movement pattern reached at least three identical locations on identical times. At least 35 snapshots of every trajectory were taken during each day.

5.2 Detailed Results

We preprocessed the data using our MBB-based algorithm for building trajectories. In order to evaluate the proposed similarity measure, we clustered the trajectories once using the *"minimal distances"* similarity measure [1] and once using our proposed *"data-amounts-based"* similarity measure.

For each simulation K-Means iterations amount was set to 20, and we examined the following amounts of trajectory clusters k: 2, 3, 4, 5, 6, 7. For each value of k, we examined 6 different options for the segmentation thresholds (using the following values of b substituted in equation 6: 1, 1.5, 2, 2.5, 3, 3.5). We compared the clustering running times, the Dunn index, and the Rand index.

After the experiments, we evaluated the results using a oneway analysis of variance (ANOVA), for the independent variable: similarity measure type (*minimal-distance* or *data-amounts-based*). Our dependent variables (tested separately) are Rand index, Dunn index, and run time.

The ANOVA results in Tables 1 and 2 show that the new *"data-amounts-based"* similarity measure significantly outperforms the *"minimal distances"* similarity measure with respect to the Dunn index. In Figure 4 we can see that our *"data-amounts-based"* similarity measure gets higher rand index results (Y axis), and in Figure 5 we can see that it gets significantly higher Dunn index results (Y axis).

Table 1. ANOVA for the Rand index target variable

source	Sum of squares	df	Mean Squares	F	Sig.
Corrected Model	10.189	11	.926	37.9	.0
Intercept	419.829	1	419.829	17177.77	.0
Clusters Amount	10.081	5	2.016	82.495	.0
Similarity Measure	.081	1	.081	3.322	.069
Clusters Amount*Similarity	.27	5	.005	.221	.954
Error	34.901	1428	.024		
Total	464.918	1440			
Corrected Total	45.09	1439			

Table 2. ANOVA for the Dunn index target variable

source	Sum of squares	df	Mean Squares	F	Sig.
Corrected Model	4.28	11	.389	61.944	.0
Intercept	1071.336	1	1071.336	170549.8	.0
Clusters Amount	.013	5	.003	.407	.844
Similarity Measure	4.265	1	4.265	678.905	.0
Clusters Amount*Similarity	.003	5	.001	.09	.994
Error	8.97	1428	.006		
Total	1084.587	1440			
Corrected Total	13.25	1439			

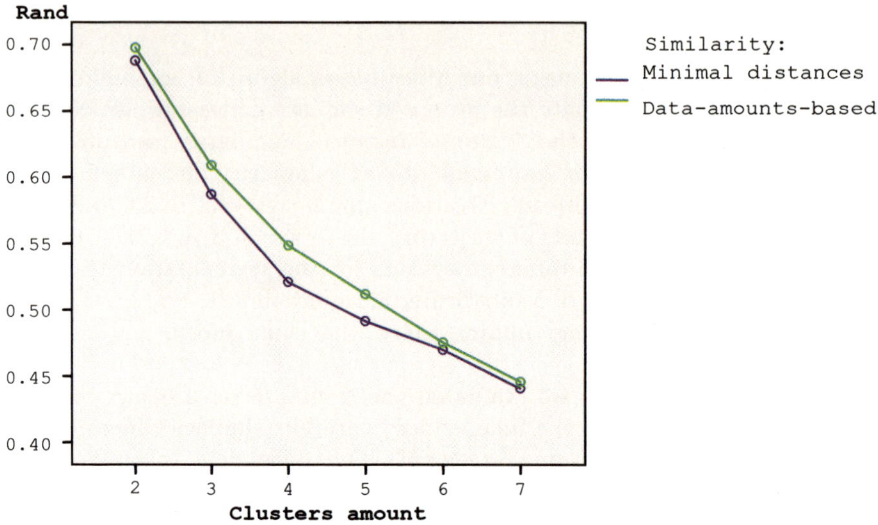

Fig. 4. Rand index vs. different cluster amounts

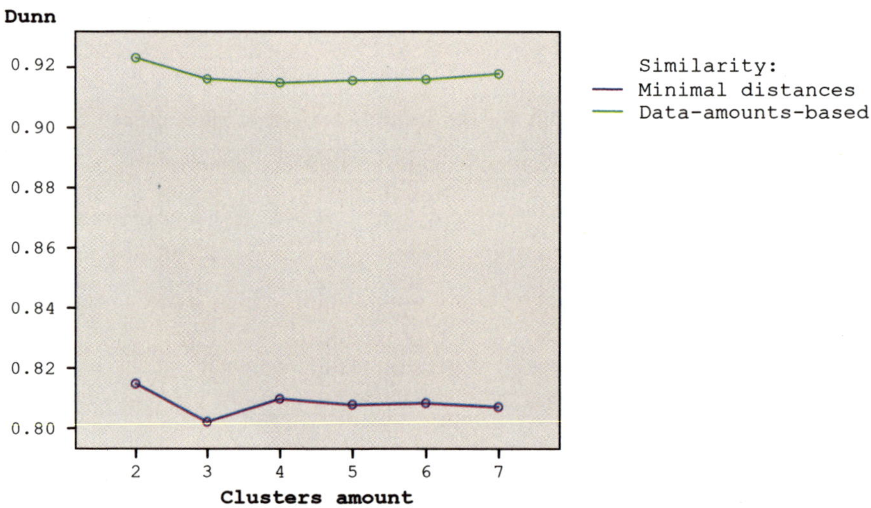

Fig. 5. Dunn index vs. different cluster amounts

We can see in Figure 6 that run durations (Y axis) are a few seconds higher using our suggested "data-amounts-based" similarity measure. As can be seen in Table 3 the two methods are not significantly different in terms of the running time variable.

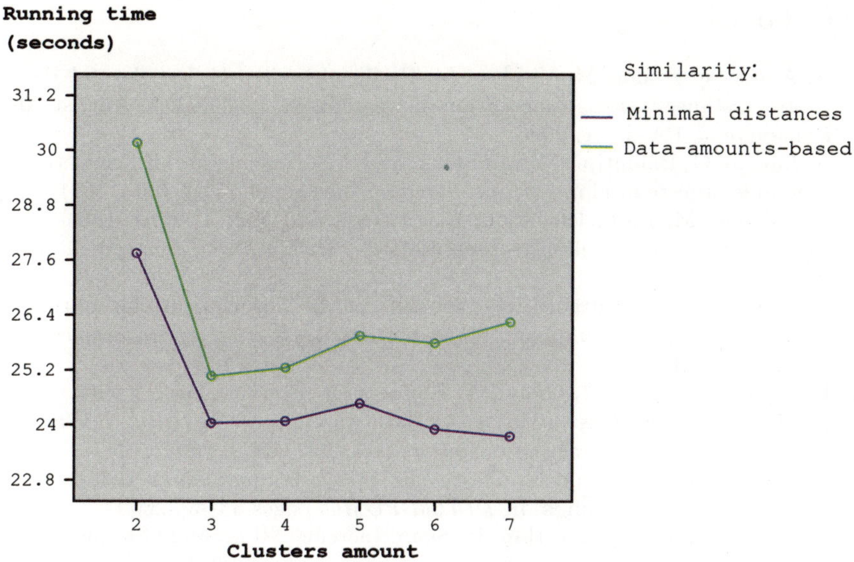

Fig. 6. Run durations index vs. different cluster amounts

Table 3. ANOVA for the running time target variable

source	Sum of squares	df	Mean Squares	F	Sig.
Corrected Model	1.285	11	.117	.646	.79
Intercept	260.735	1	260.735	1440.63	.0
Clusters Amount	.949	5	.19	1.049	.387
Similarity Measure	.304	1	.304	1.68	.195
Clusters Amount*Similarity	.032	5	.006	.035	.999
Error	258.45	1428	.181		
Total	520.471	1440			
Corrected Total	259.735	1439			

6 Conclusions

In this chapter, we presented a new method for measuring similarity between trajectories for finding clusters of periodic trajectories characterizing a given mobile object. The new similarity measure is based on the minimal distances similarity [1] and it was shown to outperform existing similarity measures with respect to two cluster validity indices.

Further research is needed for exploring the effect of summarization thresholds on the efficiency and effectiveness of spatio-temporal data mining algorithms using the proposed similarity measure.

References

1. A. Anagnostopoulos, M. Vlachos, M. Hadjieleftheriou, E. Keogh, and P.s. Yu. Global distance-based segmentation of trajectories. In *KDD'06*, August 20–23, Philadelphia, PA, USA, 2006
2. A. Baggio, G. Ballintijn, M. van Steen, and A.S. Tanenbaum. Efficient tracking of mobile objects in globe. *The Computer Journal*, 44(5):340–353, 2001
3. P. Bakalov, M. Hadjieleftheriou, E.J. Keogh, and V.J. Tsotras. Efficient trajectory joins using symbolic representations. *Mobile Data Management*, 86–93, 2005
4. M. Bennewitz, W. Burgard, G. Cielniak, and S. Thrun. Learning motion patterns of people for compliant robot motion, I. *Journal of Robotic Research*, 24(1):31–48, 2005
5. F. Braz, S. Orlando, R. Orsini, A. Raffaeta, A. Roncato, and C. Silvestri. Approximate aggregations in trajectory data warehouses. In *IEEE, ICDE Workshop on Spatio-Temporal Data Mining 2007 (STDM07)*, 2007
6. H. Cao, O. Wolfson, and G. Trajcevski. Spatio-temporal data reduction with deterministic error bounds. In *DIALM-POMC*, pages 33–42, 2003
7. H. Cheng, X. Yan, and J. Han. IncSpan: Incremental mining of sequential patterns in large database. In *KDD 2004*, pages 527–532, 2004
8. I. Cohen and G.G. Medioni. Detecting and tracking moving objects for video surveillance. In *CVPR 1999*, 2319–2325, 1999
9. M. D'Auria, M. Nanni, and D. Pedreschi. Time-focused density-based clustering of trajectories of moving objects. *JIIS Special Issue on Mining Spatio-Temporal Data*, 27(3):267–268, 2006
10. S. Elnekave, M. Last, and O. Maimon. Incremental clustering of mobile objects. In *IEEE, ICDE Workshop on Spatio-Temporal Data Mining (STDM07)*, 2007
11. C. Faloutsos, M. Ranganathan, and Y. Manolopoulos. Fast sub-sequence matching in time-series databases. *ACM SIGMOD*, 419–429, 1994
12. F. Giannotti, A. Mazzoni, S. Puntoni, and C. Renso. Synthetic generation of cellular network positioning data. In *GIS'05: Proceedings of the 13th ACM International Workshop on Geographic Information Systems*, pages 12–20, 2005
13. J.Y. Goh and D. Taniar. *An Efficient Mobile Data Mining Model*, volume 3358, LNCS, pages 54–58. Springer, Berlin Heidelberg New York, 2004
14. S.Y. Hwang, Y.H. Liu, J.K. Chiu, and F.P. Lim. Mining mobile group patterns: A trajectory-based approach. In *Proceedings of the PAKDD 2005*, pages 713–718, 2005
15. A.K. Jain, M.N. Murty, and P.J. Flynn. Data clustering: A review. *ACM Computing Surveys (CSUR)*, 31(3):264–323, 1999
16. J. Krumm and E. Horvitz. Predestination: Inferring destinations from partial trajectories. In *Eighth International Conference on Ubiquitous Computing (UbiComp 06)*, 2006
17. J.W. Lee, O.H. Paek, and K.H. Ryu. Temporal moving pattern mining for location-based service. *Journal of Systems and Software* 73(3), 481–490, 2004
18. Y. Li, J. Han, and J. Yang. Clustering moving objects. In *KDD-2004 – Proceedings of the Tenth ACM SIGKDD International Conference on Knowledge Discovery and Data Mining*, pages 617–622, 2004
19. S. Ma, S. Tang, D. Yang, T. Wang, and J. Han. Combining Clustering with moving sequential pattern mining: A novel and efficient technique. In *PAKDD 2004*, pages 419–423, 2004

20. S. Ma, S. Tang, D. Yang, T. Wang, and C. Yang. *Incremental Maintenance of Discovered Mobile User Maximal Moving Sequential Patterns*, volume 2973, NCS, pages 824–830, Springer, Berlin Heidelberg New York, 2004
21. M. Mamoulis, H. Cao, G. Kollios, M. Hadjieleftheriou, Y. Tao, and D.W. Cheung. Mining, indexing, and querying historical spatiotemporal data. In *Proceedings of the 10th ACM SIGKDD International Conference on Knowledge Discovery and Data Mining*, pages 236–245, 2004
22. R. Nehme and E.A. Rundensteiner. SCUBA: Scalable cluster-based algorithm for evaluating continuous spatio-temporal queries on moving objects. In *10th International Conference on Extending Database Technology*, Munich, Germany, March 26–30, 2006
23. New York Times site, http://www.nytimes.com/2007/07/09/nyregion/09ring.html, 6/8/2007
24. D. Niculescu and B. Nath. Trajectory based forwarding and its applications. In *MOBICOM 03*, 2003
25. A. Odoi, S.W. Martin, P. Michel, J. Holt, D. Middleton, and J. Nath. Geographical and temporal distribution of human giardiasis in Ontario, Canada. *International Journal of Health Geographics*, 2:5, 2003
26. N. Pelekis, I. Kopanakis, I. Ntoutsi, G. Marketos, and Y. Theodoridis. Mining trajectory databases via a suite of distance operators. In *IEEE, Workshop on Spatio-Temporal Data Mining (STDM07)*, 2007
27. W. Peng and M. Chen. Developing data allocation schemes by incremental mining of user moving patterns in a mobile computing system. *IEEE Transactions on Knowledge and Data Engineering*, 15(1):70–85, 2003
28. D. Pfoser. Indexing the trajectories of moving objects. *IEEE Data Engineering Bulletin*, 2002
29. D. Pfoser, C.S. Jensen, and Y. Theodoridis. Novel approaches in query processing for moving object trajectories. In *Proceedings of the 26th International Conference on Very Large Data Bases*, pages 395–406, 2000
30. K. Porkaew, I. Lazaridis, and S. Mehrotra. Querying mobile objects in spatio-temporal databases. In *Proceedings of the 7th International Symposium on Advances in Spatial and Temporal Databases*, pages 55–78, 2001
31. S. Rasetic, J. Sander, J. Elding, and M.A. Nascimento. A trajectory splitting model for efficient spatio-temporal indexing. In *VLDB 2005*, pages 934–945, 2005
32. A. Schenker, H. Bunke, M. Last, and A. Kandel. *Graph-Theoretic Techniques for Web content Mining*, volume 62, Series in Machine Perception and Artificial Intelligence. World Scientific, Singapore, 2005
33. A. Solmo and A.E. Howe. incremental clustering for profile maintenance in information gathering web agents. In *Proceedings of the International Conference on Autonomous Agents*, pages 262–269, 2001
34. Y. Tao, C. Faloutsos, D. Papadias, and B. Liu. Prediction and indexing of moving objects with unknown motion patterns. In *Proceedings of the International Conference on Management of Data (SIGMOD)*, pages 611–622, ACM Press, USA, 2004
35. Technical News site. http://technews.acm.org/archives.cfm?fo=2007-07-jul/jul-06-2007.html, 6/8/07
36. M. Vlachos, G. Kollios, and D. Gunopulos. Discovering similar multidimensional trajectories. In *Proceedings – International Conference on Data Engineering*, pages 673–684, 2002

37. Y. Wang, E.P. Lim, and S.Y. Gunopulos. Mining group patterns of mobile users. In *Proceedings of the 14th International Conference on Database and Expert Systems Applications, DEXA*, pages 287–296, 2003
38. J. Yang, and M. Hu. *TrajPattern: Mining Sequential Patterns from Imprecise Trajectories pf Mobile Objects*, volume 3896, LNCS, Springer, Berlin Heidelberg New York, pages 664–681, 2006
39. J. Zhang, and L. Gruenwald. Spatial and temporal aware, trajectory mobility profile based location management for mobile computing. In *DEXA Workshops*, pages 716–720, 2002

Part III

Graph-Based Methods

Matching of Hypergraphs — Algorithms, Applications, and Experiments

Horst Bunke,[1] Peter Dickinson,[2] Miro Kraetzl,[3] Michel Neuhaus[4] and Marc Stettler[1]

[1] Institute of Computer Science and Applied Mathematics
University of Bern, Neubrückstrasse 10, CH-3012 Bern, Switzerland
bunke@iam.unibe.ch, stettler@iam.unibe.ch

[2] C3ID, Defence Science and Technology Organisation
Edinburgh SA 5111, Australia
peter.dickinson@dsto.defence.gov.au

[3] C3ID, Defence Science and Technology Organisation
Canberra ACT 2600, Australia
miro.kraetzl@defence.gov.au

[4] Laboratoire d'Informatique LIP6, University Pierre et Marie Curie
104 avenue du Président Kennedy, F-75016 Paris, France
mneuhaus@iam.unibe.ch

Summary. In this chapter we introduce hypergraphs as a generalisation of graphs for object representation in structural pattern recognition. We also propose extensions of algorithms for the matching and error-tolerant matching of graphs to the case of hypergraphs, including the edit distance of hypergraphs. In a series of experiments, we demonstrate the practical applicability of the proposed hypergraph matching algorithms and show some of the advantages of hypergraphs over conventional graphs.

Keywords. Structural pattern recognition, graph, graph matching, hypergraph, hypergraph matching

1 Introduction

In the field of pattern recognition, one can distinguish between the statistical and the structural approach. Statistical pattern recognition is characterised by the use of feature vectors for pattern representation, while symbolic data structures, such as strings, trees, and graphs are predominant in the structural approach. From the representational point of view, structural pattern recognition is more powerful than statistical pattern recognition because each feature vector can be understood as a special case of a string, tree, or graph. On the other hand, statistical pattern recognition provides us with a much

H. Bunke et al.: *Matching of Hypergraphs — Algorithms, Applications, and Experiments*, Studies in Computational Intelligence (SCI) **91**, 131–154 (2008)
www.springerlink.com

richer repository of tools than the structural approach, i.e. algorithms for classification, clustering, pattern dimensionality reduction, etc.

The most general representation formalism in structural pattern recognition is graphs. Graphs provide us with a very convenient way for object representation when relational properties among individual parts of patterns need to be modelled. Another advantage of graphs is that they can be of arbitrary size, while the dimensionality of a feature vector is always fixed for a particular problem. A large number of successful applications involving graphs for pattern representation have been reported [1]. Despite all these advantages, graphs are restricted in the sense that only binary relations between nodes can be represented, through graph edges. An extension is provided by hypergraphs, where each edge is a subset of the set of nodes [2]. Hence higher-order relations between nodes can be directly modelled in a hypergraph, by means of hyperedges. A large body of theoretical work on hypergraphs has been published. However, not many applications in the field of image analysis and pattern recognition involving hypergraphs have been reported. Refs. [3, 4] list a number of applications of hypergraphs in low level image processing, and [5] describes a 3-D object recognition system using hypergraphs. We notice in particular that not much attention has been payed to the problem of high-level matching of hypergraphs. In [6] the problem of maximum common sub-hypergraph, and in [5] the problem of hypergraph monomorphism computation has been considered. However, the problem of error-tolerant hypergraph matching has not yet been addressed in the literature.

In an earlier paper by the authors [7], formal definitions and algorithms for hypergraph matching, including hypergraph isomorphism, sub-hypergraph isomorphism, and maximum common sub-hypergraph computation, were introduced. In the work described in this chapter we extend the framework introduced in [7] by dealing with the problem of error-tolerant matching and edit distance computation of hypergraphs. In the current chapter we also present some experimental results on error-tolerant hypergraph matching. In the experiments described in this paper, there are two major objectives. First, we want to study how the run time of error-tolerant matching algorithms behaves when one goes from conventional graphs to hypergraphs and increases the order of hyperedges. Secondly we are interested to see if through the use of hypergraphs, rather than graphs, a higher performance in certain graph classification tasks can be achieved.

The remainder of this chapter is structured as follows. In the next section we introduce basic definitions and concepts. Then in Section 3, we generalise a number of matching tasks, such as isomorphism, subgraph isomorphism, maximum common subgraph and graph edit distance, from graphs to hypergraphs. Computational procedures to actually perform theses matching tasks are discussed in Section 4. Experimental results obtained with these procedures on various data sets are presented in Section 5. Finally, conclusions are drawn and suggestions for future work presented in Section 6.

2 Preliminaries

Let L_V and L_E be finite sets of labels for the nodes and edges of a graph, or the nodes and hyperedges of a hypergraph, respectively.

Def. 2.1: A graph is a 4-tuple $g = (V, E, \alpha, \beta)$, where

- V is the finite set of nodes (also called vertices)
- $E \subseteq V \times V$ is the set of edges
- $\alpha : V \to L_V$ is a function assigning labels to nodes
- $\beta : E \to L_E$ is a function assigning labels to edges □

Edge (x, y) originates at node x and terminates at node y. An undirected graph is obtained as a special case if there exists an edge $(y, x) \in E$ for any edge $(x, y) \in E$ with $\beta(x, y) = \beta(y, x)$. If $V = \emptyset$ then g is called the *empty graph*. The notation $|g|$ will be used to denote the number of nodes of graph g.

The edges of a graph can be interpreted as unary or binary relations. Unary relations correspond to loops, i.e. edges $(x, x) \in E$, while binary relations correspond to directed edges of the form $(x, y) \in E$, $x \neq y$. Hypergraphs are a generalization of ordinary graphs in the sense that higher-order relations between nodes can be modeled.

Def. 2.2: Let $N \geq 1$ be an integer. A *hypergraph* of order N is a 4-tuple $h = (V, \mathcal{E}, \alpha, \mathcal{B})$, where

- V is the finite set of nodes (also called vertices)
- $\mathcal{E} = \cup_{i=1}^{N} E_i$ with $E_i \subseteq V^i$ is the set of hyperedges; $i = 1, \ldots, N$
- $\alpha : V \to L_V$ is a function assigning labels to nodes
- $\mathcal{B} = \{\beta_1, \ldots, \beta_N\}$ is the set of hyperedge labeling functions with $\beta_i : E_i \to L_E$ □

Each E_i is a subset of the set of hyperedges. It consists of i-tuples $(x_1, \ldots, x_i) \in V^i$, where each i-tuple is a hyperedge of hypergraph h. We call i the order of hyperedge $e = (x_1, \ldots, x_i)$. The elements of E_1 are the loops of the hypergraph and the elements of E_2 correspond to the edges of a (normal) graph. A hyperedge of degree higher than two can be used to model higher-order relations among the nodes. Note that graphs according to Def. 2.1 are a special case of Def. 2.2 if $N = 2$.

Hyperedges can be ordered or unordered. In the former case, the order of the individual elements of a hyperedge, $e = (x_1, \ldots, x_i) \in E_i$, is important while in the latter case it doesn't matter, i.e. for unordered hyperedges we have $(x_1, \ldots, x_i) = (x_{l_1}, \ldots, x_{l_i})$ where $(l_1, \ldots, l_i)$ is a permutation of $(1, \ldots, i)$. Hyperedges are labeled by means of functions from set $\mathcal{B}$. There is a separate hyperedge labeling function β_i, for each hyperedge order i. Note that, according to Def. 2.2, there can exist at most one hyperedge $e = (x_1, \ldots, x_i) \in E_i$ for any given i-tuple of nodes $(x_1, \ldots, x_i)$.

There are several possibilities to graphically represent a hypergraph. In [2] it was proposed to draw a node as a point, an edge from subset E_1 as a loop, an edge from subset E_2 as a line segment connecting the pair of nodes involved, and edges of higher degree as simple closed curves that enclose the corresponding nodes. An example is shown in Fig. 1 (from [2]). The underlying hypergraph is defined by $V = \{x_1, \ldots, x_8\}$, $\mathcal{E} = E_1 \cup E_2 \cup E_3$, $E_1 = \{x_7\}$, $E_2 = \{(x_1, x_2), (x_5, x_8)\}$, $E_3 = \{(x_6, x_7, x_8), (x_3, x_4, x_5), (x_2, x_3, x_7)\}$. There are no labels in this example.

In this chapter we adopt a different graphical notation, where rectangles are used to graphically represent nodes, and circles or ellipses are used to represent hyperedges of degree three and higher. If $e = (x_1, \ldots, x_i) \in E_i$, $i \geq 3$, then we draw i arrows pointing from the hyperedge symbol to $x_1, \ldots, x_i$, respectively. Labels are written next to nodes or next to hyperedge symbols. A similar notation has been used in [6]. Fig. 2 shows the hypergraph depicted in Fig. 1 using our graphical notation.

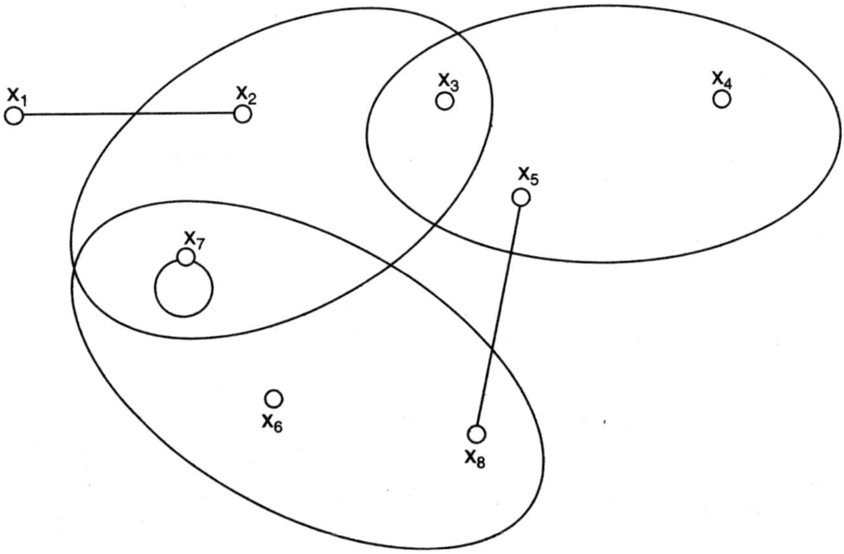

Fig. 1. Example of a hypergraph (from [2])

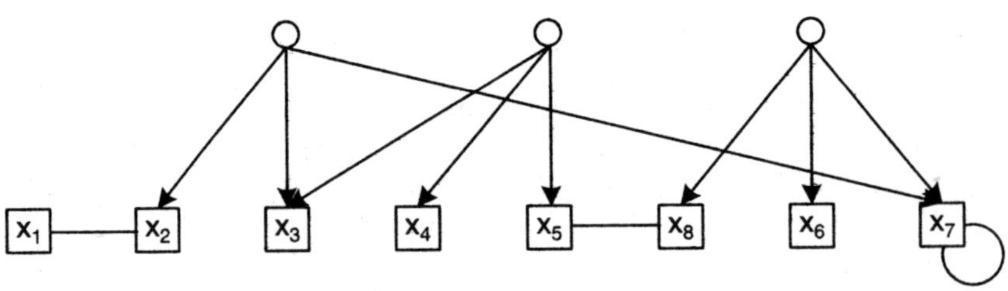

Fig. 2. Alternative representation of the hypergraph shown in Fig. 1

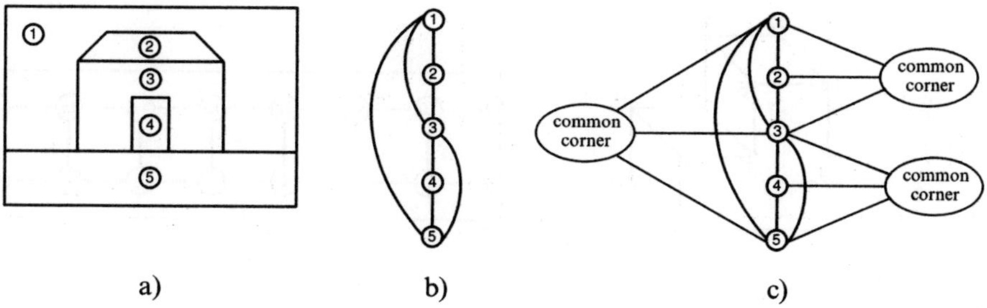

Fig. 3. a) image after segmentation; b) region adjacency graph; c) hypergraph with hyperedges representing region adjacency of order two and three

We conclude this section with a few examples that illustrate how hypergraphs can be used in various applications. These examples are also intended to show that, from the practical point of view, hypergraphs have a higher representational power than normal graphs.

Example 2.1: In pattern recognition and computer vision, *region adjacency graph* (rag) is a popular data structure to formally represent the contents of an image after it has been segmented into homogeneous regions. In a *rag* nodes represent image regions and edges between nodes indicate whether two regions are adjacent to each other or not. Fig. 3a shows an image that has been segmented into homogeneous regions, and Fig. 3b shows the corresponding *rag*. For certain applications it may be interesting to know whether three (or more) regions meet at a common corner in the image. Such a relation among three (or more) regions can't be directly represented in a normal graph. But in a hypergraph it is straightforward to model relations of this kind by means of hyperedges. Fig. 3c shows a hypergraph that corresponds to Fig. 3a. It includes hyperedges that represent region adjacency of degree three. □

Example 2.2: Wireframe models are a common representation in 3-D object modeling and computer graphics. Fig. 4a shows a polyhedral object and Fig. 4b the corresponding wireframe model, given in terms of a graph. In this representation graph nodes represent the vertices of the polyhedron, and graph edges correspond to the edges on the polyhedron. Note that the graph in Fig. 4b only captures the topology of the object, and doesn't include any quantitative, metric information.[1] In a graph, such as the one depicted in Fig. 4b, only binary relations between the vertices of an object can be represented. It is not possible to directly represent properties such as the collinearity or coplanarity of $n \geq 3$ vertices. However, in a hypergraph it is easy to directly model relations of this kind by means of hyperedges. Fig. 4c shows an extended version of Fig. 4b, where the relation of collinearity has been added. □

[1] However, metric information can be easily incorporated by means of node and edge labels.

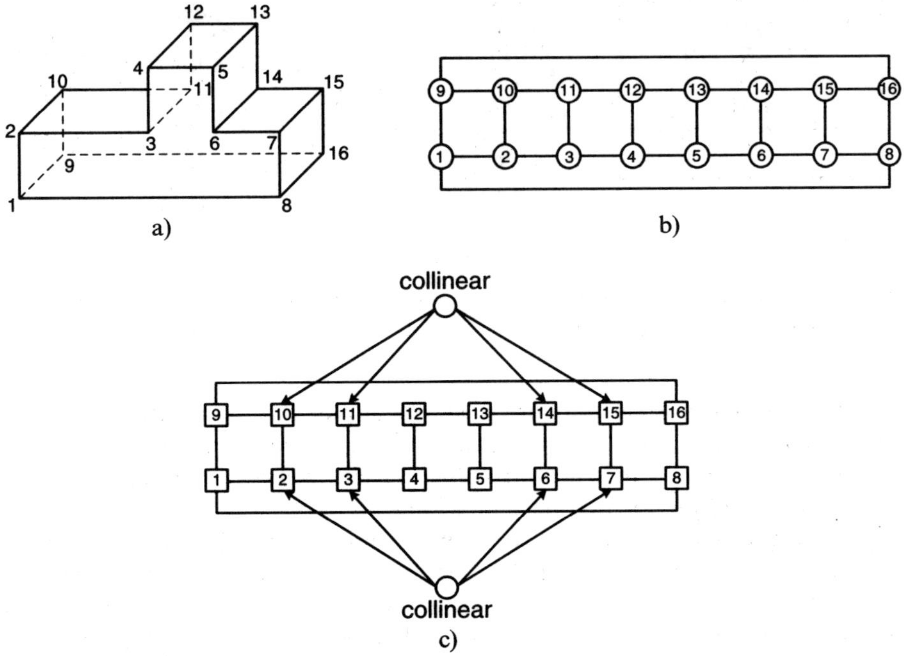

Fig. 4. a) a polyhedral object; b) graph representation; c) hypergraph showing collinear vertices

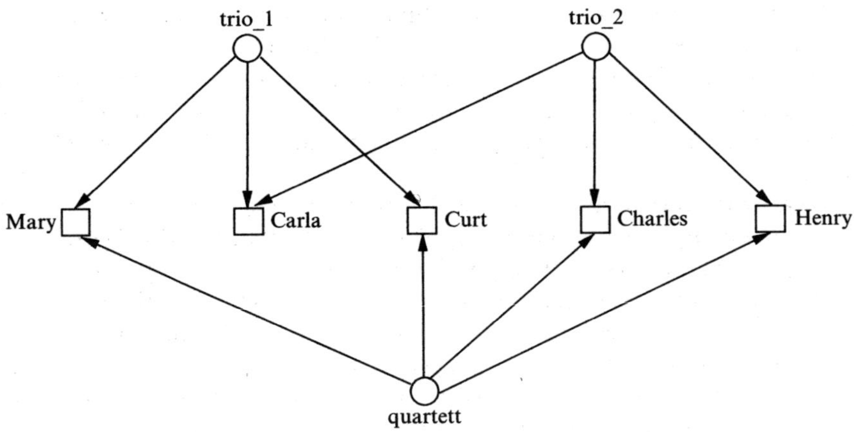

Fig. 5. Hypergraph with directed hyperedges, modeling social network

Example 2.3: This example involves a social network that represents relationships among musicians. Fig. 5 shows a hypergraph where each node represents a person and hyperedges indicate subgroups of persons who perform together in a trio or in a quartet. Such relations can't be directly modeled in a normal graph. □

3 Hypergraph Matching

Graph matching is the task of establishing a correspondence between the nodes and edges of two given graphs such that some constraints are satisfied. Well-known instances of the graph matching problem include graph and subgraph isomorphism [8], maximum common subgraph computation [9, 10] and graph edit distance [11, 12]. In this section we'll extend these matching problems from graphs to the case of hypergraphs.

Def. 3.1: Let $h = (V, \mathcal{E}, \alpha, \mathcal{B})$ and $h' = (V', \mathcal{E}', \alpha', \mathcal{B}')$ be two hypergraphs of order N and N', respectively. We call h a *sub-hypergraph* of h' if

- $V \subseteq V'$
- $E_i \subseteq E'_i$ for $i = 1, \ldots, N$
- $\alpha(x) = \alpha'(x)$ for all $x \in V$
- $\beta_i(e) = \beta'_i(e)$ for all $e \in E_i$ and $i = 1, \ldots, N$ □

According to this definition, h' may include a hyperedge $e = (x_1, \ldots, x_i)$ that is not included in h although nodes $x_1, \ldots, x_i$ are included in h. A more restricted version of this definition is obtained if we require that h' must not include any hyperedge of this kind. Formally, we replace the second condition in Def. 3.1 by $E_i = E'_i \cap V^i$ for $i = 1, \ldots, N$, and call the resulting type of sub-hypergraph an *induced sub-hypergraph.*

Def. 3.2: Let h and h' be two hypergraphs with $N = N'$. A *hypergraph isomorphism* between h and h' is a bijective mapping $f : V \rightarrow V'$ such that

- $\alpha(x) = \alpha'(f(x))$ for all $x \in V$
- for $i = 1, \ldots, N$ and any hyperedge $e = (x_1, \ldots, x_i) \in E_i$ there exists a hyperedge $e' = (f(x_1), \ldots, f(x_i)) \in E'_i$ such that $\beta_i(e) = \beta'_i(e')$ and for any hyperedge $e' = (f(x_1), \ldots, f(x_i)) \in E'_i$ there exists a hyperedge $e = (f^{-1}(x'_1), \ldots, f^{-1}(x'_i)) \in E_i$ such that $\beta'_i(e') = \beta_i(e)$ □

If $f : V \rightarrow V'$ is a hypergraph isomorphism between two hypergraphs h and h', and h' is an (induced) sub-hypergraph of another hypergraph h'', then f is called an (induced) *sub-hypergraph isomorphism* from h to h''.

Def. 3.3: Let h and h' be two hypergraphs. A *common sub-hypergraph* of h and h' is a hypergraph h'' such that there exist sub-hypergraph isomorphisms from h'' to h and from h'' to h'. We call h'' a *maximum common sub-hypergraph* of h and h', *mcsh(h,h')*, if there exists no other common sub-hypergraph of h and h' that has more nodes and, for a fixed number of nodes, more hyperedges than h''. □

This definition is a generalization of the concept of maximum common subgraph. As the maximum common subgraph of a pair of graphs is in general not unique, also the maximum common sub-hypergraph of a pair of hypergraphs is not unique. Def. 3.3 is based on sub-hypergraph isomorphism. Alternatively, one can also use the relation of induced sub-hypergraph isomorphism to define maximum common induced sub-hypergraph.

Def. 3.4: Let h_1 and h_2 be two hypergraphs. An *error-correcting hypergraph matching* (echgm) from h_1 to h_2 is a bijective function $f : \hat{V}_1 \rightarrow \hat{V}_2$ where $\hat{V}_1 \subseteq V_1$ and $\hat{V}_2 \subseteq V_2$. □

We say that node $x \in \hat{V}_1$ is *substituted* by node $y \in \hat{V}_2$ if $f(x) = y$. If $\alpha_1(x) = \alpha_2(x)$ then the substitution is called an *identical* substitution. Otherwise it is termed a *non-identical* substitution. Furthermore, any node from $V_1 - \hat{V}_1$ is *deleted* from h_1, and any node from $V_2 - \hat{V}_2$ is *inserted* in h_2 under f. We will use $\hat{h}_1$ and $\hat{h}_2$ to denote the sub-hypergraphs of h_1 and h_2 that are induced by the sets $\hat{V}_1$ and $\hat{V}_2$, respectively.

The mapping f *directly* implies an edit operation on each node in h_1 and h_2. That is, nodes are substituted, deleted, or inserted, as described above. Additionally, the mapping f *indirectly* implies edit operations on the hyperedges of h_1 and h_2. If $f(x_1) = y_1$, $f(x_2) = y_2$, ..., $f(x_i) = y_i$ and there exist hyperedges $e_1 = (x_1, \ldots, x_i) \in E_{i1}$ and $e_2 = (y_1, \ldots, y_i) \in E_{i2}$ then hyperedge e_1 is substituted by hyperedge e_2 under f. Note that e_1 and e_2 may have different labels. If there exists no hyperedge $e_1 = (x_1, \ldots, x_i) \in E_{i1}$ but a hyperedge $e_2 = (y_1, \ldots, y_i) \in E_{i2}$ then hyperedge e_2 is inserted in h_2. Similarly, if $e_1 \in E_{i1}$ but there exists no hyperedge e_2 then hyperedge e_1 is deleted from h_1 under f. If a node $x \in V_1$ is deleted from h_1, then any hyperedge that includes node x is deleted, too. Similarly, if $x \in V_2$ is inserted in h_2, then any hyperedge that includes x is inserted, too. In our framework, hyperedge substitutions are only possible among hyperedges of the same order. However, via hyperedge deletions and insertions, any subset of the hyperedges of h_1 can be transformed into any subset of the hyperedges of h_2. In other words, insertions and deletions may be used to simulate the substitution of hyperedges of different order. Obviously, any *echgm* f can be understood as a set of edit operations (substitutions, deletions, and insertions of both nodes and hyperedges) that transform a given hypergraph h_1 into another hypergraph h_2.

Example 3.1: Assume that Curt leaves the group of musicians represented in Fig. 5 and is replaced by Colette. Colette takes Curt's part in the quartet, but Curt's part in *trio_1* is taken over by Charles. The new situation is depicted in Fig. 6. Obviously, the transition from Fig. 5 to Fig. 6 can be modeled by one node substitution, replacing label 'Curt' by 'Colette', one hyperedge deletion (the hyperedge containing Mary, Carla and Curt in Fig. 5), and one hyperedge insertion (the hyperedge containing Mary, Carla and Charles in Fig. 6). □

Def. 3.5: The *cost* of an *echgm* $f : \hat{V}_1 \rightarrow \hat{V}_2$ from hypergraph h_1 to a hypergraph h_2 is given by

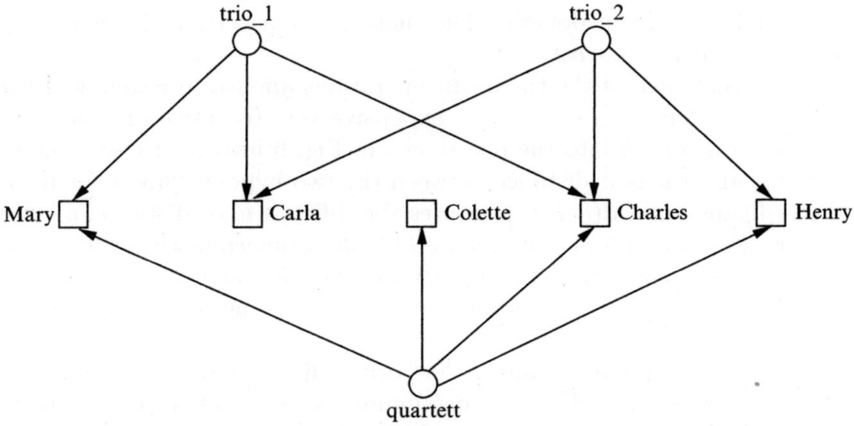

Fig. 6. Illustration of edit operations

$$c(f) = \sum_{x\in\hat{V}_1} c_{ns}(x) + \sum_{x\in V_1-\hat{V}_1} c_{nd}(x) + \sum_{x\in V_2-\hat{V}_2} c_{ni}(x) + \sum_{j=1}^{N}\sum_{e\in E_j^s} c_{es}(e) + \sum_{j=1}^{N}\sum_{e\in E_j^d} c_{ed}(e) + \sum_{j=1}^{N}\sum_{e\in E_j^i} c_{ei}(e)\,,$$

where

- $c_{ns}(x)$ is the cost of substituting node $x \in \hat{V}_1$ by $f(x) \in \hat{V}_2$,
- $c_{nd}(x)$ is the cost of deleting node $x \in V_1 - \hat{V}_1$ from h_1,
- $c_{ni}(x)$ is the cost of inserting node $x \in V_2 - \hat{V}_2$ in h_2,
- $c_{es}(e)$ is the cost of substituting hyperedge e,
- $c_{ed}(e)$ is the cost of deleting hyperedge e,
- $c_{ei}(e)$ is the cost of inserting hyperedge e,

and E_j^s, E_j^d, and E_j^i are the sets of hyperedges of order j that are substituted, deleted, and inserted, respectively. All costs are non-negative real numbers. □

Notice that the sets E_j^s, E_j^d, and E_j^i are implied by the mapping f. A particular set of costs $c_{ns}, c_{nd}, \ldots, c_{ei}$ according to Def. 3.5 will be called a *cost-function.*

Def. 3.6: Let f be an *echgm* from a hypergraph h_1 to a hypergraph h_2 under a particular cost function. We call f an *optimal echgm* if there exists no other *echgm* f' from h_1 to h_2 with $c(f') < c(f)$. □

The cost of an optimal *echgm* from a hypergraph h_1 to a hypergraph h_2 is also called the *edit distance* of h_1 and h_2, and denoted by $d(h_1, h_2)$.

In practice, the costs $c_{ns}, \ldots, c_{ei}$ introduced in Def. 3.5 are used to model the likelihood of errors or distortions that may corrupt ideal hypergraphs of the underlying problem domain. The more likely a certain distortion is to occur, the smaller is its cost.

Assume that each of the three edit operations applied in Example 3.1 has a cost equal to one. Then the least expensive way to transform the hypergraph shown in Fig. 5 into the one shown in Fig. 6 involves those three edit operations, and the edit distance between the two hypergraphs is equal to 3.

The edit distance, $d(h, h')$, measures the difference, or distance, of a pair of hypergraphs, h and h'. Clearly, if h and h' are isomorphic then $d(h, h') = 0$. In general, the greater the dissimilarity between h and h' is, the more edit operations are needed to transform h into h' and the larger the edit distance becomes.

In ordinary graph matching, a few other distance measures have been proposed. They are based on the maximum common subgraph of a pair of graphs. One of those measures [13] is defined as

$$d(g, g') = 1 - \frac{|mcs(g, g')|}{max(|g|, |g'|)} \tag{3.1}$$

In this definition, $|g|$ denotes the size of graph g, for example, the number of nodes. Clearly this definition can be applied to hypergraphs as well if we replace maximum common subgraph by maximum common sub-hypergraph.

Other graph distance measures proposed in the literature are [14] and [15]. These measures are similar to Eq. (3.1) in the sense that they are based on the maximum common subgraph of g and g'. However, they use different quantities for normalization. In [14] the size of the union of g and g' serves as a normalization factor, while in [15] the minimum common supergraph is used. Similarly to Eq. (3.1) both measure can be directly adapted to the case of hypergraphs.

4 Algorithms for Hypergraph Matching

In the previous section, a number of theoretical concepts have been introduced. However, no algorithmic procedures were considered. In the current section we'll discuss possible algorithms for hypergraph matching. For the matching of normal graphs, many algorithms have been proposed in the literature. They are based on various computational paradigms, including combinatorial search, neural networks, genetic algorithms, graph eigenvalue decomposition, and others. For a recent survey we refer to [1].

We start with the problem of extending graph and subgraph isomorphism computation to the case of hypergraphs. One of the best known graph matching algorithms, which can be used for both graph and subgraph isomorphism detection, is the one by Ullman [8]. It is a combinatorial search procedure that explores all possible mappings between the nodes of the two graphs under consideration. In order to avoid the exploration of partial mappings that can't lead to a correct solution, a look-ahead strategy is used. In this section, we'll discuss a generalization to of this algorithm to hypergraph matching.

Given two graphs g_1 and g_2 that need to be tested for subgraph isomorphism, Ullman's algorithm sequentially maps each node x of g_1 to a node y of g_2 and checks a number of constraints. Let $x \in V_1$, $y \in V_2$ and $f : V_1 \to V_2$ be the mapping being constructed. The partial mapping constructed up to a certain point in time is augmented by $f(x) = y$ if the following three constraints are satisfied:

1. There exists no node $x' \in V_1$, $x' \neq x$, with $f(x') = y$, i.e. no other node of g_1 has already been assigned to y
2. Nodes x and y have the same label
3. The assignment $f(x) = y$ is compatible with all previous node assignments under function f, i.e. if the assignment $f(u) = v$ has been made before and there is an edge (x, u) or (u, x) in g_1, there must be an edge $(f(x), f(u))$ or $(f(u), f(x))$ in g_2 with the same label

To extend Ullman's algorithm to the case of hypergraphs, we adopt constraints 1 and 2 without any changes. Only constraint 3 needs to be generalized in the sense that not only compatibility with respect to all edges, but w.r.t. all hyperedges is checked. Clearly such an extension is straightforward to implement.

A significant speedup in Ullman's algorithm is achieved through the use of a lookahead technique. The basic idea is to maintain a *future match* table where all possible future node assignments are recorded. Initially, all pairs $(x, y) \in V_1 \times V_2$ where x and y have identical node labels are possible. During the course of the search, pairs that violate Constraint 3 are eliminated. Consequently the number of assignments to be investigated in the tree search procedure is reduced. Obviously this lookahead procedure can be integrated in our sub-hypergraph matching schema. Assume that $x_1, \ldots, x_n \in V_1$ have been mapped to $f(x_1), \ldots, f(x_n) \in V_2$. Once a new assignment $f(x) = y$ has been made, we inspect all future (i.e. remaining) nodes $u_1, \ldots, u_m \in V_1$ and $v_1, \ldots, v_k \in V_2$ and delete any pair (u_i, v_j) from the future match table if nodes $x_1, \ldots, x_n, x, u_i \in V_1$ are part of a hyperedge in h_1 but $f(x_1), \ldots, f(x_n), f(x), v_j$ are not part of a hyperedge with the same label and order in h'.

The computational paradigm underlying Ullman's algorithm is tree search. Also the problem of maximum common subgraph computation can be solved by means of tree search [10]. An isomorphism between part of the first graph g_1 and part of the second graph g_2 is constructed that has the maximum possible size. This procedure can be extended from graphs to hypergraphs by extending the consistency checks in the common subgraph from edges to hyperedges.

Finally, we turn to the problem of edit distance computation. Tree search is a standard procedure to not only find the maximum common subgraph of a

Algorithm 1 Pseudo-code of algorithm for computing an optimal *echgm*

Input: Hypergraphs $h_1 = (V_1, \alpha_1, \mathcal{E}_1, \mathcal{B}_1)$ and $h_2 = (V_2, \alpha_2, \mathcal{E}_2, \mathcal{B}_2)$, where $V_1 = \{u_1, \ldots, u_{|V_1|}\}$ and $V_2 = \{v_1, \ldots, v_{|V_2|}\}$; cost function c

Output: An optimal *echgm* f from h_1 to h_2

OPEN=$\emptyset$
For each node $w \in V_2$ insert the partial *echgm* $\{f(u_1) = w\}$ into OPEN
Insert $\{f(u_1) = \varepsilon\}$ into OPEN
loop
    Remove the partial *echgm* $p_{min} = \arg\min_{p \in \text{OPEN}}\{c(p)\}$ from OPEN
    if p_{min} is a complete *echgm* **then**
        Return p_{min} as the solution and terminate
    else
        Let $p_{min} = \{f(u_1) = W_1, \ldots, f(u_k) = W_k\}$, where $W_1, \ldots, W_k \in V_2 \cup \{\varepsilon\}$
        if $k < |V_1|$ **then**
            For each $w \in V_2 \setminus \{W_1, \ldots, W_k\}$ insert $p_{min} \cup \{f(u_{k+1}) = w\}$ into OPEN
            Insert $p_{min} \cup \{f(u_{k+1}) = \varepsilon\}$ into OPEN
        else
            Insert $p_{min} \cup \bigcup_{w \in V_2 \setminus \{W_1, \ldots, W_k\}} \{f(\varepsilon) = w\}$ into OPEN
        end if
    end if
end loop

pair of graphs, but also to compute their edit distance [16, 17]. Next we introduce an extended tree search procedure for the edit distance of hypergraphs. Our algorithm is a special version of the well-known A*-algorithm [18].

A pseudo-code description of the algorithm is shown in Alg. 1. The algorithm takes, as input, a pair of hypergraphs $h_1 = (V_1, \mathcal{E}_1, \alpha_1, \mathcal{B}_1)$ and $h_2 = (V_2, \mathcal{E}_2, \alpha_2, \mathcal{B}_2)$ together with a cost function c. Upon termination, the algorithm outputs an optimal *echgm* from h_1 to h_2, the cost of which is equal to the edit distance, $d(h_1, h_2)$, of h_1 and h_2. The optimal *echgm* to be output is given in the form $\{f(v_1) = w_1, f(v_2) = w_2, \ldots, f(v_{|V_1|}) = w_{|V_1|}\}$, where $w_i \in V_2 \cup \{\varepsilon\}$. If $w_i \in V_2$ then element v_i is substituted by node $w_i \in V_2$. Otherwise, if $w_i = \varepsilon$ then v_i is deleted from h_1. An optimal *echgm* to be output by the algorithm may contain elements of the form $f(\varepsilon) = v$ for $v \in V_2$, which indicates that node v (and all its incident edges) are inserted in h_2. Note that we require function f to be complete, i.e. there must be an element $w \in V_2 \cup \{\varepsilon\}$ with $f(v) = w$ for every node $v \in V_1$. Conversely, there must be an element $v \in V_1 \cup \{\varepsilon\}$ with $f(u) = v$ for every node $u \in V_1$.

We assume without loss of generality that the nodes of V_1 are ordered as $v_1, \ldots, v_{|V_1|}$. Our algorithm maintains a list called OPEN, where several partial matches of the form $p = \{f(u_1) = W_1, \ldots, f(u_k) = W_k\}$ are stored, $W_i \in V_2 \cup \{\varepsilon\}$; $i = 1, \ldots, |V_1|$. Each partial match, p, has a cost, $c(p)$, associated with it (for details see below).

The algorithm starts with an empty list OPEN (step 1). It then constructs all partial matches that are possible for the first node $u_1 \in V_1$ (steps 2 and 3). In the course of the main loop (steps 4 to 17), the algorithm follows a best-first strategy, i.e. it selects the partial match p_{min} with the lowest cost among all partial matches stored in OPEN (step 5). Once p_{min} is retrieved and removed from OPEN, it is checked whether p_{min} represents a complete *echgm*, i.e. an *echgm* that involves all nodes of V_1 and all nodes of V_2 (step 6). If this is he case, p_{min} is returned as the solution and the algorithm terminates (step 7). In this case one can be sure that the optimal solution, i.e. the *echgm* from h_1 to h_2 with minimum cost has been found. Otherwise the algorithm examines whether $p_{min} = \{f(u_1) = W_1, \ldots, f(u_k) = W_k\}$ includes all nodes of V_1 (step 10). If this is not the case, partial mapping p_{min} is expanded. This means that new partial matches $p_{min} \cup \{f(u_{i+1}) = W\}$ are inserted into OPEN for all $W \in V_2 \setminus \{W_1, \ldots, W_k\}$ (step 11). Additionally the new partial match $\{f(u_1) = W_1, \ldots, f(u_k) = W_k\} \cup \{f(u_{k+1}) = \varepsilon\}$ is inserted in OPEN (step 12). If p_{min} includes all nodes of V_1, i.e. if $k = |V_1|$ in line 10, the algorithm continues with line 14, where all missing nodes of V_2 — in case there are any — are added to p_{min}. Then the updated version of p_{min} is added to OPEN.

The overall structure of the algorithm shown in Alg. 1 is similar to the computation of the edit distance of ordinary graphs. The most significant difference is in the way the costs $c(p)$ of a partial match are computed (line 5). Assume we have retrieved, in step 5, partial match $p = \{f(u_1) = W_1 \ldots, f(u_k) = W_k\}$, have turned it into $p' = p \cup \{f(u_{k+1}) = W_{k+1}\}$ where $W_{k+1} \in V_2$, and have inserted p' into OPEN in step 11. Then the cost of p', $c(p')$, is computed as follows. First we need to retrieve all hyperedges $e = (x_1, \ldots, x_j)$ in h_1 that include node u_{k+1} and no node that doesn't belong to set $\{u_1, \ldots, u_{k+1}\}$. Next we need to find out whether a hyperedge $e' = (f(x_1) \ldots, f(x_j))$ exists in h_2 with the same label as e. If such a hyperedge exists, no additional cost is added to the cost of p, $c(p)$. If a hyperedge $e' = (f(x1), \ldots, f(x_j))$ exists in h_2, but with a different label or with a different order of nodes, we need to increment $c(p)$ by the cost of substituting the label of e by the label of e'. Thirdly, if no hyperedge $e' = (f(x_1), \ldots, f(x_j))$ exists in h_2, we need to delete e and increment $c(p)$ by the cost of deleting hyperedge e. Furthermore, we need to retrieve all hyperedges e' from h_2 that include node $f(x_{k+1}) = W_{k+1}$ and no node that doesn't belong to $\{W_1 \ldots, W_k\}$. For each such hyperedge, e', we have to find out whether there exists a corresponding hyperedge in h_1. If no such hyperedge exists, an insertion of e' in h_2 is implied through the partial mapping p', and $c(p)$ needs to be incremented accordingly. We need to add that, whenever p_{min} is augmented by an element of the form $f(u_{k+1}) = w$ (step 11), $f(u_{k+1}) = \varepsilon$ (step 12), or $f(\varepsilon) = w$ (step 14), the cost of a node substitution, node deletion, or node insertion, respectively, is to be added to $c(p_{min})$ in addition to the costs of the hyperedge edit operations discussed above. To complete the discussion of how to compute the cost $c(p)$ for partial matches p in OPEN we need to consider two more cases. First, for partial matches of the form $p_{min} \cup \{f(u_{k+1}) = \varepsilon\}$ (step 12), we need to add the cost of deleting all hyperedges in h_1 that are incident to node u_{k+1}. Secondly, for all partial matches defined by line 14, we need to add the cost of inserting all hyperedges incident to a node from $V_2 \setminus \{W_1, \ldots, W_k\}$.

Finally we like to mention that the search procedure described in this section can be enhanced by including a lookahead mechanism that returns a lower estimate of the future cost. Such a mechanism has been used in the matching of ordinary graphs [16, 17]. It can potentially avoid the exploration of useless branches in the search tree and can speed up the computation. However, our current implementation does not provide for such a lookahead and we leave the design of an efficient lookahead mechanism for future work.

5 Experimental Results

In this section we present the results of two different sets of experiments. The experiments described in Section 5.1 are conducted on synthetic data. Here we are interested to study the computational complexity of hypergraph matching. Our special interest is to see how the computation time behaves when one goes from ordinary graphs to hypergraphs of comparable size, and when one increases the order of the hyperedges in a hypergraph. In the second set of experiments we use pseudo-real and real data. Here our objective is to see whether the performance of graph classification can be improved by using hypergraphs rather than graphs for object representation. All experiments reported in this chapter are conducted on a 3 GHz Pentium 4 computer with 1024 MB memory under the Windows 2000 operating system.

5.1 Experiments on Synthetic Data

The hypergraphs used in the first series of experiments are derived from the graph database described in [19].[2] This database was designed to test graph and subgraph isomorphism algorithms. It includes various types of graphs, for example, regular and irregular meshes, and graphs with bounded valence. For the experiments described below, graphs with randomly generated edges are considered. These graphs consist of n nodes with labels and unlabelled edges that are randomly inserted between pairs of nodes. There is a parameter $0 < r \leq 1$ that controls the edge density. That is, the total number of edges is given by $n(n-1)r$.

As the database described in [19] consists of graphs rather than hypergraphs, the first task in the experimental evaluation is to convert all graphs to hypergraphs. This conversion is accomplished according to the following rules:

1. Each node of a given graph g becomes a node of the corresponding hypergraph h; the label of the node remains unchanged.

[2] The database is available under `http://amalfi.dis.unina.it`.

2. The hyperedges of a hypergraph h are all of the same order N. This condition is imposed in order to make the analysis of the computational complexity, as a function of the order of the hyperedges, as transparent as possible.
3. Let m be the number of edges in g. Then there should be $2m/N$ hyperedges in h. If $2m/N$ is not an integer number, we choose the largest integer smaller than $2m/N$ to be the number of hyperedges in h. This condition ensures that the amount of data to be processed by the matching algorithm remains approximately the same when we increase the order of the hypergraph. If $N = 2$, i.e. for an ordinary graph, there are $2m$ node instances incident to an edge. All these node instances need to be checked during graph matching. In a hypergraph with m' hyperedges of order N we have $m'N$ node instances incident to a hyperedge. Hence if we let $m' = 2m/N$ the amount of data to be dealt with by the matching algorithm remains approximately the same independently of the order of the hypergraph. The hyperedges of h are randomly generated in such a way that no node can participate more than once in the same hyperedge. For $N = 2$, the original graphs from [19] are taken.

Our first experiment is concerned with the computational complexity of the proposed hypergraph isomorphism algorithm. In particular we are interested to see how the computation time of the algorithm behaves when we increase the order of the underlying hypergraphs. In this experiment, pairs of isomorphic hypergraphs with $n = 40, 60, 80, 100$ nodes are used. For each value of n, three different values of parameter r are chosen, i.e., $r = 0.01, 0.05, 0.1$. The order of the hypergraphs is increased in steps of one from $N = 2$ up to $N = 10$. For each parameter configuration, 100 tests are run and the average computation time is recorded.

Figs. 7, a–d show the average computation time needed for testing a pair of hypergraphs for isomorphism. The computation time is shown as a function of the order N of the hyperedges. In each plot, three different functions are displayed, corresponding to the three different values of parameter r. We observe a similar behaviour of each individual function in each plot. When the order N is increased, the computation time increases up to a maximum value and then decreases again. The strongest (weakest) increase occurs for the largest (smallest) number of hyperedges, i.e. for the largest (smallest) value of parameter r. This behaviour can be explained as follows. Our algorithm for hypergraph isomorphism uses a future match table as described in [8]. This future match table is the basis of a forward-checking procedure that eliminates all future node correspondences that don't comply with constraints 1 to 3 mentioned in Section 4. If the order of the underlying hypergraph increases, the handling of the future match table requires substantially more time. On the other hand, with an increasing order of the hypergraph, the topological constraints imposed by the future match table become stronger and thus the search space is reduced. The combination of these two effects leads to the behaviour shown

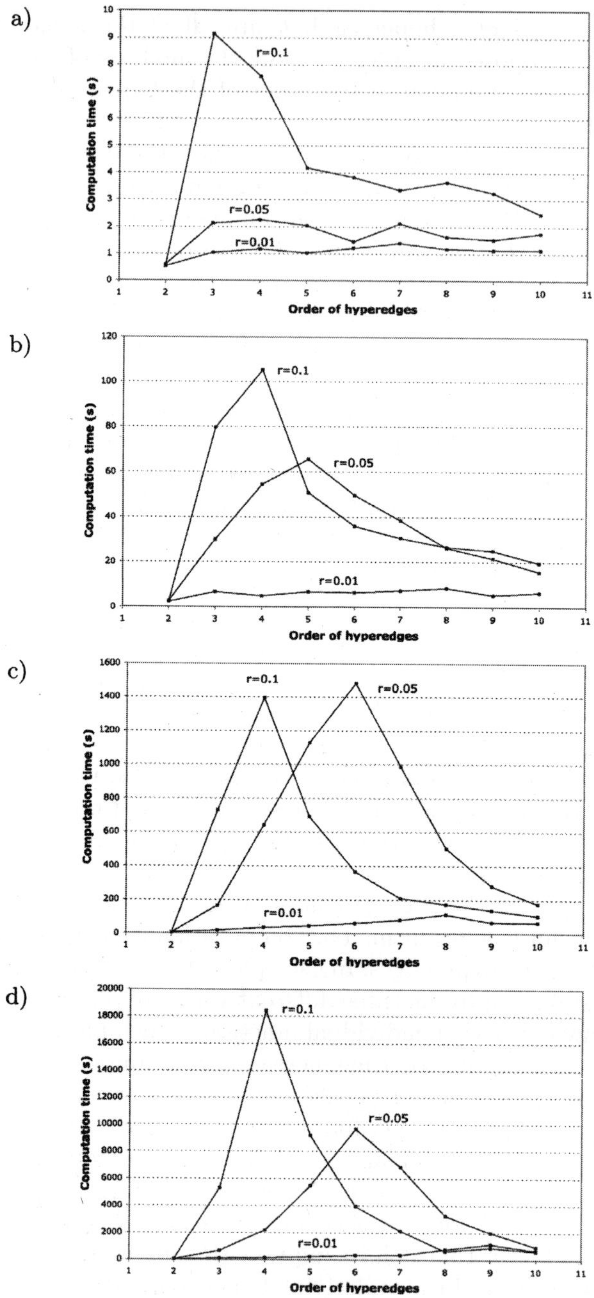

Fig. 7. Computation time of hypergraph isomorphism as a function of the order of the hyperedges; hypergraph size is a) 40 nodes, b) 60 nodes, c) 80 nodes, d) 100 nodes

in Figs. 7, a–d. The two effects become more pronounced as the number of hyperedges increases, i.e. with an increasing value of parameter r.

From Figs. 7, a–d one can draw some first conclusions. As the order of the hyperedges grows, testing two hypergraphs for isomorphism becomes computationally more expensive. However, the additional overhead does not monotonically increase with the order of the hyperedges. The computation time assumes a local maximum, and then decreases again. Hence if a graph isomorphism problem of a certain size is computationally tractable, it can be expected that a hypergraph isomorphism problem involving the same amount of data is tractable as well. Note that our algorithm for hypergraph isomorphism can also be applied to testing two graphs for sub-hypergraph isomorphism.

Next we analyse the complexity of computing the maximum common sub-hypergraph. In this experiment, again pairs of graphs from the database described in [19] are used. Their conversion into hypergraphs is done in the same way as in the first experiment. The size of the graphs varies from 35 up to 100 nodes. Two different values of parameter r for controlling the hyperedge density are used. The size of the maximum common sub-hypergraph is between 10% and 90% of the size of the two underlying hypergraphs. Again, the algorithm is run on 100 pairs of hypergraphs for each parameter configuration and the average computation time is recorded.

Fig. 8 shows the time needed to compute the maximum common sub-hypergraph as a function of the order of the hyperedges. In contrast with Figs. 7, a–d we observe that there is no obvious dependency of the computation time on the hyperegde order. The plot shown in Fig. 8 is based on hypergraphs with 35 nodes, edge densitities corresponding to $r = 0.1$ and $r = 0.2$ and an average size of the maximum common sub-hypergraph of 3.5 nodes. Similar results are obtained for other parameter configurations. The independency of the computational complexity of the hyergraph order can be explained by the fact that, in contrast to the algorithm for sub-hypergraph and hypergraph isomorphism, there is no future match table used for maximum common sub-hypergraph computation. The algorithm follows a best-first tree-search strategy. The computational complexity of the search depends on the number of nodes and the number of node instances incident to a hyperedge, but not on the order of the hyperedges. From this experiment we can conclude that going from ordinary graphs to hypergraphs and increasing the order of the hyperedges does not lead to any substantial computational overhead in maximum common sub-hypergraph computation.

5.2 Experiments on Pseudo-real and Real Data

The aim of the experiments described in Section 5.2 is to investigate whether hypergraphs have a higher representational power than ordinary graphs. That is, it is attempted to find out whether a higher recognition rate can be achieved

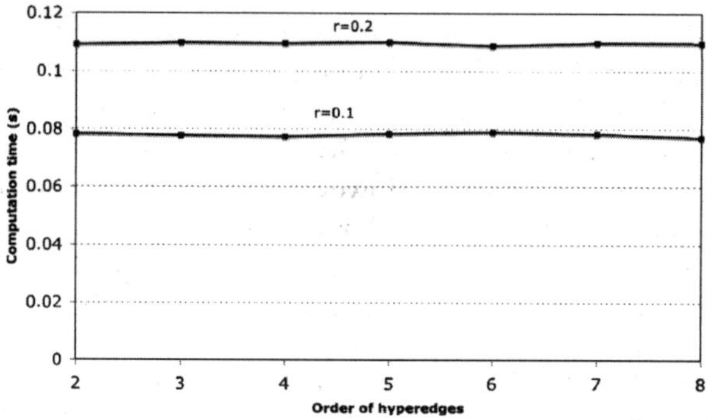

Fig. 8. Computation time of maximum common sub-hypergraph as a function of the order of the hyperedges; hypergraph size is 35 nodes and the maximum common sub-hypergraph includes 3.5 nodes on the average

in certain graph classification tasks when one uses hypergraphs instead of ordinary graphs for object representation. Simple k-nearest-neighbour classifiers based on graph and hypergraph edit distance are used in the experiments described below.

Two data sets have been used in the experiments described in this chapter. The first, the so-called *letter data set*, consists of graphs representing line drawings of capital letters. First, 15 prototype line drawings, one for each class, are constructed in a manual fashion. The considered classes correspond to those 15 capital letters of the Roman alphabet that can be drawn with straight lines only (A, E, F, H, I, K, L, M, N, T, V, W, X, Y, Z). An illustration of the prototypes is provided in Fig. 9. A noisy sample set of graphs is then created by repeatedly applying random distortions to the clean prototypes. The distortions include the translation of ending points of line segments as well as the insertion of spurious line segments and the deletion of existing line segments. There is a single parameter that controls the distortion strength. Several examples of distorted line drawings representing letter A are shown in Fig. 10. Line drawings are then converted into graphs by representing ending points of lines by nodes and lines by unlabeled edges. Each node is provided with a two-dimensional label specifying its position with respect to an underlying coordinate system.

In the graph representation there is no explicit information included as to whether two line segments cross each other. However, it can be expected that knowledge of this kind is potentially valuable for better recognizing the individual characters. Obviously there is no direct way to include this kind of information in an ordinary graph. But it is straightforward to represent information about crossing line segments in a hypergraph. We just need to define a special type of hyperedge of order four. Hyperedges of this kind connect a quadruple of nodes if and only if the nodes represent ending points of lines that cross each other. The hypergraph representation obtained thus

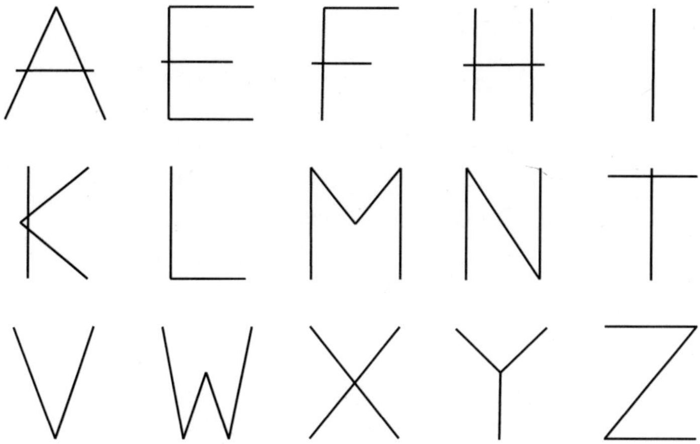

Fig. 9. Prototype letters

Fig. 10. Samples of distorted letters

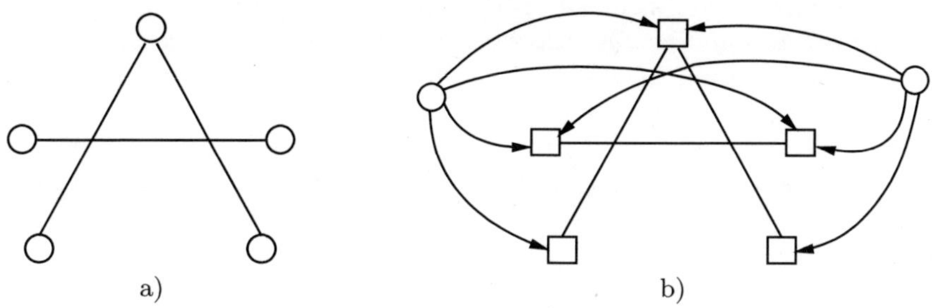

Fig. 11. a) Graph representation of letter A; b) Hypergraph representation of the same letter

includes all nodes and edges contained in the ordinary graph representation of the letters plus all possible hyperedges of order four. An example is shown in Fig. 11.

For both the graph and the hypergraph representation, the data set is split into a training, a validation, and a test set of size 750 each. Thus there are 75 instances of each letter in each of these sets. Training, validation, and

Table 1. Experimental results

Database	Graphs	Hypergraphs
Letters		
weak distortion	98.0%	99.73%*
medium distortion	98.0%	98.67%
strong distortion	94.66%	94.13%
Fingerprints	76.2%	78.03%*

* Improvement statistically significant ($\alpha = 5\%$)

test set are pairwise disjoint. The graphs and hypergraphs in the training set serve as the prototypes of the nearest neighbour classifier. The validation set is used in order to find the optimal number of nearest neighbours to be used for classification. That is, the number k of nearest neighbours in the k-nearest-neighbour classifier that leads to the smallest error on the validation set is chosen for classifying the test set. The optimization of the number of nearest neighbours is done independently for the graph and the hypergraph representation for each considered distortion level.

The classification results on the test set for three different degrees of distortion are given in the first three rows of Table 1. We observe that for distortions of small and medium strength the hypergraph representation leads to a higher recognition rate. The improvement is statistically significant at the 95% level for small distortions. For severe distortions (see Fig. 10) the graph representation is superior to the hypergraphs. This can be explained by the fact that the additional information included in the hypergraphs becomes rather adverse if the line drawings are heavily distorted.

The final experiment described in this chapter is based on the so-called *fingerprint database*. This data set consists of graphs representing fingerprint images labelled according to the five standard Henry-Galton fingerprint classes *arch, tented arch, left loop, right loop*, and *whorl* [20, 21]. Example fingerprint images are shown in Fig. 12. In our experiments, we use the 3,300 fingerprints from the NIST-4 database [22] that are labeled with a single class only. First, the foreground of a fingerprint image is segmented from the background, the direction of ridge lines is locally estimated using a Sobel operator [23], and a characteristic singular region is extracted from the resulting orientation field. These regions are then skeletonized and finally turned into a graph consisting of nodes with a position label and edges with an orientation label [24]. The data set consists of a training and a validation set each of size 150, and a test set of size 3,000.

A closer examination reveals that the graphs extracted from the fingerprint images contain many chains of nodes with degree 2. It can be conjectured that the interior nodes along such a chain do not convey salient information for discriminating between the different fingerprint classes. For this reason the

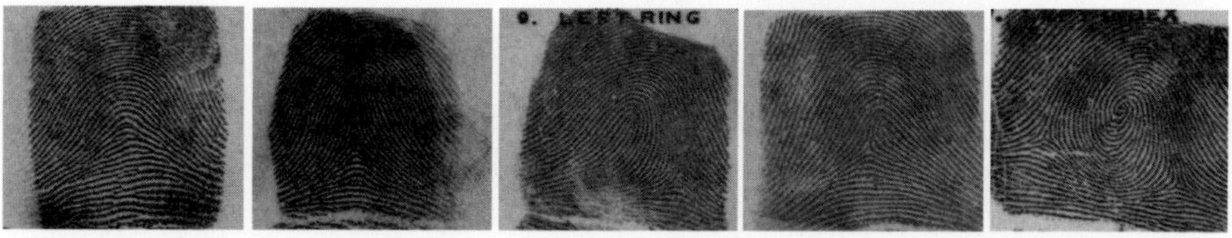

a) b) c) d) e)

Fig. 12. Fingerprint data set: Examples from the five classes a) *arch*, b) *tented arch*, c) *left loop*, d) *right loop*, and e) *whorl*

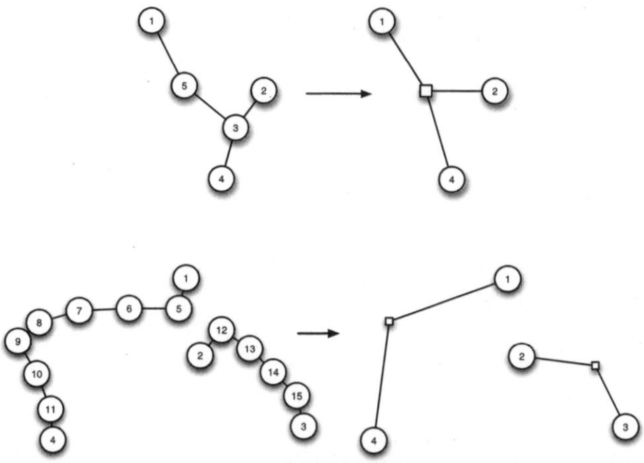

Fig. 13. a) Example of converting a graph representing a fingerprint into a hypergraph: The resulting hypergraph is of order three. b) Another example where the resulting hypergraph is of order two

graphs extracted from the images are converted into hypergraphs by only keeping the extremal nodes of connected components (i.e. the nodes of degree 1 at the end of a chain) and connecting these nodes by a hyperedge. If a connected component contains n extremal nodes, then those extremal nodes are connected by a hyperedge of degree n. All other nodes and edges are deleted from the connected component. Fig. 13,a shows an example with one connected component containing three extremal nodes, and Fig. 13,b shows a similar example with two connected components, each having two extremal nodes.

The classification results on the test set are given in the last row of Table 1. Using the ordinary graph representation a correct recognition rate of 76.2% is obtained. A higher correct recognition rate of 78.03% is achieved with the hypergraph representation. This improvement is statistically significant at the 95% level.

The experimental results achieved on both the letter and the fingerprint database clearly indicate that hypergraphs have the potential to improve the recognition accuracy in various graph classification tasks.

6 Conclusions

Graphs have become a well-established representation formalism in pattern recognition. There is a large number of applications where graphs and graph matching algorithms have been used successfully [1]. Nevertheless, graphs are restricted in the sense that only two-dimensional relations can be modelled. In this paper we have investigated a more general framework that is based on hypergraphs. Hypergraphs allow us to represent not only binary relations, but relations of any finite order, and include graphs as a special case.

A fundamental requirement for any graph-based formalism in pattern recognition is the availability of related graph matching algorithms. For the case of normal graphs, such algorithms exist. Examples are isomorphism, subgraph isomorphism, maximum common subgraph, and graph edit distance computation. On top of such algorithms, classification and clustering procedures can be implemented. In this paper we show that similar matching algorithms can be designed for the domain of hypergraphs. This makes the enhanced representational power of hypergraphs available for a potentially large number of practical pattern recognition applications.

In a number of experiments with artificial data, we have studied the computational complexity of the proposed algorithms for hypergraph matching. We were particularly interested in the question whether the computational overhead increases when one goes from ordinary graphs to hypergraphs, and increases the order of hyperedges. For the case of hypergraph isomorphism it turns out that there is in fact an increase in computational complexity when the order of the hyperedges grows. However, the complexity is not monotonically increasing with a growing order of the hyperedges. It increases only up to a local maximum and then decreases again. Hence if a graph isomorphism problem of a certain size is computationally tractable, it can be expected that a hypergraph isomorphism problem of comparable size is tractable as well. In case of maximum common sub-hypergraph computation, our results indicate that there is no dependency of the computational complexity on the order of the hyperedges at all. Hence hypergraphs of any order may be used without introducing any additional computational overhead.

In a few other experiments, using pseudo-real and real data, it has been shown that hypergraphs have more representational power than ordinary graphs. This increase in representation power leads to higher correct recognition rates in certain graph classification tasks.

The algorithms considered in this chapter are characterized by their rather high computational complexity, which is due to the fact that the underlying graph matching tasks are NP complete.[3] Therefore their applicability is limited to rather small graphs. Recently, some novel approximate algorithms for graph edit distance computation have been proposed [25, 26]. These algorithms run quite fast in practice and are able to cope with fairly large graphs. At the same time they do not necessarily lead to a decrease of the recognition rate when used in conjunction with a nearest neighbour classifier. One of our future research goals consists in extending the scope of these approximate algorithms from graphs to hypergraphs.

Another challenging problem left to future research is to identify application areas and classification tasks that are particularly suitable for hypergraphs. In the applications considered in Section 5.2 the objects were originally given in terms of graphs and turned into hypergraphs 'a posteriori'. It would interesting to find other applications where the primary information is given in form of a hypergraph, or in such a way that converting it into a hypergraph is more straightforward than converting it into an ordinary graph. To analyse the recognition accuracy obtained for such problems could lead to further insight into the suitability of hypergraphs as a representation tool in pattern recognition.

References

1. D. Conte, P. Foggia, C. Sansone, and M. Vento. Thirty years of graph matching in pattern recognition. *International Journal of Pattern Recognition and Artificial Intelligence* 18(3):265–298, 2004
2. C. Berge, *Hypergraphs.* North-Holland, Amsterdam, 1989
3. A. Bretto, H. Cherifi, and D. Aboutajdine. Hypergraph imaging: An overview. *Pattern Recognition*, 35(3):651–658, 2002
4. C. Bretto and L. Gillibert. Hypergraph-based image representation. In *Proceedings of the 5th International Workshop on Graph-Based Representations in Pattern Recognition*, volume 3434, LNCS, pages 1–11. Springer, Berlin Heidelberg New York, 2005
5. A. Wong, S. Lu, and M. Rioux. Recognition and shape synthesis of 3-D objects based on attributed hypergraphs. *IEEE Transactions on Pattern Analysis and Machine Intelligence.* 11(3):279–290, 1989
6. D. Demko. Generalization of two hypergraphs. Algorithm of calculation of the greatest sub-hypergraph common to two hypergraphs annotated by semantic information. In *Graph Based Representations in Pattern Recognition, Computing.* Supplement 12, pages 1–10. Springer, Berlin Heidelberg New York, 1998
7. H. Bunke, P. Dickinson, M. Kraetzl. Theoretical and algorithmic framework for hypergraph matching. In *Proceedings of the 13th International Conference on Image Analysis and Processing*, volume 3617, LNCS, pages 463–470. Springer, Berlin Heidelberg New York, 2005
8. J. Ullman. An algorithm for subgraph isomorphism. *Journal of the Association for Computing Machinery*, 23(1):31–42, 1976

[3] For graph isomorphism it is actually not known whether it is in P or NP.

9. G. Levi. A note on the derivation of maximal common subgraphs of two directed or undirected graphs. *Calcolo*, 9:341–354, 1972
10. J. McGregor. Backtrack search algorithm and the maximal common subgraph problem. *Software Practice and Experience*, 12(1):23–34, 1982
11. A. Sanfeliu, K. Fu. A distance measure between attributed relational graphs for pattern recognition. *IEEE Transactions on Systems, Man, and Cybernetics (Part B)*, 13(3):353–363, 1983
12. B. Messmer, H. Bunke. A new algorithm for error-tolerant subgraph isomorphism detection. *IEEE Transactions on Pattern Analysis and Machine Intelligence* 20(5):493–504, 1998
13. H. Bunke, K. Shearer. A graph distance metric based on the maximal common subgraph. *Pattern Recognition Letters*, 19(3):255–259, 1998
14. W. Wallis, P. Shoubridge, M. Kraetzl, D. Ray. Graph distances using graph union. *Pattern Recognition Letters*, 22(6):701–704, 2001
15. M.L. Fernandez, G. Valiente. A graph distance metric combining maximum common subgraph and minimum common supergraph. *Pattern Recognition Letters*, 22(6–7):753–758, 2001
16. H. Bunke, G. Allermann. Inexact graph matching for structural pattern recognition. *Pattern Recognition Letters*, 1:245–253, 1983
17. A. Wong, M. You, S. Chan. An algorithm for graph optimal monomorphism. *IEEE Transactions on Systems, Man, and Cybernetics*, 20(3):628–638, 1990
18. P. Hart, N. Nilsson, B. Raphael. A formal basis for the heuristic determination of minimum cost paths. *IEEE Transactions of Systems, Science, and Cybernetics*, 4(2):100–107, 1968
19. P. Foggia, C. Sansone, M. Vento. A database of graphs for isomorphism and sub-graph isomorphism benchmarking. In *Proceedings of the 3rd International Workshop on Graph-Based Representations in Pattern Recognition*, 176–188, 2001
20. M. Kawagoe, A. Tojo. Fingerprint pattern classification. *Pattern Recognition*, 17:295–303, 1984
21. K. Karu, A. Jain. Fingerprint classification. *Pattern Recognition*, 29(3):389–404, 1996
22. C. Watson, C. Wilson. NIST special database 4, fingerprint database, 1992
23. D. Maltoni, D. Maio, A. Jain, S. Prabhakar. *Handbook of Fingerprint Recognition.* Springer, Berlin Heidelberg New York, 2003
24. M. Neuhaus, H. Bunke. A graph matching based approach to fingerprint classification using directional variance. In *Proceedings of 5th International Conference on Audio- and Video-Based Biometric Person Authentication*, volume 3546, LNCS, pages 191–200. Springer, Berlin, Heidleberg, New York, 2005
25. M. Neuhaus, K. Riesen, H. Bunke. Fast suboptimal algorithms for the computation of graph edit distance. In *Proceedings of the Institute Workshop on Structural, and Syntactical Pattern Recognition*, volume 4109, LNCS, pages 163–172. Springer, Berlin Heidelberg New York, 2006
26. K. Riesen, M. Neuhaus, H. Bunke. Bipartite graph matching for computing the edit distance of graphs. In F. Escolano and M. Vento, editors, *Graph-Based Representations in Pattern Recognition*, LNCS 4538, Springer, 2007, 1–12.

Feature-Driven Emergence of Model Graphs for Object Recognition and Categorization

Günter Westphal,[1] Christoph von der Malsburg[2,3] and Rolf P. Würtz[1]

[1] Institut für Neuroinformatik
Ruhr-Universität Bochum
D-44780 Bochum, Germany
westphal|wuertz@neuroinformatik.rub.de

[2] Frankfurt Institute for Advanced Studies
Johann Wolfgang Goethe-Universität Frankfurt am Main
D-60438 Frankfurt am Main, Germany
malsburg@fias.uni-frankfurt.de

[3] Laboratoy for Computational and Biological Vision
University of Southern California
Los Angeles, 90089-2520, USA
malsburg@organic.usc.edu

Summary. An important requirement for the expression of cognitive structures is the ability to form mental objects by rapidly binding together constituent parts. In this sense, one may conceive the brain's data structure to have the form of graphs whose nodes are labeled with elementary features. These provide a versatile data format with the ability to render the structure of any mental object. Because of the multitude of possible object variations the graphs are required to be dynamic. Upon presentation of an image a so-called model graph should rapidly emerge by binding together memorized subgraphs derived from earlier learning examples driven by the image features. In this model, the richness and flexibility of the mind is made possible by a combinatorial game of immense complexity. Consequently, emergence of model graphs is a laborious task which, in computer vision, has most often been disregarded in favor of employing model graphs tailored to specific object categories like faces in frontal pose. Invariant recognition or categorization of arbitrary objects, however, demands dynamic graphs.

In this work we propose a form of graph dynamics which proceeds in three steps. In the first step position-invariant feature detectors, which decide whether a feature is present in an image, are set up from training images. For processing arbitrary objects these features are small regular graphs, termed parquet graphs, whose nodes are attributed with Gabor amplitudes. Through combination of these classifiers into a linear discriminant that conforms to Linsker's infomax principle a weighted majority voting scheme is implemented. This network, termed the preselection network, is well suited to quickly rule out most irrelevant matches and only leaves the ambiguous cases, so-called model candidates, to be processed in a third step using a rudimentary version of elastic graph matching, a standard correspondence-based technique

G. Westphal et al.: *Feature-Driven Emergence of Model Graphs for Object Recognition and Categorization*, Studies in Computational Intelligence (SCI) **91**, 155–199 (2008)
www.springerlink.com

for face and object recognition. To further differentiate between model candidates with similar features it is asserted that the features be in similar spatial arrangement for the model to be selected. Model graphs are constructed dynamically by assembling model features into larger graphs according to their spatial arrangement. The model candidate whose model graph attains the best similarity to the input image is chosen as the recognized model.

We report the results of experiments on standard databases for object recognition and categorization. The method achieved high recognition rates on identity, object category, and pose, provided that individual object variations are sufficiently covered by learning examples. Unlike many other models the presented technique can also cope with varying background, multiple objects, and partial occlusion.

Keywords: compositionality, model graphs, parquet graphs, position-invariant feature detectors, infomax principle, preselection network, model candidates, emergence of model graphs, elastic graph matching, feature- vs. correspondence-based object recognition

1 Introduction

An important requirement for the expression of cognitive structures is the ability to form mental objects by rapidly binding together constituent parts [2, 3]. In this sense, one may conceive the brain's data structure to have the form of graphs whose nodes are labeled with elementary features.

This data format has been used for visual object recognition [4, 5, 19, 30], and in the Dynamic Link Matching approach [11, 37–39, 47]. In all these approaches the data structure of stored objects has the form of graphs whose nodes are labeled with elementary features. These are called *model graphs* and provide a view-tuned representation [23, 25] of the object contained in the presented image. They provide a versatile data format with the capability to render the structure of any object. Because of the multitude of possible object variations like changes in identity, pose, or illumination, the graphs are required to be dynamic with respect both to shape and attributed features.

Upon presentation of an image a so-called model graph should rapidly emerge by binding together memorized subgraphs derived from earlier learning examples driven by the image features. Emergence of model graphs is a laborious task which, in computer vision, has most often been disregarded in favor of employing model graphs tailored to specific object categories like faces in frontal pose [11, 47, 49]. Recognition or categorization of arbitrary objects, however, demands dynamic graphs, i.e., more emphasis must be laid on the question of how model graphs are created from raw image data.

Relatively little work has been done on the dynamic creation of model graphs. The object recognition system proposed in [40] is based on Dynamic Link Matching supplied with object memory. While learning novel objects a so-called fusion graph is created through iteratively matching image graphs with the fusion graph and grafting non-matched parts of image graphs into the fusion graph. When an object is to be recognized, one or more image graphs are compared against model memory via graph matching, implemented by dynamic links. The matching parts of the fusion graph thus constitute the model graph for the object contained in the input image. The system has proven to perform well for a small number of object views. During both learning and recognition the objects are required to be placed in front of a plain background.

A different approach is the creation of model graphs with minimal user-assistance [16]. In that method, a growing neural gas [8] is used to determine shape and topology of a model graph. Binarized difference images derived from two consecutive images of the same moving object are used as an input to a growing neural gas, whose nodes are attracted to superthreshold frame differences. Upon a user-initiated event, Gabor jets are extracted at the node positions and the produced model graph is stored in a model database. During recognition, model graphs are matched in succession with the input image. The compositional aspect is thus prominent while learning novel objects but is absent during recognition. A rudimentary version of model graph dynamics is also present in [49], where model graphs are adapted to segmentation masks in order to ignore background influences.

In [41] a system is proposed that creates an object model in a probabilistic framework. The technique uses mixtures of collaborating probabilistic object models, termed *components*. Highly textured regions, so-called *parts*, are employed as local features. They are automatically extracted from earlier learning images. Each component is an expert for a small ensemble of object parts. In order to describe an object in an image several components need to be active. Model parameters, the parameters of the incorporated probability densities, are iteratively learned using expectation maximization (EM). Categorization of an object is based on the maximum a posteriori (MAP) decision rule: the object in the input image is supposed to belong to the category whose object model attained maximal a posteriori probability. In [6] a similar method is proposed which is able to categorize objects from few learning examples.

In [31] a graph dynamics is employed for object tracking. It is formulated in a maximum a posteriori framework using a hidden Markov model: the tracker estimates the object's state, expressed by a model graph, through maximization of a posterior probability. New features are added to the model graph if they can reliably be observed in the hidden Markov model's time window. Similarly, repeatedly non-matching features are removed from the model graph.

Recognition methods relying on graph matching are *correspondence-based* in the sense that image point correspondences are estimated before recognition is attempted. This estimation is usually only possible on the basis of the spatial arrangement of elementary features. There is also a class of recognition algorithms which are purely *feature-based* and completely disregard feature arrangement. A prominent example is SEEMORE [18]. There it is shown that a simple neural network can distinguish objects in a purely feature-based way if enough feature types are employed. As a model for recognition and categorization in the brain feature-based methods can be implemented as feedforward networks, which would account for the amazing speed with which these processes can be carried out, relative to the slow processing speed of the underlying neurons [33, 34]. These methods, however, encounter problems in the case of multiple objects and highly structured backgrounds. From the point of view of pattern recognition, feature-based methods are *discriminative* while graph matching is *generative* [35].

It is reasonable to assume that feedforward processing is applied as far as it goes by excluding as many objects as possible and that only ambiguous cases are subjected to correspondence-based processing, which is more time-consuming.

In this chapter we propose a form of graph dynamics, which proceeds in three steps. In the first step *position-invariant feature detectors*, which decide whether a feature is present in an image, are set up from training images. For processing arbitrary objects, features are small localized grid graphs, so-called *parquet graphs*, whose nodes are attributed with Gabor amplitudes. Through combination of these classifiers into a single layer perceptron that conforms to Linsker's infomax principle [14], the so-called *preselection network*, a weighted majority voting scheme [12] is implemented. It allows for preselection of salient learning examples, so-called *model candidates*, and likewise for preselection of salient categories the object in the presented image supposedly belongs to. Each model candidate is verified in a third step using a rudimentary version of elastic graph matching. To further differentiate between model candidates with similar features it is asserted that the features be in similar spatial arrangement for the model to be selected. In this way model graphs are constructed dynamically by assembling model features into larger graphs according to their spatial arrangement (fig. 1). Finally, the resulting model graphs are matched with a rudimentary version of elastic graph matching, and the model candidate that yields the best similarity to the input image is chosen as the recognized model (fig. 2).

The description of the method is accompanied by a *case study*, which exemplifies the various steps on an example, in which only two images of two objects are learned and distinguished.

2 Learning Set, Partitionings, and Categories

There are many different classifications that can be made on image data. For object recognition, all instances of the same object under different pose and/or illumination are to be put into the same class. An alternative learning problem may be the classification of illumination or pose regardless of object identity. A hallmark of human visual cognition is the classification into *categories*: we group together images of cats, dogs, insects, and reptiles into the

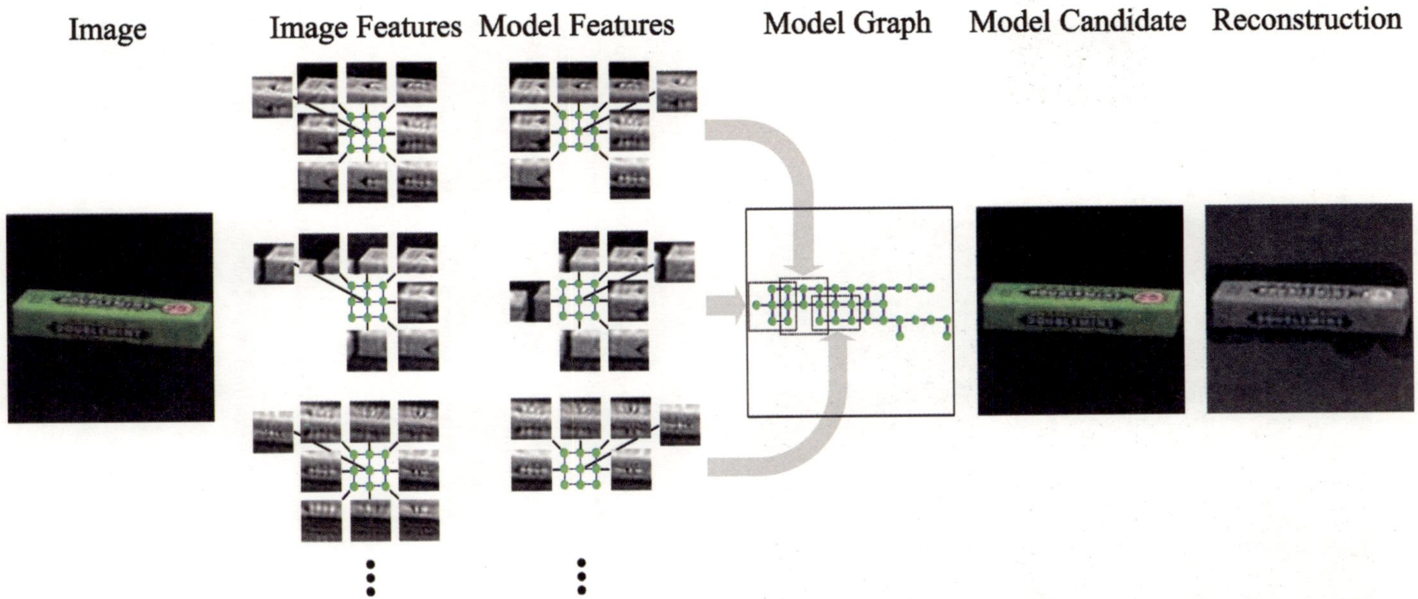

Fig. 1. Feature-Driven Emergence of Model Graphs — *Upon presentation of an image (first column) a model graph (fourth column) should rapidly emerge by binding together (arrows) memorized subgraphs, termed parquet graphs in this work, derived from earlier learning examples (third column) that match with the image features (second column). Column six shows the reconstruction from the model graph. The graph dynamics itself proceeds in three steps. In the first step position-invariant feature detectors, are learned from training images. Through combination of these classifiers into a single-layer perceptron, a weighted majority voting scheme is implemented. It allows for preselection of salient learning examples, so-called model candidates (fifth column), and likewise for preselection of salient categories the object in the presented image hypothetically belongs to. Each model candidate is verified in a third step using a variant of elastic graph matching. To further differentiate between model candidates with similar features similar spatial arrangement for the model features is asserted. Reconstruction and the model candidate contain the same object in the same pose, which is slightly different from the one in the input image*

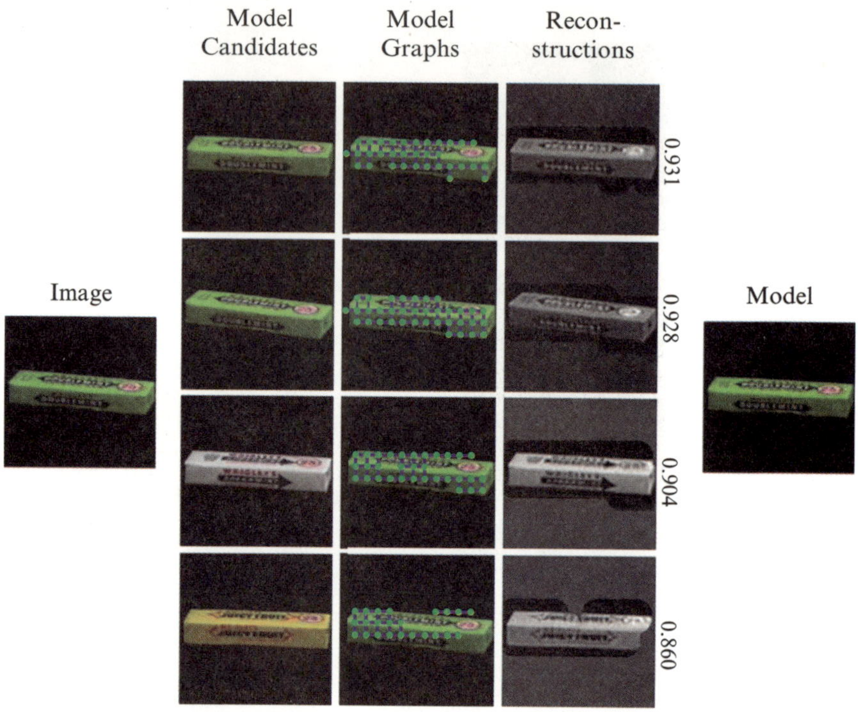

Fig. 2. Selection of the Model — *Given the input image in the first column, the preselection network selects four model candidates (second column). As has been illustrated in fig. 1, a model graph is dynamically constructed for each model candidate by assembling matching model features into larger graphs according to their spatial arrangement (third column). The fourth column shows the reconstruction from each model graph. Each model candidate is verified using a rudimentary version of elastic graph matching. Model graphs are optimally placed on the object contained in the input image in terms of maximizing the measure of similarity (third column). The attained similarities between the model candidates, represented by their model graphs, and the input image are annotated to the reconstructions. The model candidate that yields the best similarity to the input image is chosen as the recognized model (fifth column)*

category 'animal' and are able to differentiate animals from non-animals with impressive speed [33].

Following [22] we use the term *recognition* for a decision about an object's unique identity. Recognition thus requires subjects to discriminate between similar objects and involves generalization across some shape changes as well as physical translation, rotation and so forth. The term *categorization* refers to a decision about an object's kind. Categorization thus requires generalization across members of a class of objects with different shapes. Especially, the system has to generalize over object identity.

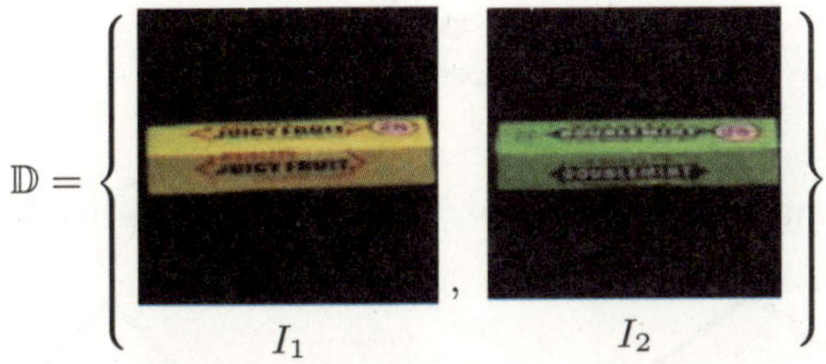

Fig. 3. Case Study: Learning Set — *The learning set comprises two images of different chewing gum packages in approximately the same pose. The images are taken from the COIL-100 database [21]. In the following these images are referred to as I_1 and I_2*

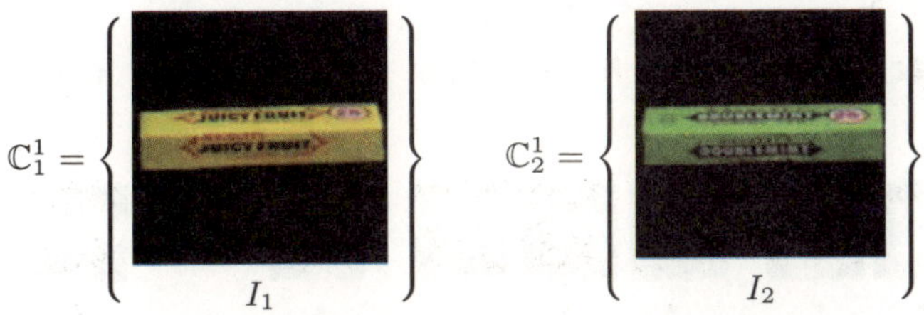

Fig. 4. Case Study: Partitioning of the Learning Set — *In our case study there exists only $K = 1$ partitioning Π^1 of the learning set (fig. 3). The partitioning consists of $C^1 = 2$ single-element categories $\mathbb{C}_1^1 = \{I_1\}$ and $\mathbb{C}_2^1 = \{I_2\}$*

We start by considering some finite set of images $\mathbb{I}$ and a subset $\mathbb{D}$, which we call the *learning set*. In our case study the learning set comprises two images of different chewing gum packages in approximately the same pose (fig. 3).

In order to accommodate the various learning tasks that can be imposed on a single image set we consider that there exist K *partitionings* Π^k of the learning set (1). A partitioning Π^k consists of C^k pairwise disjoint partitions $\mathbb{C}_c^k$.

$$\Pi^k = \left\{ \mathbb{C}_c^k \subseteq \mathbb{D} \middle| 1 \leq c \leq C^k \right\}$$
$$\text{with } \forall c \neq c' : \mathbb{C}_c^k \cap \mathbb{C}_{c'}^k = \emptyset \quad \text{and} \quad \bigcup_{c=1}^{C^k} \mathbb{C}_c^k = \mathbb{D} \tag{1}$$

The objects in the images of a particular partition are conceived to share a common semantic property, for instance, being images of animals, or having the same illumination direction. Accordingly, partitions in the following are termed *categories*. *Category labels* c range between 1 and C^k; their range implicitly depends on the number of categories in the underlying partitioning Π^k. For simultaneous recognition of the object's identity and the object's pose the learning set is subdivided into single-element categories while for object categorization purposes the learning set is usually organized in a hierarchy of categories. In fig. 4 the single partitioning of the learning set in our case study is shown.

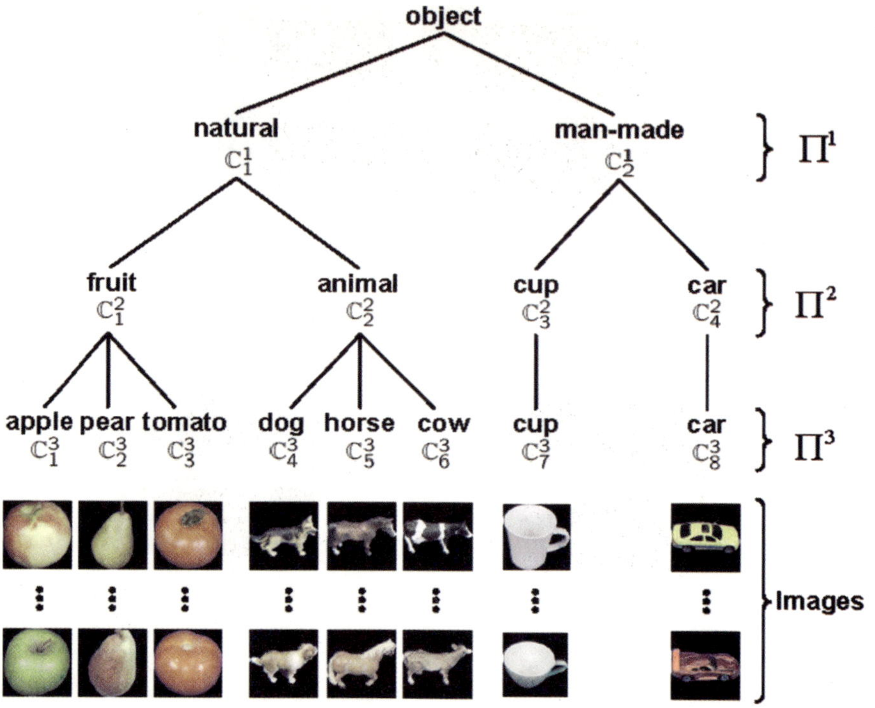

Fig. 5. Hierarchical Organization of Categories — *A hierarchy of categories on the ETH-80 image database [13], which contains images of apples, pears, tomatoes, dogs, horses, cows, cups, and cars in varying poses and identities. We created $K = 3$ partitionings Π^1, Π^2, and Π^3. Partitioning Π^1 comprises $C^1 = 2$ categories of natural ($\mathbb{C}_1^1$) and man-made objects ($\mathbb{C}_2^1$). Partitioning Π^2 comprises $C^2 = 4$ categories of fruits ($\mathbb{C}_1^2$), animals ($\mathbb{C}_2^2$), cups ($\mathbb{C}_3^2$), and cars ($\mathbb{C}_4^2$). Finally, partitioning Π^3 comprises $C^3 = 8$ categories of apples ($\mathbb{C}_1^3$), pears ($\mathbb{C}_2^3$), tomatoes ($\mathbb{C}_3^3$), dogs ($\mathbb{C}_4^3$), horses ($\mathbb{C}_5^3$), cows ($\mathbb{C}_6^3$), cups ($\mathbb{C}_7^3$), and cars ($\mathbb{C}_8^3$)*

A hierarchical categorization task can be exemplified with the ETH-80 image database [13]. That database comprises images of apples, pears, tomatoes, dogs, horses, cows, cups, and cars in varying poses and identities and has been used for the categorization experiments in sect. 7.2. For those experiments we created $K = 3$ partitionings of the learning set as shown in fig. 5.

3 Parquet Graphs

The feature-based part of the technique described in this paper can work with any convenient feature type. A successful application employing color and multiresolution image information is presented in [45]. For the current combination of feature- and correspondence-based methods we chose small regular graphs labeled with Gabor features. We call them *parquet graphs* inspired by the look of ready-to-lay parquet tiles. These can work as simple feature detectors for preselection and be aggregated to larger graph entities for correspondence-based processing.

Throughout this paper, parquet graphs are constituted out of $V = 9$ nodes. In the following, a parquet graph f is described with a finite set of node attributes: Each node v is labeled with a triple $(\mathbf{x}_v, \mathcal{J}_v, b_v)$ where $\mathcal{J}_v$ is a Gabor jet derived from an image at an absolute node position $\mathbf{x}_v$. Computation and parameters of the Gabor features is the same as in [11, 47]. In order to make use of segmentation information it is convenient to mark certain nodes as *invalid* and exclude them from further calculation in that way. For this purpose the node attributes comprise the validity flag b_v that can take the values 0 and 1, meaning 'invalid' and 'valid'. The horizontal and vertical node distances Δx and Δy are set to 10 pixels in this work.

$$f = \{\, (\mathbf{x}_v, \mathcal{J}_v, b_v) | \, 1 \leq v \leq V\} \tag{2}$$

In fig. 6 an example of a parquet graph that has been placed on the object in learning image I_1 is shown. Where appropriate, instances of parquet graphs are, more generally, called features or feature instances.

For selection of salient categories and model candidates, the feature-based part of the proposed system, a parquet graph describes a patch of texture derived from an image regardless of its position in the image plane. Particularly, this means that the node positions are irrelevant for the decision whether two images contain a similar patch of texture. Later, for verification of the selected model candidates, i.e., learning images that may serve as models for the input image, larger graphs are constructed dynamically by assembling

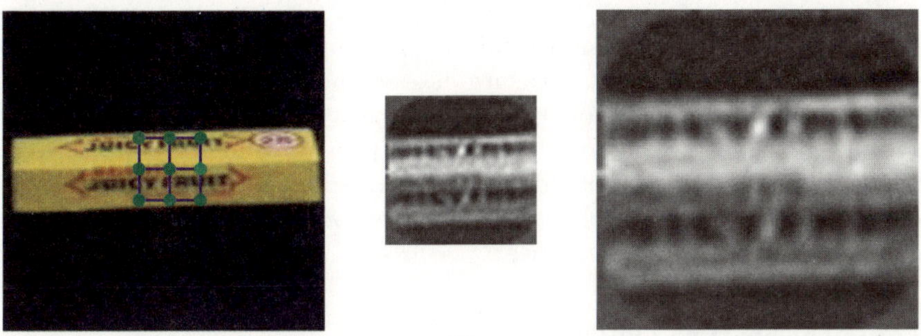

(a) (b) (c)

Fig. 6. Example of a Parquet Graph — *Figure* **(a)** *shows a parquet graph that has been placed on the object in learning image I_1. Each node of a parquet graph is attributed with Gabor amplitudes derived from an image at the node's position. Figure* **(b)** *shows the reconstruction from the parquet graph. Figure* **(c)** *is an enlarged version of fig.* **(b)**. *The reconstruction is computed with the algorithm from [24]*

parquet graphs derived from earlier learning images according to their spatial arrangement. Thus, within the correspondence-based part, the node positions will become important.

3.1 Similarity Function

The measure of similarity between two parquet graphs f and f' is defined as the normalized sum of the similarities between valid Gabor jets [28, 49] attached to nodes with the same index that stem from the given parquet graphs (4). Throughout this paper, the similarity between two Gabor jets is given by the normalized scalar product between the absolute values of the complex components of the two jets (3). Let a_n denote the absolute value of n-th filter response.

$$s_{abs}\left(\mathcal{J},\mathcal{J}'\right)=\frac{\sum_n a_n a'_n}{\sqrt{\sum_n a_n^2 \sum_n {a'_n}^2}} \tag{3}$$

By definition, the factors $(b_v b'_v)$ are 1 if the respective jets $\mathcal{J}_v$ and $\mathcal{J}'_v$ have both been marked as valid, and 0 otherwise. Thus, these factors assert that only similarities between jets that have both been marked as valid are taken into account. If all products become 0, the similarity between the two parquet graphs yields 0.

$$s_{graph}\left(f,f'\right)=\begin{cases}\left(\sum\limits_{v=1}^{V} b_v b'_v\right)^{-1} \sum\limits_{v=1}^{V} (b_v b'_v) s_{abs}\left(\mathcal{J}_v,\mathcal{J}'_v\right) & \text{if } \sum\limits_{v=1}^{V} b_v b'_v > 0 \\ 0 & \text{otherwise}\end{cases} \tag{4}$$

From the viewpoint of the correspondence problem, two parquet graphs in different images establish a local *array* of contiguous point-to-point correspondences. The similarity measure assesses how well points in two images specified by the given parquet graphs actually correspond to each other. It is well worth noting that parquet graphs provide a means to protect from accidentally establishing point-to-point correspondences in that contiguous, topographically smooth fields of good correspondences are favored over good but topographically isolated ones.

3.2 Local Feature Detectors

For the assessment whether two parquet graphs f and f' convey similar patches of texture with respect to a given sensitivity profile we introduce local feature detectors that return 1 if the similarity between the given parquet graphs is greater or equal than a given similarity threshold $0 < \vartheta \leq 1$, and 0 otherwise (5). We say that two parquet graphs *match* with respect to a given similarity threshold if the local feature detector returns 1.

$$\varepsilon(f, f', \vartheta) = \begin{cases} 1 \text{ if } s_{graph}(f, f') \geq \vartheta \\ 0 \text{ otherwise} \end{cases} \tag{5}$$

Matching features are one argument for point-to-point *correspondences*, which needs to be backed up by the spatial arrangement of several matching features.

4 Learning a Visual Dictionary

Our goal is to formulate a graph dynamics that, upon image presentation, lets a model graph rapidly emerge by binding together memorized subgraphs derived from earlier learning examples. To this end we need to compute a repertoire of parquet graphs from learning examples in advance. These play the role of a visual dictionary. Parquet graphs derived from an input image during classification are looked up in the dictionary to find out which image and model features match. Each coincidence of a matching feature in the image and model domain may then be accounted as a piece of evidence that the input image belongs to the same categories as the learning image which contains the model feature.

4.1 Feature Calculators

In (6) we define R functions f^r capable of extracting a set of features out of an image. In this work parquet graphs are exclusively used as local image features. Let $\mathbb{F}$ denote the set of all possible features and let $\wp(\mathbb{F})$ denote the power set of $\mathbb{F}$. In the following these functions will be called *feature calculators*. The index r implicitly specifies the parameterization of the parquet graphs returned from the respective feature calculator f^r, like the similarity threshold ϑ^r, which is employed in the local feature detectors (5). Generally, feature calculators are not restricted to parquet graphs; other feature types have been used in [1, 27, 43, 45].

$$f^r : \mathbb{I} \to \wp(\mathbb{F}) \text{ with } r \in \{1, \ldots, R\} \tag{6}$$

For extraction of parquet graphs, the inter-node distances Δx and Δy are also used to specify a grid in the image plane. At each grid position allowing for placement of a whole parquet graph, a parquet graph is extracted. Scanning of the image starts in the upper left corner from left to right to the lower right corner. If the image is known to be figure-ground segmented, parquet graphs with the majority of nodes residing in the background will be disregarded, the others have background points marked as invalid.

In the case study, we employ only $R = 1$ feature calculator f^1. The feature calculator returns a set of parquet graphs with ten pixels distance between two neighbored nodes in horizontal and in vertical direction, respectively. In fig. 7 the result of consecutively applying this feature calculator to both learning examples is shown.

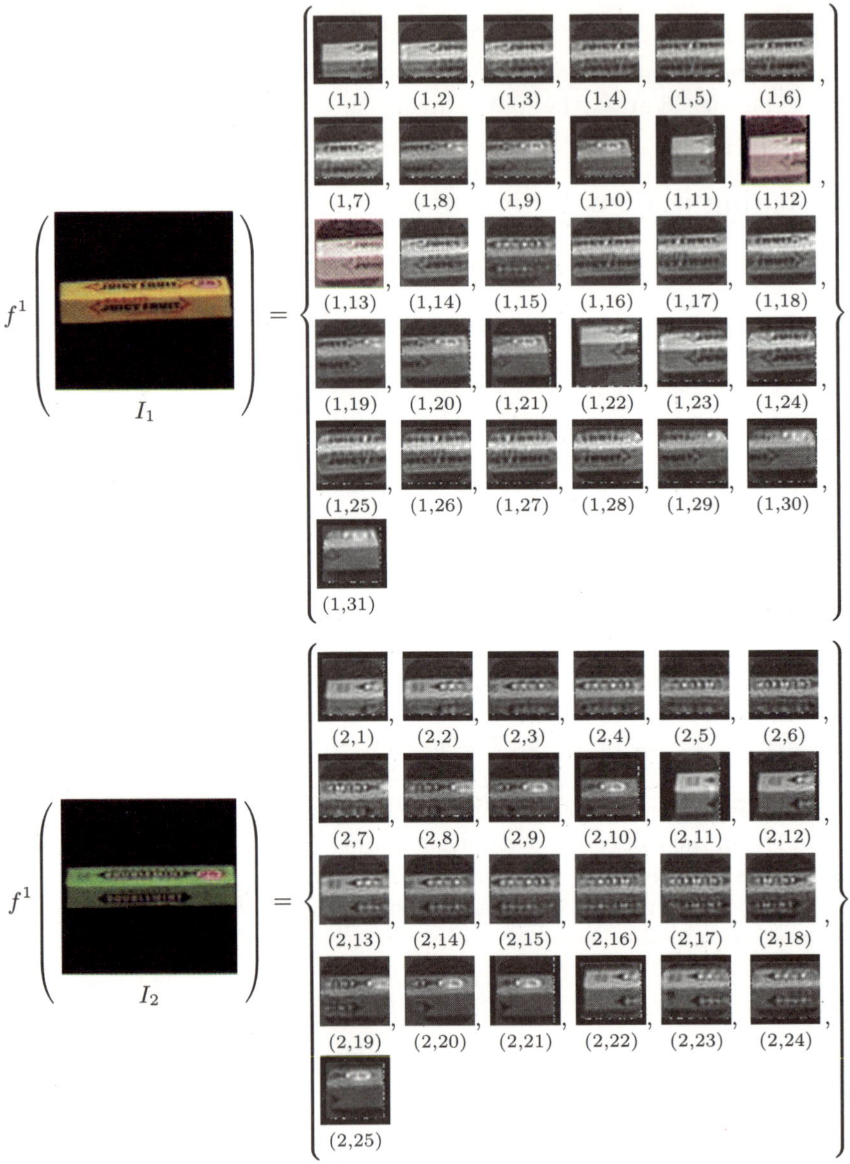

Fig. 7. Case Study: Application of the Feature Calculator to the Learning Images — *The thumbnail images in the returned sets on the right hand side are reconstructions from the extracted parquet graphs. Each reconstruction is uniquely labeled with a tuple. The first component addresses the learning image the parquet graph stems from while the second component is a sequential number*

4.2 Feature Vectors

Looking at the number of parquet graphs that have been extracted from just two images (fig. 7), it is clear that for learning sets with thousands or even ten thousands of images the total number of features would grow into astronomical dimensions. Consequently, we have to limit the total number of features to a tractable number. For this task we employ a simple variant of vector quantization [9] given as pseudo code in fig. 8. A vector quantizer maps data vectors in some vector space into a finite set of *codewords*, which are supposed to represent the original set of input vectors well. A collection of codewords that purposefully represent the set of input vectors is termed *codebook*. The design of an optimal codebook is NP-hard.

Algorithm 1: *vectorQuantization*

Parameter : Learning Set: $\mathbb{D}$
Parameter : Feature Calculator: $f^r : \mathbb{I} \rightarrow \wp(\mathbb{F})$
Parameter : Similarity Threshold: ϑ^r; $0 < \vartheta^r \leq 1$
Result : Feature Vector of Length T^r: $\mathbf{f}^r$

$\mathbb{F}^r \leftarrow \emptyset$
$T^r \leftarrow 0$

forall $I \in \mathbb{D}$ **do**
  forall $f \in f^r(I)$ **do**
    if $\forall f' \in \mathbb{F}^r : \varepsilon(f, f', \vartheta^r) = 0$ **then**
      $\mathbb{F}^r \leftarrow \mathbb{F}^r \cup \{f\}$
      $T^r \leftarrow T^r + 1$
    end
  end
end

$\mathbf{f}^r =: (f_t^r)_{1 \leq t \leq T^r} \leftarrow (0)_{1 \leq t \leq T^r}$

$t \leftarrow 0$
forall $f \in \mathbb{F}^r$ **do**
  $f_t^r \leftarrow f$
  $t \leftarrow t + 1$
end

return $\mathbf{f}^r$

Fig. 8. Vector Quantization Method — *The algorithm computes a codebook of codewords. In this work parquet graphs become employed as codewords while the codebook is a set of these parquet graphs. The size of the feature set depends considerably on the value of the similarity threshold ϑ^r. For lower values of ϑ^r many features will be disregarded and the final feature set will become rather small. Conversely, higher values of ϑ^r close to one lead to low compression rates and large feature sets*

Using the vector quantization given in fig. 8, each of the R feature calculators is used to compute a feature vector $\mathbf{f}^r$ with $r \in \{1, \ldots, R\}$. In the following T^r denotes the number of features in feature vector $\mathbf{f}^r$. All R feature vectors constitute the visual dictionary. Let, as a shorthand, f_t^r address the feature with index t in the feature vector with index r, throughout.

In our case study, application of the vector quantization algorithm using feature calculator f^1 with a similarity threshold of $\vartheta^1 = 0.92$ yields the result presented in table 1. The table's left column comprises parquet graphs that have been chosen as codewords while in the right column lists the disregarded parquet graphs. The lower labels have been introduced in fig. 7, the upper labels are the similarities between the disregarded parquet graph and the respective codeword. The final feature vector $\mathbf{f}^1 = \left(f_t^1\right)_{1 \leq t \leq 8}$ comprises $T^1 = 8$ parquet graphs.

5 Preselection Network

In this section we will present the second step of the proposed form of graph dynamics: a feedforward neural network that allows for preselection of salient learning examples, so-called *model candidates*, and likewise for preselection of *salient categories* the object in the presented image supposedly belongs to. This network will be called the *preselection network*. Its design is motivated by the well-established finding that individual object-selective neurons tend to preferentially respond to particular object views [15, 23]. The preselection network's output neurons take the part of these view-tuned units.

The preselection network is a fully-connected single layer perceptron [26] that implements a weighted majority voting scheme [12]. In the network's input layer *position-invariant feature detectors* submit their assessments whether their reference feature is present in an image to dedicated input neurons while the output layer comprises one neuron for each predefined category. Synaptic weights are chosen such that the network conforms to Linsker's *infomax principle* [14]. That principle implies that the synaptic weights in a multilayer network with feedforward connections between layers develop, using a Hebbian-style update rule [10], such that the output of each cell preserves maximum information [29] about its input. Subject to constraints, the infomax principle thus allows to directly assign synaptic weights. The time-consuming adaption of synaptic weights becomes unnecessary at the expense of having to set up the preselection network in batch mode, i.e., the complete learning set has to be presented. This network setup in conjunction with the application of the winner-take-most or winner-take-all nonlinearity as decision function [25] implements a weighted majority voting scheme that allows for the desired preselection of salient categories and model candidates.

Here, the selection of salient categories and model candidates is only based on feature coincidences in image and model domain. As their spatial arrangement is disregarded, false positives are frequent among the selected model

Table 1. Case Study: Computation of Feature Vector $\mathbf{f}^1$

Codewords	Disregarded Features
$f_1^1 =$ (1,1)	0.96 (1,2), 0.93 (1,3), 0.97 (1,12), 0.95 (1,13), 0.92 (1,14), 0.93 (1,22), 0.95 (2,1), 0.93 (2,12)
$f_2^1 =$ (1,4)	0.97 (1,5), 0.97 (1,6), 0.96 (1,7), 0.94 (1,8), 0.95 (1,15), 0.93 (1,16), 0.93 (2,2), 0.94 (2,3), 0.94 (2,4), 0.93 (2,5), 0.93 (2,6), 0.93 (2,7), 0.92 (2,8), 0.92 (2,14), 0.93 (2,15), 0.93 (2,16), 0.92 (2,17), 0.92 (2,18)
$f_3^1 =$ (1,9)	0.95 (1,10), 0.94 (1,19), 0.96 (1,20), 0.96 (2,9), 0.94 (2,10), 0.94 (2,19), 0.95 (2,20)
$f_4^1 =$ (1,11)	0.94 (1,21), 0.94 (1,31), 0.97 (2,11)
$f_5^1 =$ (1,17)	0.97 (1,18), 0.94 (1,24), 0.93 (1,25), 0.93 (1,26), 0.95 (1,27), 0.93 (1,28)
$f_6^1 =$ (1,23)	0.93 (1,29), 0.92 (1,30), 0.93 (2,22), 0.94 (2,23)
$f_7^1 =$ (2,13)	0.93 (2,24)
$f_8^1 =$ (2,21)	0.96 (2,25)

candidates. To rule them out similar spatial arrangement of features will be asserted for the model to be selected in the correspondence-based verification part (sect. 6).

5.1 Neural Model

In the preselection network we employ two types of generalized McCulloch & Pitts neurons [17], variant A with identity and variant B with a Heaviside threshold function $H(\cdot)$ as output function. The output of a neuron of type A is equal to the weighted sum of its inputs $\sum_{n=1}^{N} x_n w_n$ with x_n being the presynaptic neurons' outputs and the w_n being synaptic weights. The output of a neuron of type B is 1, if the weighted sum of its inputs is greater than 0, and 0 otherwise.

5.2 Position-Invariant Feature Detectors

To test the presence of a particular feature from the visual dictionary, in the following called *reference feature*, in an image we construct a *position-invariant feature detector* out of local feature detectors (sect. 3.2). For this task, we distribute instances of local feature detectors uniformly over the image plane. For a given reference feature, combining the local feature detectors in a linear discriminant yields a position-invariant feature detector that returns 1 if the reference feature is observed at at least one position, and 0 otherwise (7). In fig. 9 it is shown how a position-invariant feature detector is constructed for a feature f_t^r from the visual dictionary. For a given feature f_t^r, the symbol τ_t^r denotes the respective position-invariant feature detector and $\tau_t^r(I)$ its result. We will say that a position-invariant feature detector τ_t^r has *found* or *observed* its feature f_t^r in input image I if $\tau_t^r(I) = 1$. From now on, we use the term *feature detector* only for the position-invariant version.

$$\tau_t^r : \mathbb{I} \rightarrow \{0,1\} \, ; \tau_t^r(I) = H\left(\sum_{f \in f^r(I)} \varepsilon\left(f, f_t^r, \vartheta^r\right)\right) \tag{7}$$

For the sake of simplicity we regard the feature detectors as the perceptron's processing elements [26], rather than an additional layer.

Each time a feature detector has found its reference feature f_t^r in the input image, we add pairs of matching features (f, f_t^r) to a table, where f stems from the input image. That table is used for efficient construction of image and model graphs in the correspondence-based verification part (sect. 6). The table is cleared before each image presentation.

$$\mathcal{F}_{match}(I) \leftarrow \mathcal{F}_{match}(I) \cup \bigcup_{f \in f^r(I)} \left\{ \left(f, f_t^r\right) \middle| \, \varepsilon\left(f, f_t^r, \vartheta^r\right) = 1 \right\} \tag{8}$$

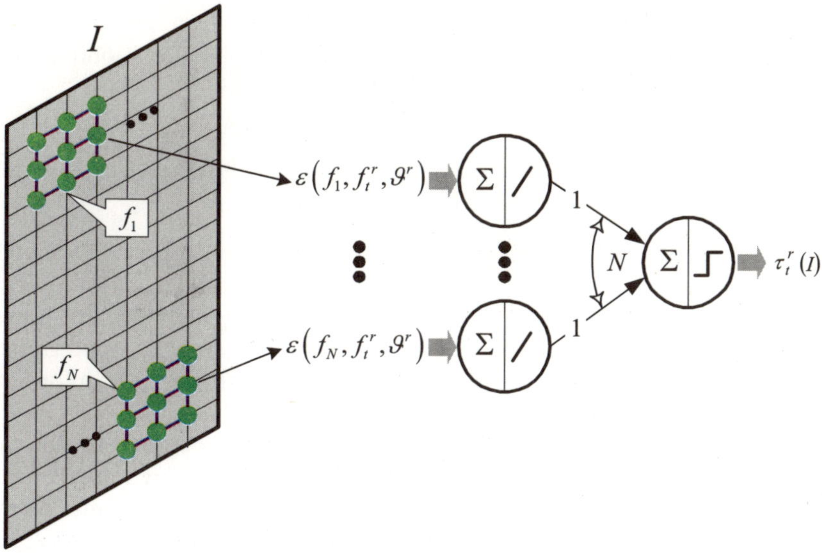

Fig. 9. Position-Invariant Feature Detector — *The position-invariant feature detector returns* 1 *if a given feature* f_t^r *is present in image* I*, and* 0 *otherwise. At each grid position allowing for placement of a whole parquet graph a local feature detector is installed that compares the local graph with the reference feature* f_t^r*. Technically, this has been implemented by applying feature calculator* f^r *to the given image* I*. If the feature calculator returns a set of* N *parquet graphs* $\{f_n | 1 \leq n \leq N\}$*, each local feature detector compares its feature* f_n *with the reference feature* f_t^r *with respect to similarity threshold* ϑ^r*. Then, each local feature detector passes its result into a single layer perceptron with* N *input units of type A, one output unit of type B, and feedforward connections of strength* 1 *between each input unit and the output neuron. The net's output is* 1 *if at least one of the local feature detectors has found its reference feature in the given image, and* 0 *otherwise. In this fashion, a position-invariant feature detector is instantiated for each feature in the visual dictionary*

5.3 Weighting of Feature Detectors

From the example in table 1 it becomes clear that the feature detectors have varying relevance for the selection of salient categories. In the following the contributions of feature detectors to choosing salient categories are described through *measures of information.* Shannon has defined information as the decrease of uncertainty [29]. In this sense, a natural definition of the measures of information is presented in (9). For a given feature detector τ_t^r that has found its reference feature f_t^r in the input image and for a given partitioning Π^k, the information $i_t^{r,k}$ that feature detector contributes to the decision about choosing categories of partitioning Π^k is defined by the difference between the largest possible amount of uncertainty and the feature detector's amount of uncertainty encoded by the Shannon entropy $\mathcal{H}_t^{r,k}$. $\mathcal{P}\left[\mathbb{C}_c^k \middle| f_t^r\right]$ describes the conditional probability that the genuine category is $\mathbb{C}_c^k$ given that feature f_t^r

has been observed. In this fashion measures of information are calculated for all features in the visual dictionary with respect to all partitionings of the learning set. Similar approaches are proposed in [7, 36].

$$i_t^{r,k} = \ln C^k - \mathcal{H}_t^{r,k} = \ln C^k + \sum_{c=1}^{C^k} \mathcal{P}\left[\mathbb{C}_c^k \,|f_t^r\right] \ln \mathcal{P}\left[\mathbb{C}_c^k \,|f_t^r\right] \tag{9}$$

For a given partitioning Π^k, the measures of information range between 0 and $\ln C^k$. If a feature occurs in all categories of that partitioning, the respective feature detector cannot make a contribution and, accordingly, its measure of information is 0. Conversely, if a feature occurs in only one category, the respective feature detector contributes maximally; its measure of information is $\ln C^k$.

Assuming that all prior probabilities for choosing a category are the same, the conditional probabilities $\mathcal{P}\left[\mathbb{C}_c^k \middle| f_t^r\right]$ are calculated through application of Bayes' rule (10). The $n_t^r(\mathbb{C})$ denote the total number of observations of feature f_t^r in the images of the parameterized category. For a given category $\mathbb{C}_c^k$ and a given feature f_t^r, we may interpret this probability as the frequency of that feature among the categories of partitioning Π^k. In table 2 the calculation of measures of information in our case study is demonstrated.

$$\mathcal{P}\left[\mathbb{C}_c^k \middle| f_t^r\right] = \frac{n_t^r\left(\mathbb{C}_c^k\right)}{\sum_{c'=1}^{C^k} n_t^r\left(\mathbb{C}_{c'}^k\right)} \tag{10}$$

5.4 Neurons, Connectivity, and Synaptic Weights

The preselection network is a single-layer perceptron comprising a layer of input and a layer of output neurons. In the network's input layer, we assign neurons of type A to the feature detectors. Thus, the network comprises $V_{\text{in}} = \sum_{r=1}^{R} T^r$ input neurons. By definition, each input neuron passes the result of its feature detector into the network. In the network's output layer, we assign neurons of type A to the predefined categories. Accordingly, the network contains $V_{\text{out}} = \sum_{k=1}^{K} C^k$ output neurons.

For fulfillment of the infomax principle, we define the synaptic weight $w_{t,c}^{r,k}$ between the presynaptic neuron assigned to a feature detector τ_t^r and the postsynaptic neuron assigned to a category $\mathbb{C}_c^k$ as follows. Imagine that feature f_t^r can both be observed in the input image and in at least one image of that category. Then, this may be considered as a piece of evidence that the input image belongs to that category. Consequently, feature detector τ_t^r should contribute its quantitative amount of information $i_t^{r,k}$ to the output of the postsynaptic neuron assigned to that category $\mathbb{C}_c^k$. Conversely, if that category contains only images in which that feature cannot be observed, the feature detector should never be allowed to make a contribution at all.

Table 2. Case Study: Calculation of Measures of Information

FEATURE INDEX (t)	FEATURE (f_t^1)	$n_t^1(\mathbb{C}_1^1)$	$n_t^1(\mathbb{C}_2^1)$	$\mathcal{P}\left[\mathbb{C}_1^1 \middle\| f_t^1\right]$	$\mathcal{P}\left[\mathbb{C}_2^1 \middle\| f_t^1\right]$	$i_t^{1,1}$
1	 (1, 1)	7	2	$\frac{7}{9}$	$\frac{2}{9}$	0.1634
2	 (1,4)	7	12	$\frac{7}{19}$	$\frac{12}{19}$	0.035
3	 (1,9)	4	4	$\frac{1}{2}$	$\frac{1}{2}$	0
4	 (1,11)	3	1	$\frac{3}{4}$	$\frac{1}{4}$	0.1307
5	 (1,17)	7	0	1	0	0.6931
6	 (1,23)	3	2	$\frac{3}{5}$	$\frac{2}{5}$	0.0201
7	 (2,13)	0	2	0	1	0.6931
8	 (2,21)	0	2	0	1	0.6931

Using this construction rule of synaptic weights, we define $R \times K$ matrices of synaptic weights $\mathbf{W}^{r,k}$: one matrix per feature vector/partitioning combination. For a given feature vector $\mathbf{f}^r$ and a given partitioning Π^k, weight matrix $\mathbf{W}^{r,k}$ (11) is of dimensions $(C^k \times T^r)$. That matrix comprises the synaptic weights $w_{t,c}^{r,k}$ of the connections between the input neurons assigned to feature detectors τ_t^r and the output neurons assigned to categories $\mathbb{C}_c^k$. The indices t of the presynaptic neurons range between 1 and T^r and the indices c of the postsynaptic neurons between 1 and C^k.

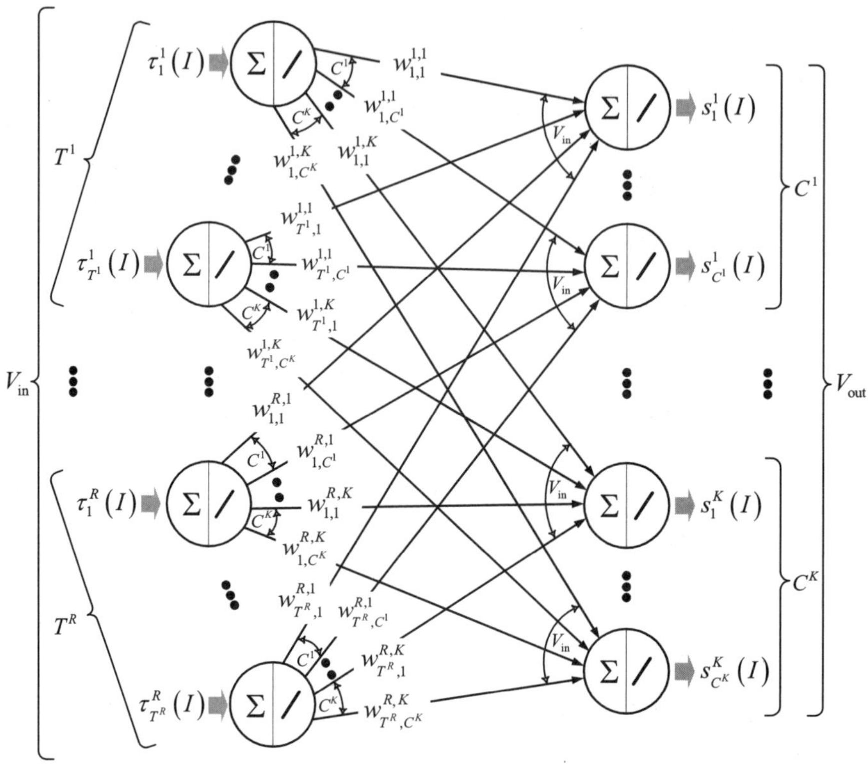

Fig. 10. Preselection Network — *The preselection network is a fully-connected single-layer perceptron. In its input layer neurons of type A have been assigned to the feature detectors. Accordingly, the network comprises* $V_{\text{in}} = \sum_{r=1}^{R} T^r$ *input neurons. Each input neuron passes the binary result of its feature detector into the network. In the network's output layer neurons of type A have been assigned to the predefined categories. Accordingly, the network contains* $V_{\text{out}} = \sum_{k=1}^{K} C^k$ *output neurons. The synaptic weights* $w_{t,c}^{r,k}$ *are chosen in a way such that the whole network conforms to Linsker's infomax principle. The output of the postsynaptic neuron that has been assigned to a given category* $\mathbb{C}_c^k$ *will be called the saliency of that category and is denoted by* $s_c^k(I)$

$$\mathbf{W}^{r,k} = \left(H\left(\sum_{I' \in \mathbb{C}_c^k} \tau_t^r(I') \right) i_t^{r,k} \right)_{\substack{1 \le c \le C^k \\ 1 \le t \le T^r}} =: \left(w_{t,c}^{r,k} \right)_{\substack{1 \le c \le C^k \\ 1 \le t \le T^r}} \tag{11}$$

In our case study, feature vector $\mathbf{f}^1$ comprises eight features and the learning set has been partitioned into two categories. Accordingly, weights matrix $\mathbf{W}^{1,1}$ is of dimensions (2×8). The matrix is shown in fig. 11.

$$\begin{aligned} \mathbf{W}^{1,1} &= \left(H\left(\sum_{I\in\mathbb{C}_c^1} \tau_t^1(I)\right) i_t^{1,1} \right)_{\substack{1\le c\le 2\\ 1\le t\le 8}} \\ &= \begin{pmatrix} 0.1634 & 0.035 & 0 & 0.1307 & 0.6931 & 0.0201 & 0 & 0 \\ 0.1634 & 0.035 & 0 & 0.1307 & 0 & 0.0201 & 0.6931 & 0.6931 \end{pmatrix} \\ &=: \left(w_{t,c}^{1,1}\right)_{\substack{1\le c\le 2\\ 1\le t\le 8}} \end{aligned}$$

Fig. 11. Case Study: Weight Matrix — *In our case study, feature vector* $\mathbf{f}^1$ *comprises eight features and the learning set has been partitioned into two categories. Accordingly, weight matrix* $\mathbf{W}^{1,1}$ *is of dimensions* (2×8). *The measures of information can be looked up in table 2*

5.5 Saliencies

The output of the postsynaptic neuron of a category $\mathbb{C}_c^k$ will be called the *saliency* of that category and is denoted by $s_c^k(I)$. With respect to an input image I, that saliency is defined as the sum of the measures of information $i_t^{r,k}$ of those feature detectors τ_t^r whose reference feature coincides in the input image and in at least one image of category $\mathbb{C}_c^k$. Thus, a saliency value is the accumulated evidence contributed by these feature detectors: the more pieces of evidence have been collected, the more likely the input image belongs to that category. For each partitioning of the learning set we can calculate a *saliency vector* $\mathbf{s}^k$ of length C^k by summing up the matrix vector products of the weight matrices $\mathbf{W}^{r,k}$ with the vector of feature detector responses $(\tau_t^r(I))_{1\le t\le T^r}$ over all feature R vectors in the visual dictionary (12). In fig. 10 the complete preselection network is shown.

$$\mathbf{s}^k : \mathbb{I} \to \mathbb{R}^{C^k};\ \mathbf{s}^k(I) = \sum_{r=1}^{R} \mathbf{W}^{r,k} \cdot (\tau_t^r(I))_{1\le t\le T^r} =: \left(s_c^k(I)\right)_{1\le c\le C^k} \tag{12}$$

5.6 Selection of Salient Categories and Model Candidates

For selection of salient categories for the input image I we apply a winner-take-most nonlinearity as a decision rule [25]. For a given partitioning Π^k the set $\Gamma^k(I)$ comprises all categories of the partitioning with super-threshold saliencies. The threshold is defined relative to the maximal saliency with a factor θ^k with $0 < \theta^k \le 1$ (13), i.e., the θ^k are relative thresholds. For $\theta^k = 1$ only the most salient category will be selected, the decision rule becomes the winner-take-all nonlinearity.

$$\Gamma^k(I) = \left\{ \mathbb{C}_c^k \in \Pi^k \middle|\, s_c^k(I) \ge \theta^k \max_{1\le c'\le C^k} \left\{ s_{c'}^k(I) \right\} \right\} \tag{13}$$

A set of *model candidates* $\mathbb{M}(I)$ for an input image I, i.e., learning images of objects that reasonably may become models for the object in the input image, are calculated by set intersection on salient categories (14). The selected model candidates will be passed to the correspondence-based verification part for further selection.

$$\mathbb{M}(I) = \bigcap_{k=1}^{K} \bigcup_{\mathbb{C} \in \Gamma^k(I)} \mathbb{C} \tag{14}$$

In fig. 12 the average numbers of model candidates in dependence on a relative threshold θ^1 are given. The experiment was carried out with the object recognition application proposed in sect. 7.1. The learning set comprised 5600 images taken from the COIL-100 database [21]. From these images $K = 1$ partitioning Π^1 with $C^1 = 5600$ single-element categories was created. We learn that, on average, the preselection network favorably rules out most irrelevant matches, i.e., the average numbers of model candidates are small relative to the total number of learning images, and that the average number of model candidates grows rapidly with decreasing relative thresholds. The average numbers of model candidates are, however, subjected to considerable mean variations, especially for small values of θ^1.

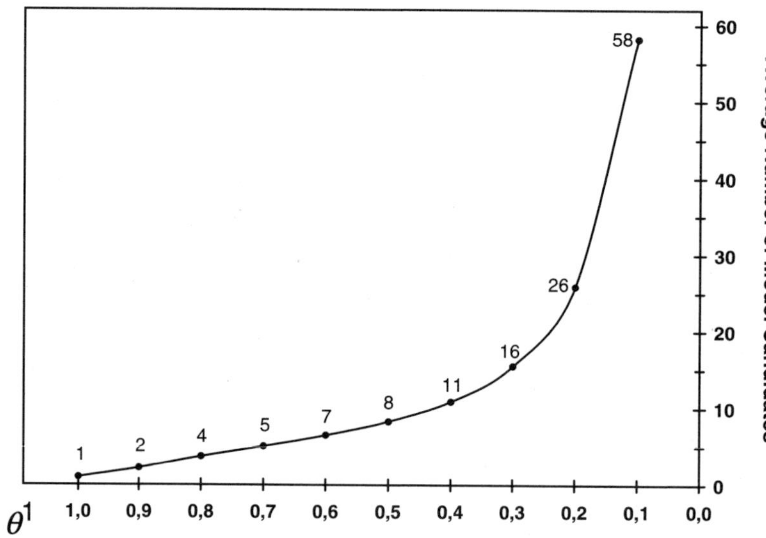

Fig. 12. Average Number of Model Candidates in Dependence on a Relative Threshold — *The average number of model candidates in dependence on the relative threshold θ^1 is given. The experiment was carried out with the object recognition application proposed in sect. 7.1*

6 Verification of Model Candidates

Up to here, model candidates have been selected by set intersection on salient categories (14). The categories' saliencies as computed by the preselection network are solely based on the detection of coincidental features in the model and image domain. The spatial arrangement of features, parquet graphs in our case, has been fully ignored, which can be particularly harmful in cases of multiple objects or structured backgrounds.

In the following model candidates are further verified through asserting that the features be in similar spatial arrangement for the model to be selected. More specifically, they are verified with a rudimentary version of elastic graph matching [11, 38, 47], a standard correspondence-based technique for face and object recognition. For each model candidate an image and a model graph are dynamically constructed through assembling corresponding features into larger graphs according to their spatial arrangement. For each model candidate the similarity between its image and model graph is computed. The model candidate whose model graph attains the best similarity is chosen as the model for the input image. Its model graph is the closest possible representation of the object in the input image with respect to the learning set.

6.1 Construction of Graphs

Construction of graphs proceeds in three steps. First, from the table of matching features (8) all feature pairs whose model feature stems from the current model candidate are transferred to a table of corresponding features. Second, templates of an image and of a model graph are instantiated with unlabeled nodes. Number and positioning of nodes is determined by the valid nodes of image and model parquet graphs. Third, at each node position, separately for image and model graph, a bunch of Gabor jets is assembled whose jets stem from node labels of valid-labeled parquet graph nodes located at that position. The respective nodes of the image or model graph become attributed with these bunches.

Table of Corresponding Features

During calculation of the categories' saliencies pairs of matching features have been collected in a table of matching features $\mathcal{F}_{match}(I)$ (8). Given a model candidate $M \in \mathbb{M}(I)$ for the input image I (14), all feature pairs whose model feature stems from M are transferred to a table of *corresponding* features $\mathcal{F}_{corr}(I, M)$, which will be used for efficient aggregation of parquet graphs into larger model and image graphs. We assume that the table comprises N feature pairs, a number that depends implicitly on the model candidate. Let f_n^I denote the image and f_n^M the model parquet graph of the n-th feature pair. Note that from now on we speak of *corresponding* rather than of *matching* parquet graphs and assume that those graphs establish local arrays of contiguous point-to-point correspondences between the input image and the model candidate.

$$\mathcal{F}_{corr}(I, M) = \left\{ \left(f_n^I, f_n^M\right) \in \mathcal{F}_{match}(I) \,\middle|\, 1 \leq n \leq N \ \wedge \ H\left(\sum_{r=1}^{R} \sum_{f \in f^r(M)} \varepsilon\left(f, f_n^M, 1\right)\right) = 1 \right\} \tag{15}$$

Nodes of parquet graphs are attributed with a triple consisting of an absolute image position, a Gabor jet derived from an image at that position, and a validity flag (sect. 3). For being able to globally address node label components, the following notation is introduced: nodes of image parquet graphs are attributed with triples $\left(\mathbf{x}_{n,v}^I, \mathcal{J}_{n,v}^I, b_{n,v}^I\right)$ where n specifies the feature pair in the table of corresponding features and v specifies the node index. The same notation is used for model parquet graphs, with a superscript M for distinction.

$$\begin{aligned} f_n^I &= \left\{ \left(\mathbf{x}_{n,v}^I, \mathcal{J}_{n,v}^I, b_{n,v}^I\right) \middle| \, 1 \leq v \leq V \right\} \\ f_n^M &= \left\{ \left(\mathbf{x}_{n,v}^M, \mathcal{J}_{n,v}^M, b_{n,v}^M\right) \middle| \, 1 \leq v \leq V \right\} \end{aligned} \tag{16}$$

Graph Templates

First, templates of an image and of a model graph are instantiated without node labels. Number and positioning of nodes are determined by the valid-labeled nodes of image and model parquet graphs. Their positions are collected in sets $\mathbb{X}^I$ and $\mathbb{X}^M$, respectively. The creation of graph templates is illustrated in fig. 13.

$$\begin{aligned} \mathbb{X}^I &= \bigcup\nolimits_{n,v} \left\{ \mathbf{x}_{n,v}^I \middle| \, b_{n,v}^I = 1 \right\} \\ \mathbb{X}^M &= \bigcup\nolimits_{n,v} \left\{ \mathbf{x}_{n,v}^M \middle| \, b_{n,v}^M = 1 \right\} \end{aligned} \tag{17}$$

Node Labels

The nodes of model and image graphs become attributed with bunches of Gabor jets: nodes of image graphs become labeled with bunches of Gabor jets that stem from node labels of valid-labeled nodes of image parquet graphs located at a given position $\mathbf{x}$ in the input image. Nodes of model graphs are just the same attributed with bunches of jets that stem from node labels of valid-labeled nodes of model parquet graphs located at a given position $\mathbf{x}$ in the model candidate. Let $\beta^I(\mathbf{x})$ denote a bunch assembled at an absolute position $\mathbf{x}$ in the input image. The same notation is used for the model graph's bunches, with a superscript M for distinction. Whenever possible we omit the position $\mathbf{x}$ and write β^I and β^M instead. The assembly of Gabor jets into bunches is also illustrated in fig. 13.

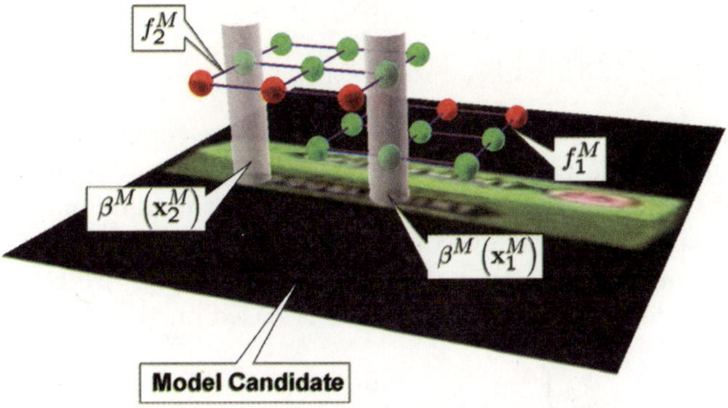

(a)

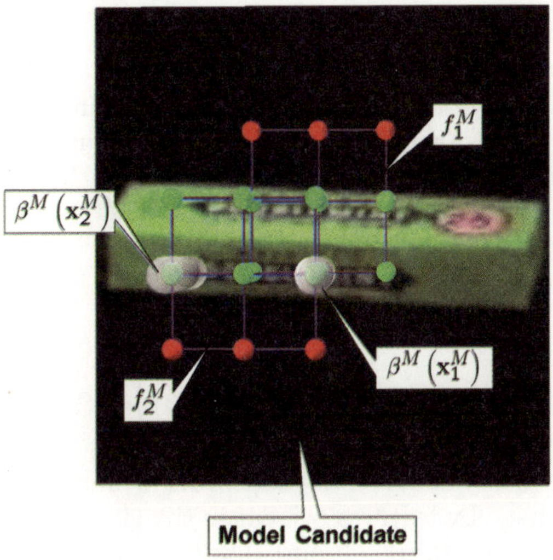

(b)

Fig. 13. Construction of Model Graphs — *Figure* **(a)** *provides a side, fig.* **(b)** *a top view of the same setup. For clarity, both figures show only two overlapping model parquet graphs* f_1^M *and* f_2^M *drawn from the table of corresponding features. For illustration of the overlap the graphs are drawn in a stacked manner. Number and position of the model graph's nodes are determined by the valid-labeled model parquet graph nodes (green nodes). Nodes that reside in the background have been marked as invalid (red nodes). In fig.* **(b)** *the shape of the emerging model graph can be foreseen. Compilation of bunches is demonstrated with two bunches only. Like stringing pearls, all valid Gabor jets at position* $\mathbf{x}_1^M$ *are collected into bunch* $\beta^M\left(\mathbf{x}_1^M\right)$ *and those at positions* $\mathbf{x}_2^M$ *become assembled into bunch* $\beta^M\left(\mathbf{x}_2^M\right)$*. From fig.* **(a)** *we learn that bunch* $\beta^M\left(\mathbf{x}_1^M\right)$ *comprises two jets while bunch* $\beta^M\left(\mathbf{x}_2^M\right)$ *contains only one jet. Image graphs are constructed in the very same fashion*

$$\begin{aligned}\beta^I(\mathbf{x}) &= \bigcup\nolimits_{n,v} \left\{ \mathcal{J}^I_{n,v} \middle| \mathbf{x}^I_{n,v} = \mathbf{x} \wedge b^I_{n,v} = 1 \right\} \\ \beta^M(\mathbf{x}) &= \bigcup\nolimits_{n,v} \left\{ \mathcal{J}^M_{n,v} \middle| \mathbf{x}^M_{n,v} = \mathbf{x} \wedge b^M_{n,v} = 1 \right\}\end{aligned} \tag{18}$$

For the assessment whether a point in the image corresponds to a point in the model candidate a measure of similarity between two bunches is needed. The similarity between two bunches is defined as the maximal similarity between the bunches' jets, which is computed in a cross run. If one of the bunches is empty the similarity between them yields 0. The jets are compared using the similarity function given in (3), which is based on the Gabor amplitudes.

$$s_{bunch}(\beta, \beta') = \begin{cases} 0 & \text{if } \beta = \emptyset \vee \beta' = \emptyset \\ \max\limits_{\mathcal{J} \in \beta, \mathcal{J}' \in \beta'} \{ s_{abs}(\mathcal{J}, \mathcal{J}') \} & \text{otherwise} \end{cases} \tag{19}$$

Graphs

Like parquet graphs, image and model graphs are specified by a set of node labels. Node labels comprise an absolute position in the input or model image drawn from the sets of node positions (17) and the bunch assembled at that position (18). The image graph is decorated with a superscript I while the model graph receives a superscript M.

$$\begin{aligned}\mathcal{G}^I &= \bigcup\nolimits_{\mathbf{x} \in \mathbb{X}^I} \left\{ \left(\mathbf{x}, \beta^I(\mathbf{x}) \right) \right\} \\ \mathcal{G}^M &= \bigcup\nolimits_{\mathbf{x} \in \mathbb{X}^M} \left\{ \left(\mathbf{x}, \beta^M(\mathbf{x}) \right) \right\}\end{aligned} \tag{20}$$

Model graphs of suited model candidates provide an approximation of the object in the input image by features present in the visual dictionary. In fig. 2 a number of model graphs (third column) that have been constructed for the input image given in the first column are given. The reconstructions from the model graphs of the first two model candidates in column four demonstrate that the emerged model graphs describe the object in the input image well.

The constructed graphs are to some extent reminiscent of bunch graphs [46, 47]. Nevertheless, since they represent single model candidates we rather speak of model instead of bunch graphs. It is, however, worthwhile mentioning that the proposed procedure may as well serve for the construction of bunch graphs. To this end the table of corresponding features has to provide feature pairs of model candidates picked from a carefully chosen subset $\tilde{\mathbb{M}}(I)$ of the set of model candidates $\mathbb{M}(I)$. The alternative computation of the table of corresponding features is given in (21). The graph construction procedure is then as well applicable to the construction of bunch graphs.

$$\mathcal{F}^{bunch}_{corr}\left(I, \tilde{\mathbb{M}}(I)\right) = \bigcup\nolimits_{M \in \tilde{\mathbb{M}}(I)} \mathcal{F}_{corr}(I, M) \tag{21}$$

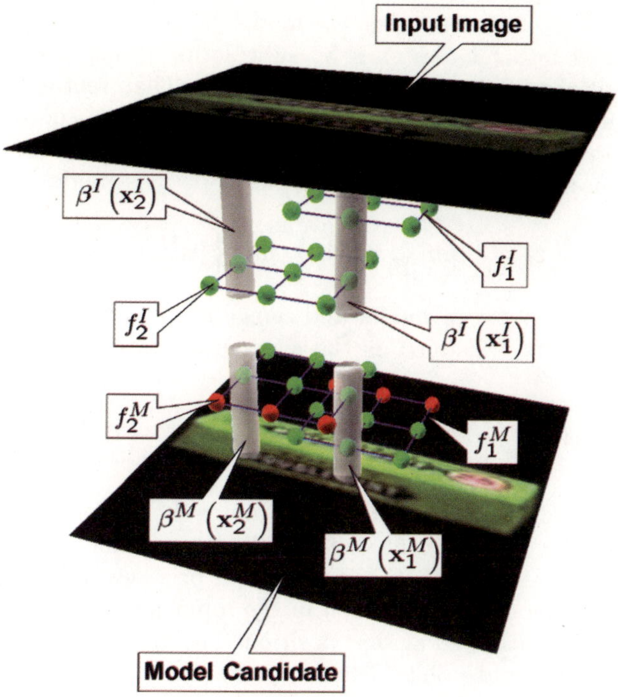

Fig. 14. Matching Setup — *The setup consists of the input image, the model candidate, and the graphs constructed using the proposed method. For clarity, only two pairs of corresponding parquet graphs have been taken from the table of corresponding features. Parquet graph f_1^I corresponds to f_1^M and f_2^I corresponds to f_2^M. Like in fig. 13, green nodes represent nodes that have been marked as valid and red nodes represent nodes that have been marked as invalid for residing in the background. Since only learning images provide figure-ground information, invalid nodes appear only in the model parquet graphs. The compilation of bunches is illustrated for two exemplary positions $\mathbf{x}_1^I$ and $\mathbf{x}_2^I$ in the input image and $\mathbf{x}_1^M$ and $\mathbf{x}_2^M$ in the model candidate. In order to find the object in the input image the model graph is iteratively moved over the entire image plane and matched with the image graph*

6.2 Matching

In order to assert that a constructed model graph represents the object in the given image well in a coherent fashion it is matched with the input image. It is moved as a template over the entire image plane in terms of maximizing the similarity between model and image graph. This action can be compared with the *scan global move* which is usually performed as the first step of elastic graph matching [11, 47]. It is also very similar to multidimensional template matching [49]. For each translation of the model graph the similarity between model and image graph is computed. The translation vector that yields the best similarity defines the optimal placement of the model graph in the image plane. In the process, the model graph's absolute node positions are transformed into relative ones by subtracting a displacement vector $\mathbf{t}_0$ from the positions of the model graph's nodes. That vector is chosen such that after subtraction the smallest x and the smallest y coordinate become zero. However, the y coordinate of the leftmost node is not necessarily 0. The same is the case for the x coordinate of the uppermost node.

$$\mathbf{t}_0 = \left(\min_{n,v} \left\{ \left(\mathbf{x}^M_{n,v}\right)_x \right\}, \min_{n,v} \left\{ \left(\mathbf{x}^M_{n,v}\right)_y \right\} \right)^\top \tag{22}$$

The similarity between model and image graph with respect to a given translation vector $\mathbf{t}$ is defined as the average similarity between image and model bunches.

$$s\left(I, M, \mathbf{t}\right) = \left|\mathcal{G}^M\right|^{-1} \sum_{\left(\mathbf{x}^M, \beta^M\right) \in \mathcal{G}^M} s_{bunch} \left(\beta^I \left(\mathbf{x}^M - \mathbf{t}_0 + \mathbf{t} \right), \beta^M \right) \tag{23}$$

In order to find the object in the input image the model graph is iteratively translated about a displacement vector in the image plane so that the measure of similarity between model and image graph becomes maximal. The model graph moves to the object's position in the input image. Let $s_{best}\left(I, M\right)$ denote the similarity attained at that position. The displacement vectors $\mathbf{t}$ stem from a set $\mathbb{G}$ of all grid points defined by the distances Δx and Δy between neighbored parquet graph nodes (sect. 4).

$$s_{best}\left(I, M\right) = \max_{\mathbf{t} \in \mathbb{G}} \left\{ s\left(I, M, \mathbf{t}\right) \right\} \tag{24}$$

6.3 Model Selection

For selection of the model, the most similar learning image for the given input image, an image and a model graph are constructed for each model candidate. The model candidate that attains the best similarity between its model and image graph is chosen as the model for the input image.

$$M_{best} = \arg\max_{M \in \mathbb{M}(I)} \left\{ s_{best}\left(I, M\right) \right\} \tag{25}$$

In fig. 2 four model candidates (column two) have been computed for the given input image (column one). The similarities attained through matching image against model graphs are annotated to the reconstructions from the model graphs (column four). Since the first model candidate yields the highest similarity, it is chosen as the model for the object in the input image.

7 Experiments

We report experimental results derived from standard databases for object recognition and categorization. The results are excerpted from [44].

7.1 Object Recognition

Object recognition experiments were conducted on the COIL-100 image database [21]. That database contains images of 100 objects in 72 poses per object, thus, 7200 image in total. We present the results of three experiments. First, we investigated the recognition performance with respect to object identity and pose for input images containing a single object, second, we analyzed the recognition performance for input images containing multiple objects, and third, recognition performance was measured for images of partially occluded objects.

Experimental results were attained in a fivefold cross-validation [48]. We thus created five pairs of disjoint learning and testing sets from all COIL-100 images. The learning sets comprise 56, the testing sets 14 views per object, thus, 5600 or 1400 images in total, respectively. The object recognition application is designed to simultaneously recognize the object's identity and pose. This is achieved by creating $K = 1$ partitioning of the learning set. That partitioning consists of single-element categories. Moreover, from each learning set a visual dictionary with $R = 2$ feature vectors of increasing length was calculated using similarity thresholds of $\vartheta^1 = 0.9$ and $\vartheta^2 = 0.95$ (Algorithm 1). Sorting feature vectors according to detailedness is harnessed in a procedure that allows for accelerated search of features in a coarse-to-fine fashion. [44]. Computation and parameters of the Gabor features are the same as in [11, 47], i.e., five scales, eight orientations, $k_{max} = \frac{\pi}{2}$, $k_{step} = \sqrt{2}$, and $\sigma = 2\pi$. For this parameterization, the horizontal and vertical node distances Δx and Δy are set to 10 pixels.

In the following we present recognition results computed within the cross-validation and their dependence on relative weighting of the feature- and correspondence-based parts. Each data point was averaged over $5 \times 1400 = 7000$ single measurements. Weighting of the feature- and correspondence-based part is controlled by the threshold scaling factor θ^1 (13) that ranges between 0.1 and 1, sampled in 0.1-steps. θ^1 determines the final number of model candidates that are passed to the correspondence-based verification part. For $\theta^1 = 1$ only one model candidates is selected while for low values the set of model candidates encompasses large portions of the learning set. That factor thus enables us to adjust the balance between the feature- and correspondence-based parts.

Recognition of Single Objects

In the first experiment we presented images containing a single object and pose. We analyzed the system's performance for each of the combinations segmented/unsegmented images and preselection network conforming/non-conforming to the infomax principle (sect. 5). The experiment was subdivided into eight test cases. In the first four test cases the recognition performance with respect to object identity was evaluated for each of these combinations

(a) (b)

Fig. 15. Input Images of a Single Object — *The figure shows an object from the COIL-100 database [21] as* **(a)** *segmented and* **(b)** *unsegmented image. Since the images of that database are perfectly segmented, the unsegmented images have been manually created by pasting the object contained in the segmented image into a cluttered background consisting of arbitrarily chosen image patches of random size derived from the other test images of the current testing set. This is the worst background for feature-based systems*

while the system's ability to recognize the objects' poses was investigated in the remaining four test cases. Since the images of the COIL-100 database are perfectly segmented, the unsegmented images have been manually created by pasting the object into a cluttered background consisting of arbitrarily chosen image patches of random size derived from the other test images of the current testing set. In fig. 15 an example of a segmented and of an unsegmented image is given. In order to asses the usefulness of the choice of synaptic weights according to (11) the preselection networks are made incompatible to the infomax principle by putting their weights out of tune using (26). Choosing the synaptic weights in this fashion the saliencies become simple counters of feature coincidences, the weighted majority voting scheme degenerates to a non-weighted one.

$$\hat{\mathbf{W}}^{r,k} = \left(H \left(\sum_{I' \in \mathbb{C}_c^k} \tau_t^r \left(I' \right) \right) \right)_{\substack{1 \leq c \leq C^k \\ 1 \leq t \leq T^r}} =: \left(\hat{w}_{t,c}^{r,k} \right)_{\substack{1 \leq c \leq C^k \\ 1 \leq t \leq T^r}} \tag{26}$$

The recognition performance with respect to object identity is shown in fig. 16(a). We considered the object in the test image to be correctly recognized if test and model image showed the same object regardless of its pose. Throughout, better recognition rates were attained if segmented images were presented. Moreover, the infomax principle always slightly improved performance where that improvement is, however, continually exhausted in gradually putting more and more emphasis on the correspondence-based part, i.e., the achieved improvement is continually used up while moving from the left to the right hand side in fig. 16. Most interestingly, a well-balanced combination of the feature- and correspondence-based parts led to optimal performance, throughout. Only for such well-balanced combinations the selection

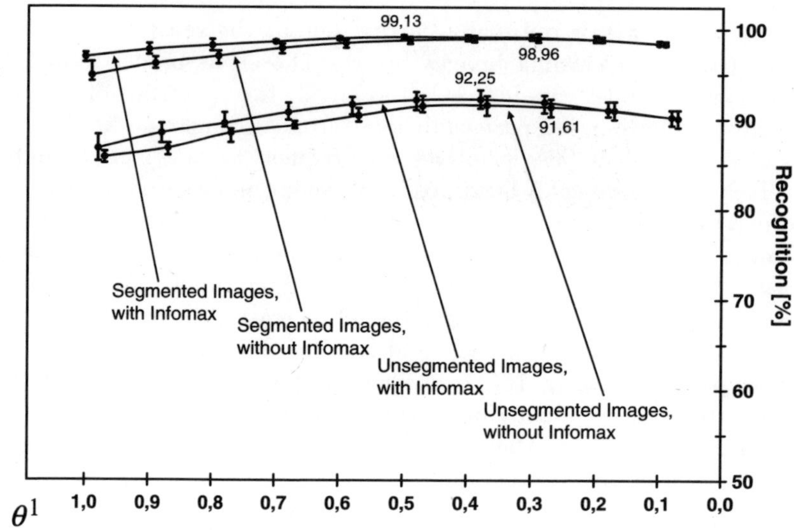

(a)

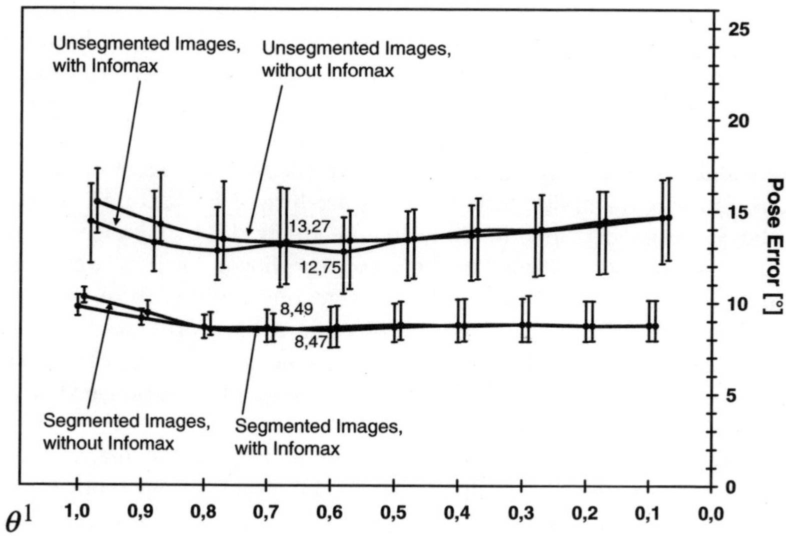

(b)

Fig. 16. Recognition of Single Objects — *The figure shows the recognition performance with respect to* **(a)** *object identity and* **(b)** *object pose depending on relative weighting of the feature- and correspondence-based parts controlled by* θ^1. *This parameter determines the final number of model candidates that are passed to the correspondence-based verification part. The best results are annotated to the respective data points. The results were better for segmented images. Optimal performance was attained by satisfying the infomax principle and for a well-balanced combination of the feature- and correspondence-based parts*

of model candidates is optimally carried out in the sense that neither too few nor too many learning images become chosen as model candidates. If the number of model candidates is too small, the spectrum of alternatives the correspondence-based part can choose from becomes too limited. This is especially harmful, if false positives are frequent among model candidates. Conversely, the number of false positives among model candidates unavoidably increases with overemphasis of the correspondence-based part: for too low values of the relative threshold even learning images of weakly salient categories become selected as model candidates. Accordingly, the probability of choosing a false positive as the final model increases, the average recognition rate decreases. The same findings hold true for the performance with respect to object pose given in fig. 16(b). The average pose errors were calculated over the absolute values of angle differences of correctly recognized, non-rotation-symmetric objects. Note that two consecutive learning images of the same object are at least five degrees apart.

Recognition of Multiple Objects

The second experiment is concerned with the recognition of multiple, simultaneously presented, non-overlapping objects, i.e., input images showed simple visual scenes. Only the recognition performance with respect to object identity was evaluated. The experiment was subdivided into six test cases. In the first three test cases we simultaneously presented $N \in \{2, 3, 4\}$ objects placed in front of a plain black background while in the last three test cases cluttered background was manually added. The procedure of background construction was the same as in the first experiment. In fig. 17 two images containing four objects with and without background are shown. Objects were randomly picked, a test image contained only different ones, and each object appeared at least once. In a test case 1400 input images were presented. The system returned the N most similar models. Each coincidence with one of the presented objects was accounted as a successful recognition response. The average recognition rates were calculated over all responses.

The result of this experiment is given in fig. 17. We learn that, compared to the single-object experiments, the point of optimal recognition performance considerably moved to the right: putting more emphasis on the correspondence-based verification part improved recognition performance. Presentation of segmented images yielded better results. For both segmented and unsegmented images the system's performance degraded smoothly with the number of simultaneously presented objects. However, overemphasis of that part caused by too small values of the relative threshold θ^1 again led to a decrease in recognition performance. This phenomenon can be observed in the test cases with unsegmented images (fig. 18(b)).

(a)

(b)

Fig. 17. Input Images of Multiple Objects — *The figure shows an example of* **(a)** *a segmented and* **(b)** *an unsegmented input image containing four objects drawn from the COIL-100 database [21]. Backgrounds were constructed in the same fashion as in the first experiment*

Recognition of Partially Occluded Objects

While in the second experiment the objects were presented in a non-overlapping manner, the third and last experiment is concerned with the recognition of partially occluded objects with respect to the same weightings of the feature- and correspondence-based parts as in the first two experiments. Again, we only evaluated recognition performance with respect to object identity. The experiment is subdivided into twelve test cases. In the first six test cases we simultaneously presented two objects where 0-50% of the object on the left was occluded by the object on the right. Occluded and occluding objects were different and randomly picked, each object appeared at least once as occluded. In the last six test cases cluttered background was added. The procedure of background construction and accounting of recognition responses was the same as in the second experiment. In fig. 19 input images of partially occluded objects are shown. In fig. 20 the average recognition rates are given.

Like in the second experiment, we learn from the results presented in fig. 20 that emphasis of the correspondence-based part improved recognition performance. Again, overemphasis of that part led to a decline. Moreover, presentation of segmented images yielded better results. For segmented (fig. 18(a)) and unsegmented images (fig. 18(b)) the system's performance smoothly degraded with the amount of occlusion.

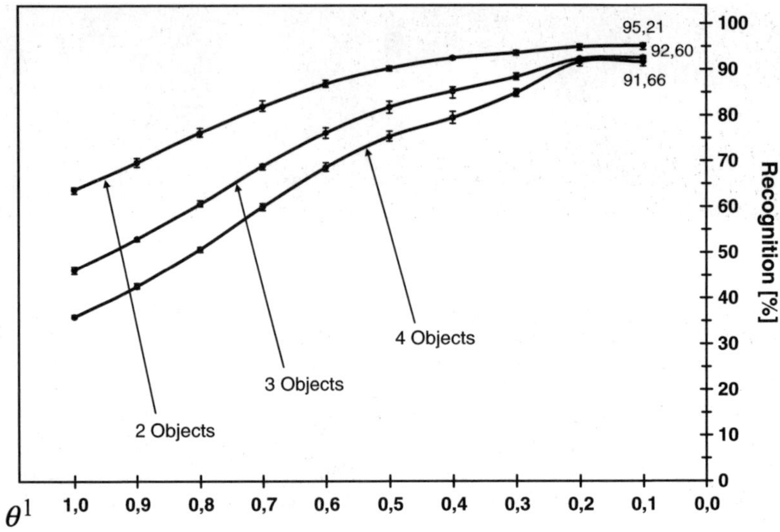

(a)

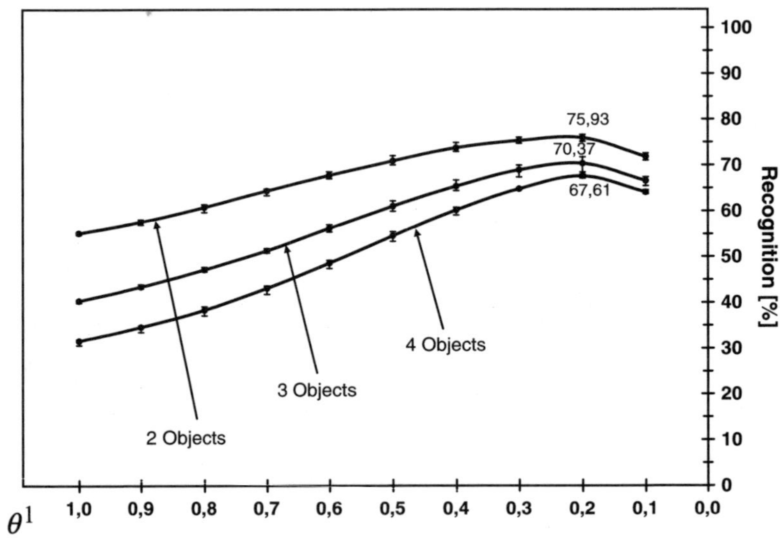

(b)

Fig. 18. Recognition of Multiple Objects — *The figure shows the recognition performance with respect to object identity in the case of multiple non-overlapping objects,* **(a)** *for segmented,* **(b)** *for unsegmented images. Compared to the first experiment, the point of optimal recognition performance has considerably moved to the right: correspondence-based verification is more important in the case of multiple objects. Overemphasis of the correspondence-based verification part, however, led to a decline. Presentation of segmented images yielded better results. Performance smoothly degraded with the number of simultaneously presented objects*

(a) (b)

Fig. 19. Input Images of Partially Occluded Object — *The figure shows* **(a)** *a segmented and* **(b)** *an unsegmented input image of partially occluded objects. The procedure of background construction was the same as in the first experiment. In this example, the occluding object covers about fifty percent of the occluded object*

Discussion

Our system performed favorably compared with other techniques. The original system of Murase & Nayar [20], that performs a nearest neighbor classification to a manifold representing a collection of objects or class views, attained a recognition rate of 100% for segmented images of single unscaled objects drawn from the COIL-100 database. Our system attained a recognition rate of 99.13% in the same test case (sect. 7.1). The recognition performance of the Murase & Nayar system is, however, unclear if it would be confronted with more sophisticated recognition tasks, for instance, images with structured backgrounds, with multiple objects, or with occluded objects.

Wersing & Körner [42] compared the performance their system of setting up the feature extraction layers in an evolutionary fashion with the Murase & Nayar system. They conducted their experiments on the COIL-100 database. In the case of segmented images their system and ours performed about equally well, see fig.4 (b) in [42] and fig. 16(a): both systems achieved recognition rates above 99%.

In the case of unsegmented images our system outperformed the system of Wersing & Körner, see fig.6 (a) in [42] and fig. 16(a): our system attained a recognition rate of 92.25% while the system of Wersing & Körner peaked slightly below 90%. It is, however, worth mentioning that the experimental setting differed considerably. Wersing & Körner performed their experiment on the first 50 objects of the COIL-100 database and constructed structured backgrounds out of fairly big patches of the remaining 50 objects. In contrast, we conducted the experiment on all objects and pasted them into a cluttered background consisting of arbitrarily chosen image patches of random size derived from the other test images.

As has been documented in the second and third experiment, our system provides a straightforward manner to analyze visual scenes with structured background. In this respect, we cannot compare our system to the ones by Murase & Nayar and Wersing & Körner, respectively.

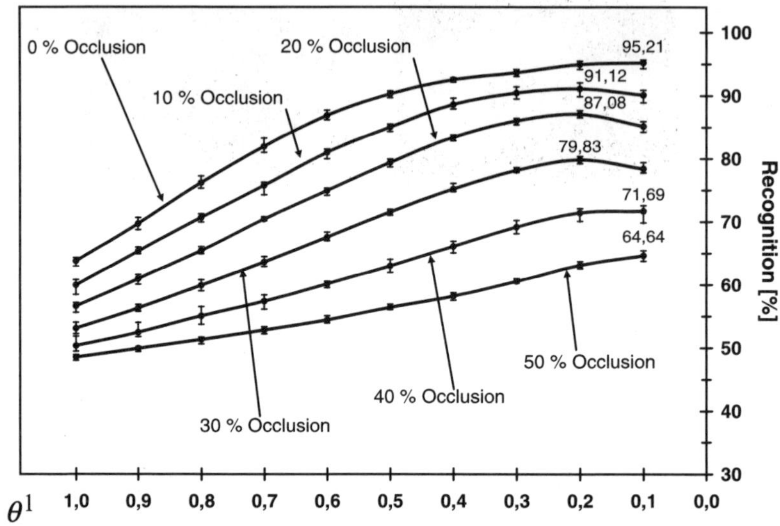

(a)

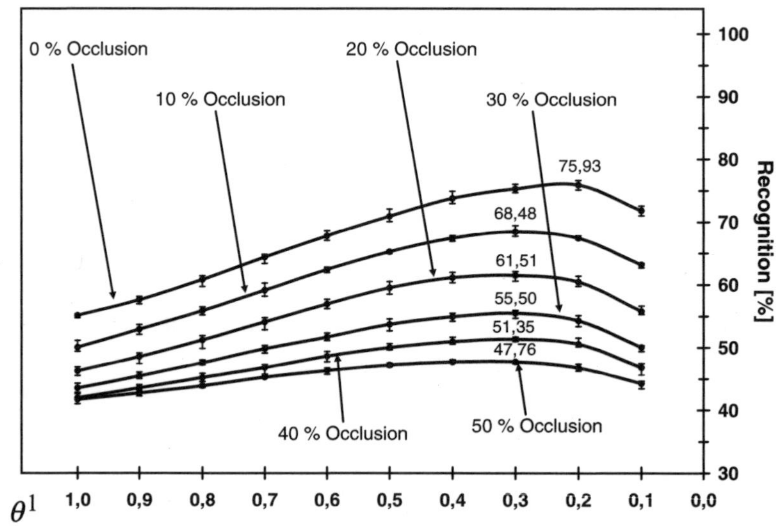

(b)

Fig. 20. Recognition of Partially Occluded Objects — *The figure shows the recognition performance with respect to object identity in the case of partially occluded objects,* **(a)** *for segmented,* **(b)** *for unsegmented images. Like in the second experiment, emphasis of the correspondence-based verification part improved recognition performance, overemphasis of that part led to a decline. Presentation of segmented images yielded better results. The system's performance smoothly degraded with the amount of occlusion*

7.2 Object Categorization

Object categorization experiments were conducted on the ETH-80 image database [13]. That database contains images of eight categories namely apples, pears, tomatoes, dogs, horses, cows, cups, and cars of ten identities per category and 41 images in different poses per identity. The databases thus consists of 3280 images in total. An interesting question with respect to object categorization is whether a given hierarchical organization of categories can be harnessed to improve categorization performance. The question how such a hierarchical organization is learned is however not addressed here. We present the results of two experiments. First, we evaluated categorization performance if the decision about the final category relies on a given hierarchical organization of categories. We employed the hierarchy given in fig. 4. Second, we evaluated categorization performance if no such hierarchy is given.

Categorization of Objects Using Hierarchically Organized Categories

Results of the first experiment were attained in a leave-one-object-out cross-validation [48]. This means that the system was trained with the images of 79 objects and tested with the images of one unknown object. We thus created 80 pairs of learning and testing sets. The learning sets contained 3239, the testing sets 41 images. We hierarchically organized the images into categories of $K = 3$ partitionings as given in fig. 4. The threshold scaling factors θ^k for selection of salient categories of partitionings Π^k, $k \in \{1, 2, 3\}$, were all set to 0.4 (13). The parameterization of parquet graph features was the same as in the object recognition experiments. For partitionings Π^1 and Π^2 we considered an object to be correctly categorized if exactly one category out of these was selected as salient and the presented object belonged to that category. For partitioning Π^3 a set of model candidates was calculated by set intersection of salient categories (14). The model candidates of that set were passed to the correspondence-based verification part. We considered the presented object to be correctly categorized if it belonged to the same of the original eight categories as the object in the model image.

In fig. 21 the averaged categorization rates computed within the leave-one-object-out cross-validation broken down into the original eight categories of apples, pears, tomatoes, dogs, horses, cows, cups, and cars are displayed. Each data point was averaged over $10 \times 41 = 410$ single measurements. Generally, categorization performance depended considerably on the sampling of categories. In this sense the system categorized apples, pears, and tomatoes well but obviously experienced difficulties in categorizing cows, dogs, horses, cars, and cups. The intra-category variations among the identities within these categories are too large. It is thus reasonable to assume that categorization performance may be improved by adding more learning examples to those

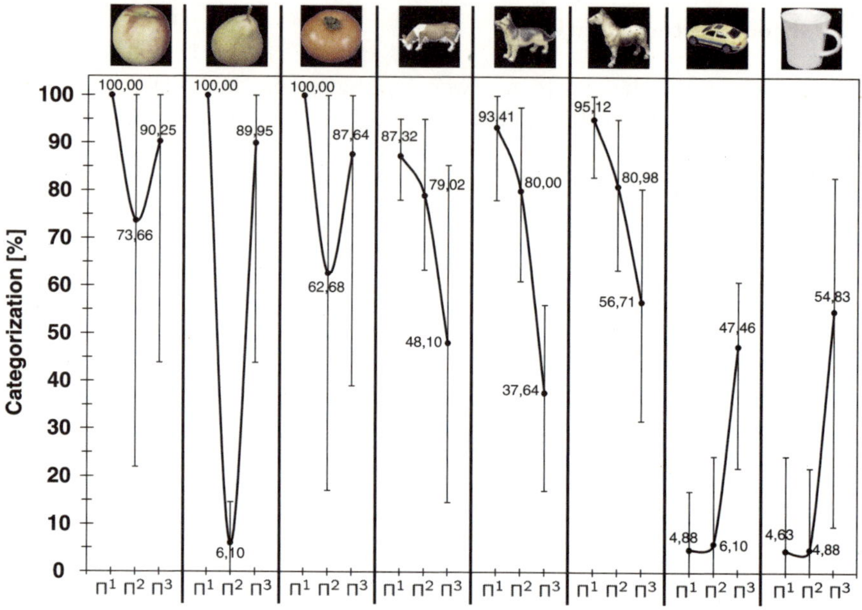

Fig. 21. Categorization of Objects Using Hierarchically Organized Categories — *The averaged categorization rates computed within the leave-one-object-out cross-validation are displayed. Each data point was averaged over* 410 *single measurements. Categorization performance depended considerably on the sampling of categories. The feature-based part's ability to unambiguously assign the object in the input image to the categories of partitionings* Π^1 *and* Π^2 *is obviously limited. For most cases, the correspondence-based verification part was able to compensate for this shortcoming, but not for the shortage of learning examples, especially in the animal categories*

categories. Moreover, the feature-based part's ability to unambiguously assign the object contained in the input image to the categories of partitionings Π^1 and Π^2 is obviously limited. This deficiency is especially prominent in the results attained for the categorization of pears, cars and cups. Due to the imbalance between natural and man-made objects, the attained results for cars and cups are even worse than those for pears. The correspondence-based verification part was to some extent able to compensate for this shortcoming and improved categorization performance for apples, pears, tomatoes, cars, and cups. However, the shortage of learning examples, especially in the animal categories, can only be cured by additional training images.

Categorization of Objects Using Single-Element Categories

For evaluation of the system's performance without predefined hierarchical organization of categories we arranged the learning set into $K = 1$ partitioning of single-element categories. We considered the object in the input image to be correctly categorized if it belonged to the same original category of apples, pears, tomatoes, cows, dogs, horses, cars, or cups as the object in the model image. The attained results depending on θ^1 are given in fig. 22. For clarity the curves are distributed over two subfigures. All other parameters were the same as above.

As in the object recognition experiments, a well-balanced combination of the feature- and the correspondence-based parts allowed for optimal categorization performance. The expectation that categorization performance would benefit from hierarchical organization of categories could not be substantiated. In the case of apples, tomatoes, cows, horses, cars, and cups average categorization performance was considerably better without hierarchy. Only for pears and dogs categorization could benefit slightly.

The categorization rates are below or close to those presented in [13]. That object categorization system, however, integrates color, texture, and shape features while our system only relies on local texture information. At least the feature-based part of the technique described in this paper can work with any convenient feature type [45]. One can thus expect to further improve categorization performance if more feature types become incorporated.

In fig. 23 a confusion matrix of the categorization performance in the case of single-element categories and optimal weightings of the feature- and correspondence-based parts is given. The optimal weightings were category-specific (fig. 22). Categorization performance depended considerably on the degree of intra-category variations: for categories with relatively small intra-category variations, for instance, the categories of fruits, cups, and cars, the system performed well while the system's performance degraded in a remarkable fashion when confronted with images of categories with larger variations among category members. This is especially prominent for the animal categories. The system performed particularly poorly for the category of dogs. However, in 75.12% (10.00% + 29.27% + 35.85%) of all cases the system assigned an input image of a dog to the category of animals vs. 80.00% in the hierarchical case (fig. 21). Images of horses and cows were assigned to that category in 84.87% and 86.10% of all cases in the non-hierarchical case vs. 80.98% and 79.02% in the hierarchical case, respectively. In sum, 82.03% of all cases input images of animals were correctly assigned to the category of animals in the non-hierarchical case while that number was $80.00\% = (79.02\% + 80.00\% + 80.98\%)/3$ with hierarchical organization of categories. These results once more confirm our statement that the data is much too sparse to make the fine distinctions between the categories of partitioning Π^3.

Discussion

Much work remains to be done on the categorization capabilities. In our experiment we have seen that the categories employed by human cognition were not helpful to improve the categorization capability when employed to structure

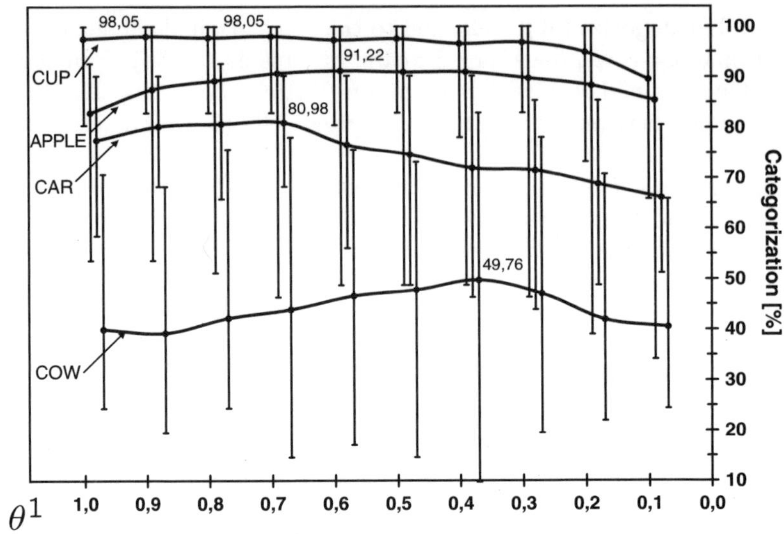

(a)

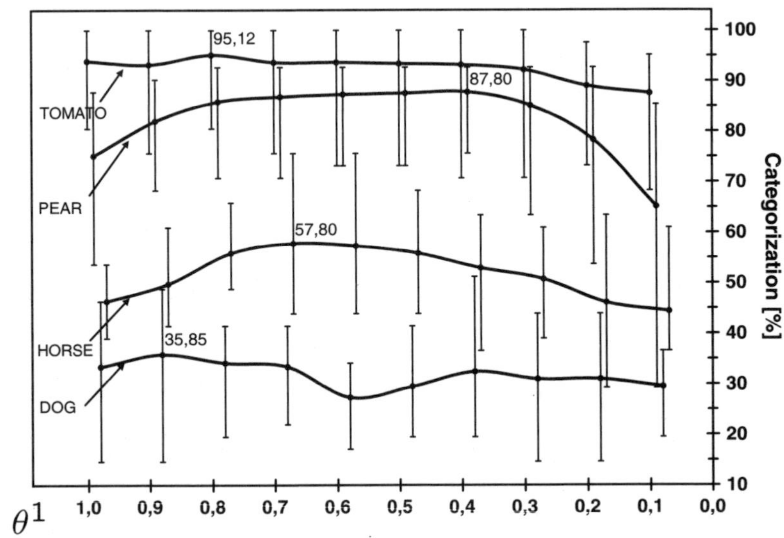

(b)

Fig. 22. Categorization of Objects Using Single-Element Categories — *The averaged categorization rates within the leave-one-object-out cross-validation are displayed. Each data point was averaged over* 410 *single measurements. Optimal categorization performance was achieved for a well-balanced combination of the feature- and correspondence-based parts. In most cases categorization performance was clearly better than in the hierarchical case*

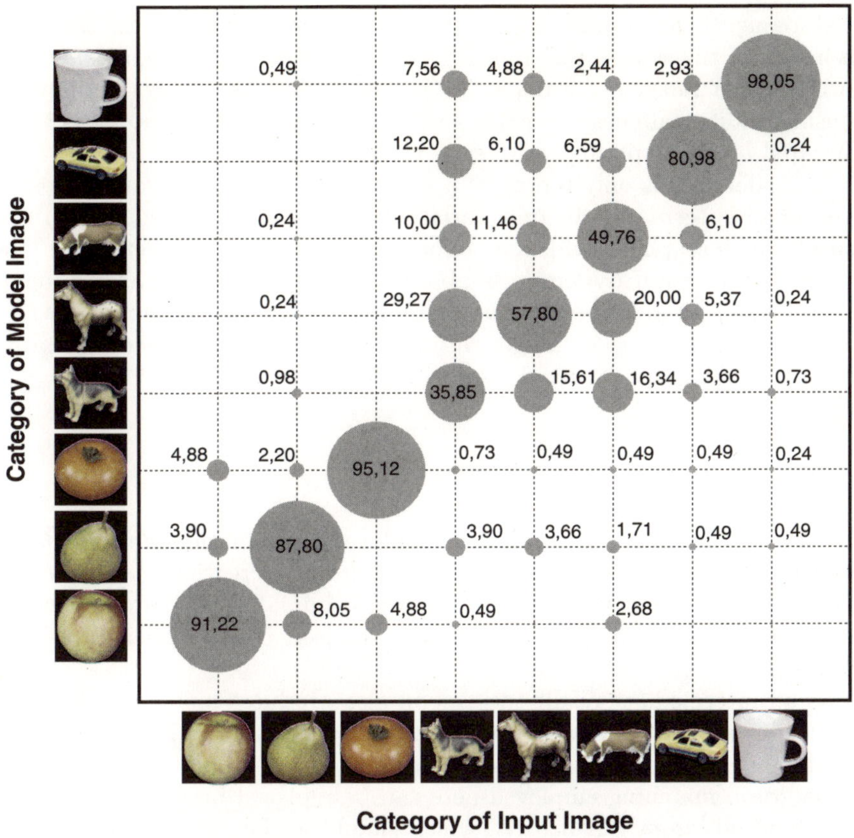

Fig. 23. Confusion Matrix of Categorization Performance — *A confusion matrix of the categorization performance in the case of single-element categories and optimal weightings of the feature- and correspondence-based parts is given. The optimal weightings were category-specific (fig. 22). The axes are labeled with the categories of the ETH-80 database [13], symbolized by images of arbitrarily chosen representants. The horizontal axis codes the categories of the object in the input images while the vertical axis codes the categories of the object in the model images. The given categorization rates are relative to the categories of the object in the input images. In each column they sum up to 100%. In order to improve readability, blobs were assigned to the categorization rates whose surface areas scale proportionally with the amount of their associated categorization rates*

the recognition process. This finding is, however, compatible with experimental results which find that in human perception recognition of a single object instance precedes categorization [22].

Another reason for the relatively poor performance is that in some cases the data was much too sparse to really cover the intra-category variations: if the variations across category members were poorly sampled, categorization failed frequently for input images supposed to be assigned to these categories. For instance, the system performed poorly for the animal categories, but categorized input images of fruits well. Categorization can always be improved by using additional cues like color and global shape. This hypothesis is substantiated by the experimental results given in [13].

As model graphs only represent a single object view they cannot possibly cover larger spectra of individual variations among category members. In this respect bunch graphs provide a more promising concept. As briefly mentioned in sect. 6, the graph dynamics is able to construct bunch graphs provided that the model features stem from carefully chosen model candidates. It is reasonable to assume that categorization performance can further be improved by using bunch graphs instead of model graphs.

8 Summary and Future Work

We have presented an algorithm that employs a combination of rapid feature-based preselection with self-organized model graph creation and subsequent correspondence-based verification of model candidates. This hybrid method outperformed both purely feature-based and purely correspondence-based approaches.

As an intermediate result the system also produces model graphs, which are the closest possible representations of a presented object in terms of memorized features. A variety of further processing can build on these graphs. The simple graph matching employed here can be replaced by the more sophisticated methods from [11, 32, 47], which should lead to increased robustness under shape and pose variations.

In the present state, the method can also be used for the purposeful initialization of sophisticated but slow techniques. For instance, it can produce a coarse pose estimation followed by refinement through correspondence-field evaluation. Another promising extension will be to use diagnostics from the classification process for novelty detection and subsequent autonomous learning.

Much work remains to be done on the categorization capabilities. In our experiment we have seen that the categories employed by human cognition were not helpful to improve the categorization capability when employed to structure the recognition process. It is, however, compatible with experimental results, which find that in human perception recognition of a single object instance precedes categorization [22].

Another reason for the relatively poor performance in categorization experiments is that the data was much too sparse to really cover the intra-category variations. Categorization can always be improved by using additional cues like color and global shape. This would, however, also require larger databases, because much more feature combinations would need to be tested. Nevertheless, the method presented here is well suited to accommodate hierarchical categories. Their impact on categorization quality as well as methods to learn the proper organization of categories from image data are subject to future studies.

References

1. M. Arentz. Integration einer merkmalsbasierten und einer korrespondenzbasierten Methode zur Klassifikation von Audiodaten. Master's thesis, Computer Science, University of Dortmund, D-44221 Dortmund, Germany, 2006
2. I. Biederman. Recognition-by-components: A theory of human image understanding. *Psychological Review*, 94:115–147, 1987
3. E. Bienenstock and S. Geman. Compositionality in neural systems. In M.A. Arbib, editor, *The Handbook of Brain Theory and Neural Networks*, pages 223–226. MIT, Cambridge, MA, London, England, 1995
4. H. Bunke. Graph grammars as a generative tool in image understanding. In M. Nagl H. Ehrig and G. Rozenberg, editors, *Graph Grammars and their Application to Computer Science*, volume 153, LNCS, pages 8–19. Springer, Berlin Heidelberg New York, 1983
5. M.A. Eshera and K.S. Fu. An image understanding system using attributed symbolic representation and inexact graph-matching. *IEEE Transactions on Pattern Analysis and Machine Intelligence*, 8(5):604–618, 1986
6. L. Fei-Fei, R. Fergus, and P. Perona. A Bayesian approach to unsupervised one-shot learning of object categories. In *Proceedings of the Ninth IEEE International Conference on Computer Vision*, volume 2, pages 1134–1141, 2003
7. G. Fritz, L. Paletta, and H. Bischof. Object recognition using local information content. In J. Kittler, M. Petrou, and M. Nixon, editors, *17th International Conference on Pattern Recognition (ICPR 2004)*, volume 2, pages 15–18, Cambridge, UK. IEEE Press, 2004
8. B. Fritzke. A self-organizing network that can follow non-stationary distributions. In *International Conference on Artificial Neural Networks (ICANN 1997)*, pages 613–618. Springer, Berlin Heidelberg New York, 1997
9. R. Gray. Vector quantization. *IEEE Signal Processing Magazine*, 1(2):4–29, April 1984
10. D.O. Hebb. *The Organization of Behavior.* Wiley, New York, 1949
11. M. Lades, J.C. Vorbrüggen, J. Buhmann, J. Lange, C. von der Malsburg, R.P. Würtz, and W. Konen. Distortion invariant object recognition in the dynamic link architecture. *IEEE Transactions on Computers*, 42(3):300–310, 1993
12. L. Lam and S.Y. Suen. Application of majority voting to pattern recognition: An analysis of its behavior and performance. *IEEE Transactions on Systems, Man, and Cybernetics – Part A: Systems and Humans*, 27(5):553–568, 1997
13. B. Leibe and B. Schiele. Analyzing appearance and contour based methods for object categorization. In *Conference on Computer Vision and Pattern Recognition (CVPR'03)*, volume 2, pages 409–415, Madison, Wisconsin, USA. IEEE Press, 2003
14. R. Linsker. Self-organization in a perceptual network. *IEEE Computer*, 105–117, 1988

15. N.K. Logothetis and J. Pauls. Psychophysical and physiological evidence for viewer-centered object representation in the primate. *Cerebral Cortex*, 3:270–288, 1995
16. H.S. Loos. *User-Assisted Learning of Visual Object Recognition.* PhD thesis, University of Bielefeld, Germany, 2002
17. W.S. McCulloch and W.H. Pitts. A logical calculus of the ideas immanent in nervous activity. *Bulletin of Mathematical Biophysics*, 5:115–133, 1943
18. B.W. Mel. SEEMORE: Combining color, shape, and texture histogramming in a neurally inspired approach to visual object recognition. *Neural Computation*, 9:777–804, 1997
19. B.T. Messmer and H. Bunke. A new algorithm for error-tolerant subgraph isomorphism detection. *IEEE Transactions on Pattern Analysis and Machine Intelligence*, 20(5):493–504, 1998
20. H. Murase and S.K. Nayar. Visual learning and recognition of 3-d objects from appearance. *International Journal of Computer Vision*, 14:5–24, 1995
21. S.A. Nene, S.K. Nayar, and H. Murase. Columbia object image library (COIL-100). Technical Report CUCS-006-96, Columbia University, 1996
22. T.J. Palmeri and I. Gauthier. Visual object understanding. *Nature Reviews Neuroscience*, 5:291–304, 2004
23. D.I. Perret, P.A.J. Smith, D.D. Potter, A.J. Mistlin, A.S. Head, and A.D. Milner. Visual cells in the temporal cortex sensitive to face view and gaze direction. *Proceedings of the Royal Society B*, 223:293–317, 1985
24. M. Pötzsch, T. Maurer, L. Wiskott, and C. von der Malsburg. Reconstruction from graphs labeled with responses of gabor filters. In C. von der Malsburg, W. von Seelen, J. Vorbrüggen, and B. Sendhoff, editors, *Proceedings of the ICANN 1996*, pages 845–850, Springer, Berlin, Heidelberg, New York, 1996
25. M. Riesenhuber and T. Poggio. Models of object recognition. *Nature Neuroscience*, 3:1199–1204, 2000
26. F. Rosenblatt. The perceptron: A probabilistic model for information storage and organization in the brain. *Psychological Review*, 65:386–408, 1958
27. P.A. Schmidt and G. Westphal. Object manipulation by integration of visual and tactile representations. In Uwe J. Ilg, Heinrich H. Bülthoff, and Hanspeter A. Mallot, editors, *Dynamic Perception*, pages 101–106. infix Verlag/IOS press, 2004
28. L.B. Shams. *Development of visual shape primitives.* PhD thesis, University of Southern California, 1999
29. C.E. Shannon. A mathematical theory of communication. *Bell Systems Technical Journal*, 27:623–656, 1948
30. L.G. Shapiro and R.M. Haralick. Structural descriptions and inexact matching. *IEEE Transactions on Pattern Analysis and Machine Intelligence*, 3(5):504–519, 1981
31. F. Tang and H. Tao. Object tracking with dynamic feature graph. In *Proceedings of the 2nd Joint IEEE International Workshop on Visual Surveillance and Performance Evaluation of Tracking and Surveillance*, pages 25–32, Beijing, China, 2005
32. A. Tewes. *A flexible object model for encoding and matching human faces.* PhD thesis, Physics Department, University of Bochum, Germany, January 2006
33. S. Thorpe, D. Fize, and C. Marlot. Speed of processing in the human visual system. *Nature*, 381:520–522, 1996

34. S. Thorpe and M.F. Thorpe. Seeking categories in the brain. *Neuroscience*, 291:260–263, 2001
35. I. Ulusoy and C.M. Bishop. Generative versus discriminative methods for object recognition. In *Proceedings of the IEEE Computer Society Conference on Computer Vision and Pattern Recognition (CVPR 2005)*, volume 2, pages 258–265,San Diego, California, USA. IEEE Press, 20–26 June 2005
36. M. Vidal-Naquet and S. Ullman. Object recognition with informative features and linear classification. In *Conference on Computer Vision and Pattern Recognition (CVPR'03)*, pages 281–288, Madison, Wisconsin, USA. IEEE Press, 2003
37. C. von der Malsburg. The Correlation theory of brain function. Internal Report 81-2, Max-Planck-Institute for Biophysical Chemistry, Department of Neurobiology, 1981
38. C. von der Malsburg. Pattern recognition by labeled graph matching. *Neural Networks*, 1:141–148, 1988
39. C. von der Malsburg. The dynamic link architecture. In M.A. Arbib, editor, *The Handbook of Brain Theory and Neural Networks*, 2nd edn., pages 1002–1005. MIT, Cambridge, MA, London, England, 2002
40. C. von der Malsburg and K. Reiser. Pose invariant object recognition in a neural system. In F. Fogelmann-Soulié, J.C. Rault, P. Gallinari, and G. Dreyfus, editors, *International Conference on Artifical Neural Networks (ICANN 1995)*, pages 127–132. EC2 & Cie, Paris, France, 1995
41. M. Weber, M. Welling, and P. Perona. Unsupervised learning of models for recognition. In *Proceedings of the 6th European Conference on Computer Vision (ECCV)*, pages 18–32, Dublin, Ireland, 2000
42. H. Wersing and E. Körner. Learning optimized features for hierarchical models of invariant object recognition. *Neural Computation*, 15:1559–1588, 2003
43. G. Westphal. Classification of molecules into classes of toxicity. Technical Report, Dr. Holthausen GmbH, Bocholt, Germany, 2004
44. G. Westphal. *Feature-driven emergence of model graphs for object recognition and categorization.* PhD thesis, University of Lübeck, Germany, 2006
45. G. Westphal and R.P. Würtz. Fast object and pose recognition through minimum entropy coding. In J. Kittler, M. Petrou, and M. Nixon, editors, *17th International Conference on Pattern Recognition (ICPR 2004)*, volume 3, pages 53–56, Cambridge, UK. IEEE Press, 2004
46. L. Wiskott. *Labeled graphs and dynamic link matching for face recognition and scene analysis.* PhD thesis, Physics Department, University of Bochum, Germany, 1995
47. L. Wiskott, J.-M. Fellous, N. Krüger, and C. von der Malsburg. Face Recognition by elastic bunch graph matching. *IEEE Transactions on Pattern Analysis and Machine Intelligence*, 19(7):775–779, 1997
48. I.H. Witten and E. Frank. *Data mining: Practical machine learning tools and techniques with java implementations.* Morgan Kaufmann, USA, 2000
49. R.P. Würtz. Object recognition robust under translations, deformations, and changes in background. *IEEE Transactions on Pattern Analysis and Machine Intelligence*, 19(7):769–775, 1997

Part IV

Special Applications

A Wavelet-based Statistical Method for Chinese Writer Identification

Zhenyu He[a] and Yuan Yan Tang[a,b]

[a] Department of Computer Science, Hong Kong Baptist University, Hong Kong
[b] Department of Electronics and Information Engineering, Huazhong University of Science and Technology, China

Summary. Writer identification is an effective solution to personal identification, which is necessary in many commercial and governmental sections of human society. In spite of continuous effort, writer identification, especially the off-line, text-independent writer identification, still remains as a challenging problem. In this paper, we propose a new method, which combines the wavelet theory and statistical model (more accurately, generalized Gaussian density (GGD) model), for off-line, text-independent writer identification. This method is based on our discovery that wavelet coefficients within each high-frequency subband of the handwritings satisfy GGD distribution. For different handwritings, the GGD parameters vary and thus can be selected as the handwriting features. Our experiments show this novel method, compared with two-dimensional Gabor model, one classic method on off-line, text-independent writer identification, not only achieves much better identification results but also greatly reduces the elapsed time on calculation.

Keywords: Writer identification, off-line, text-independent, two-dimensional Gabor, two-dimensional wavelet, generalized Gaussian model

1 Introduction

Since the beginning of civilization, identifying the statuses of uncertain persons has been crucial to the human society. Consequently, personal identification is widely used in diverse commercial and governmental sections such as financial access, health care, security control, border control and communication. In particular, personal identification is in highly increasing demand after the 9/11 terroristic attack. Traditionally, the ways for personal identification are identification cards (ID cards) and passwords. While the two ways cannot provide us an unique, secure and consistent personal identification. For example, passwords and ID cards can be shared by others and therefore are not unique. Furthermore, it is possible that we forget to bring the ID cards with us or forget the passwords and so they are not consistent. A nationwide

Z. He and Y.Y. Tang: *A Wavelet-based Statistical Method for Chinese Writer Identification*, Studies in Computational Intelligence (SCI) **91**, 203–220 (2008)
www.springerlink.com

survey in USA showed that heavy web users have an average of twenty-one passwords, and they often make confusion on their varied passwords [1]. So we need better solutions to personal identification.

Writer identification, which, speaking in a simple way, is to determine the writer from his/her handwritings (including signatures, letters, notes, etc.), is such a technique that satisfies the four requirements of personal identification: accessible, cheap, reliable and acceptable. Therefore, in spite of the existence and development of other techniques on personal identification based on DNA [2] [3], iris [4], fingerprint [5], etc., it appears that the writer identification still remains an attractive application. As a result, writer identification enjoys a huge interest from both industry and academia [6] [7] [8].

Writer identification can be classified in several ways, however the most straightforward one is to classify it into on-line and off-line writer identifications [7] [9]. The former assumes that a transducer device is connected to the computer, which can convert writing movement into a sequence of signals and then send these signals to the connected computer. The most common form of the transducer is a tablet digitizer, which consists of a plastic or electronic pen and a pressure or electrostatic-sensitive writing surface on which the user writes down his/her handwritings with the electronic pen. Since dynamic information of the writing process captured by the transducer device contains many useful writing features, on-line writer identification, compared to the off-line writer identification, is easier to achieve a high identification accuracy. On the other hand, off-line writer identification deals with handwritings scanned into a computer file in two-dimensional image representation. Despite continuous effort, off-line writer identification still remains a challenging issue [7]. In fact, off-line systems rely on more sophisticated architectures to accomplish the same identification task, but their identification results are still lower than those obtained by on-line identification systems under same testing conditions. Unfortunately, on-line systems are inapplicable in many cases. For example, on-line systems can not help us if we want to find out the writer of an existing handwriting document. Therefore, developing effective techniques on off-line writer identification is an urgent task.

Further, off-line writer identification can be divided into two types: text-dependent and text-independent writer identification [7] [9]. Text-dependent identification matches one or a small group of same characters/words and consequently requires the writer to write the same fixed text in the handwriting documents. For example, signature identification, which is well known to us, is a special case of text-dependent writer identification. Commonly, the geometric or structure features of those given characters/words are extracted as the writing features in text-dependent writer identification. But in many applications, it is impossible to find out the same text from different handwriting documents and therefore text-dependent identification is unavailable. In this case, we need text-independent identification. Text-independent identification does not use the writing features of some specific characters/words, instead, considers handwriting document layout features, text line features, etc. Gen-

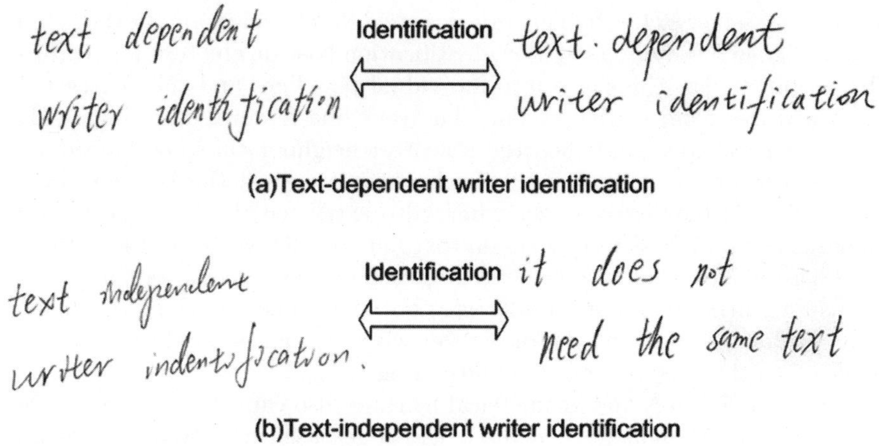

Fig. 1. Text-dependent and text-independent writer identification

erally speaking, text-dependent identification, compared to text-independent identification, has a better identification performance. However, as mentioned above, its applicability is lower than text-independent one because of its requirement on same characters/words. An example of text-independent and text-dependent writer identification is shown in Fig. 1. In this paper, we focus on the research on off-line, text-independent writer identification, which still is a changeling research topic and comparatively seldom touched by researchers. Before stating our method, we think it is wise for us to make a simple overview of other researchers' work on this field.

In [9], R. Plamondon made a summary of the early researches on writer identification. He pointed out that two general approaches had been proposed on the off-line, text-independent writer identification: transform techniques and histogram descriptions. In transform techniques, the most important variations of the writer's transfer function was reflected in the low frequency band of the Fourier spectrum of handwriting pages. In the second case, frequency distribution of different global or local properties was used.

In view of the handwritings of different people usually are visually different and inspired by the idea of multichannel spatial filtering technique, Said *et al.* proposed a texture analysis approach [7]. In this method, they regarded the handwriting as an image containing some special textures and applied a well-established two-dimensional Gabor model to extract features of such textures. In their paper, Said *et al.* also compared the two-dimensional Gabor model with the grey-scale co-occurrence matrix, and found out the two-dimensional Gabor model outperformed the grey-scale co-occurrence matrix.

Except for the global style of handwriting, some researchers found valuable features from single word or text line. In [10], Zois *et al.* morphologically processed horizontal projection profiles of single words. To increase the identification efficiency, the projections were derived and processed in segments. The

Bayesian classifier or neural network was used for classification. In [11], Hertel *et al.* designed a system for writer identification base on the text line features. They segmented a given text into individual text lines and then extracted a set of features from each text line. The text line features were regarded as the writing features. And then the k-nearest neighbor classifier was adopted for classification. In [12], A.Schlapbach *et al.* proposed the Hidden Markov Model (HMM) based recognizer which also extracted the text line features. They trained an HMM recognizer on text line for each writer in the database. The query handwriting was presented to each of these recognizers, and thus a series of log-likelihood score results were obtained. They ranked log-likelihood score results and regarded the recognizer with the highest log-likelihood score belonged to the writer of query handwriting.

Structure features and geometrical features also came into the researches' views. In [13], M.Bulacu *et al.* used the edge-based directional probability distributions as features for writer identification. They found out that the joint probability distribution of the angle combination of two "hinged" edge fragments outperformed all individual features. In addition, in [14] [15], researchers gave a definition of writer invariants for writer identification.

The rest of the paper is organized as follows. In Section 2, the two-dimensional Gabor model, a classic method for off-line, text-independent writer identification, is introduced, and will be used as a benchmark to be compared with our method in Section 4. In Section 3, we propose our method for writer identification. The experiments for writer identification using our method and the comparison between our method and two-dimensional Gabor method are offered in Section 4. Finally, the conclusion is made in Section 5.

2 A Classic Method for Writer Identification: Two-Dimensional Gabor Model

In [7], Said *et al.* firstly applied a two-dimensional Gabor model on English off-line, text-independent writer identification. Later, Zhu *et al.* also applied the same two-dimensional Gabor model on Chinese off-line, text-independent writer identification [8]. Both of the two papers said two-dimensional Gabor model achieved good results and outperformed the co-occurrence matrix in their experiments. This two-dimensional Gabor model has been widely acknowledged as one of the best methods on off-line text-independent writer identification.

The mathematical expression of the two-dimensional Gabor model used in references [7] [8] is given as follows:

$$h_e(x,y) = g(x,y)\cos[2\pi f(x\cos\theta + y\sin\theta)], \tag{1}$$

$$h_o(x,y) = g(x,y)\sin[2\pi f(x\cos\theta + y\sin\theta)], \tag{2}$$

where h_e, h_o denote the so-called even and odd symmetric Gabor filters respectively and $g(x, y)$ is an isotropic Gaussian function.

The spatial frequency responses of the Gabor filters are

$$H_e(u, v) = \frac{[H_1(u, v) + H_2(u, v)]}{2}, \tag{3}$$

$$H_o(u, v) = \frac{[H_1(u, v) - H_2(u, v)]}{2j}, \tag{4}$$

and

$$H_1(u, v) = \exp\{-2\pi^2\sigma^2[(u - f\cos\theta)^2 + (v - f\sin\theta)^2]\},$$
$$H_2(u, v) = \exp\{-2\pi^2\sigma^2[(u + f\cos\theta)^2 + (v + f\cos\theta)^2]\},$$

where $j = \sqrt{-1}$ and f, θ, σ are the spatial frequency, orientation, and space constants of the Gabor filters, respectively. For a given input image, $h_e(x, y)$ and $h_o(x, y)$ will combine to provide different Gabor subbands of the input image with different f, θ and σ.

The mean (M) and standard derivation (σ) of the Gabor subbands are selected as features to represent global writing features for writer identification.

Weighted Eucliden Distance (WED) is applied for feature matching after extracting the writing features,

$$WED(k) = \sum_{i=1}^{N} \frac{(M_i - M_i^k)^2}{\sigma_i^k}, \tag{5}$$

where M_i denotes the ith mean value of the query handwriting, M_i^k and σ_i^k denote the ith mean and standard derivation of the training handwriting of writer K respectively and N denotes the total number of mean values.

3 Our Algorithm for Writer Identification

Our algorithm for writer identification, which can be regarded as a problem of pattern recognition to some extent, contains three main steps. They are

1. **Preprocessing:** Removing the image noises and other detrimental factors which would disturb the later processings.
2. **Feature Extraction (FE):** Extracting features to fully represent the given handwriting image.
3. **Similarity Measurement (SM):** Using a certain measurement function to calculate the similarity between extracted features of the query handwriting image and the training handwriting images.

The whole procedure of our algorithm is described in Fig. 2.

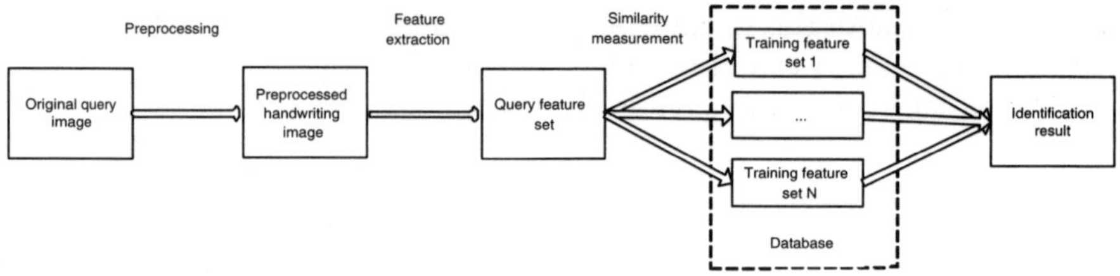

Fig. 2. The flow chart of our algorithm for writer identification

3.1 Preprocessing

As we know, the original handwriting image contains characters of different sizes, spaces between text lines and even noises. So prior to feature extraction, the original image should be preprocessed firstly. In the whole identification procedure, preprocessing plays an important role and inevitably influences the later processings and even the identification results.

Common steps adopted for preprocessing are as follows: firstly, removing the noises in the handwriting image; secondly, locating the text line and separating the single character using projection; thirdly, normalizing each character into a same size; finally, creating the preprocessed handwriting image by text padding [7] [8]. While this method only aims to the handwriting documents with a regular layout. Admittedly, automatic localization and segmentation of irregular handwriting documents are far from being successfully solved [9], and evasive in nearly all relevant papers. While we can not guarantee that all involved handwritings are written in a regular layout in practical applications. Therefore, we must find out an effective method to deal with the irregular handwriting documents. For this, in our research, we develop a software which can interactively localize and segment the characters manually from the irregular part of handwritings and generates the preprocessed handwriting images (PHI) with high-quality. In the following, we offer an example to show how our software implements the preprocessing.

Our software provides a rubber-like tool to remove the noises and needless marks. Fig. 3(a) is the original situation of the characters we want to process. Obviously, in this figure, the right character is surrounded by a circle, which may be a revising mark. Fig. 3(b) shows how to remove the outside circle, where the dotted box is the area where the rubber-like tool is working. The size of this tool can be adjusted. Therefore, one user can select large size when dealing with a large area of noise and small size when carefully dealing with the overlapping. Fig. 3(c) is the result after removing the needless mark.

Our software also provides a segment tool, which is a rectangle box with two blue "ears" at the left upper corner and right bottom corner separately. By manipulating these two "ears", the user can segment any rectangle area from the image. Fig. 4 shows the character we want is segmented and then padded

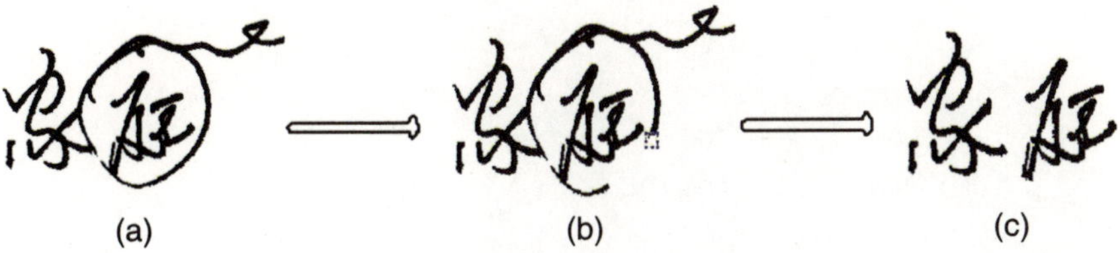

Fig. 3. Removing the noise

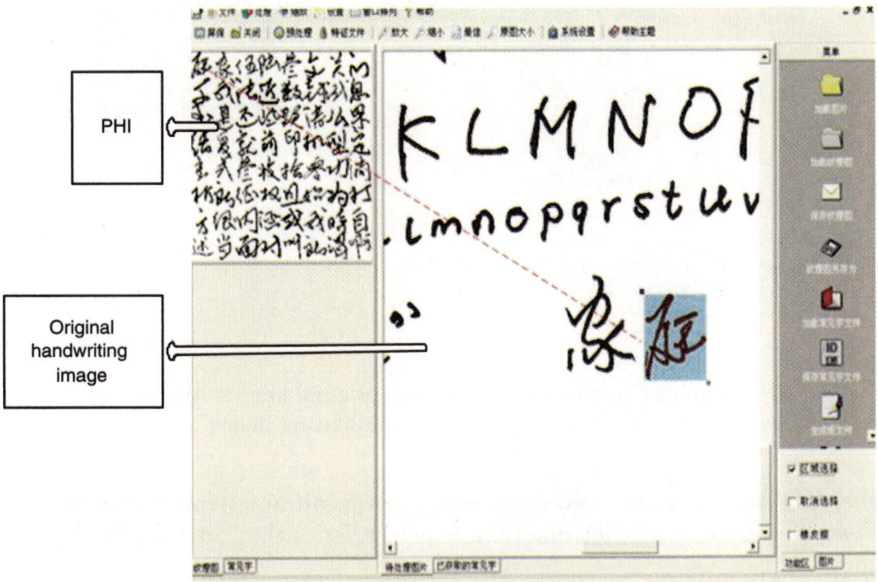

Fig. 4. Segmenting the character

into the PHI. The normalization of the character is implemented automatically by our software when the character is padded into the PHI.

Fig. 5 shows one original handwriting and the PHI generated from it using our software. [16] provides a more detailed description about the preprocessing.

3.2 Feature Extraction Based on Wavelet

Two-Dimensional Wavelet Transform

Assume that point (x, y) is a pixel in an image. It has a gray function $f(x, y)$, which indicates the gray level at this pixel. Then the wavelet transform of function $f(x, y)$ is defined as

$$W_f(x, y) = f(x, y) * \psi(x, y), \tag{6}$$

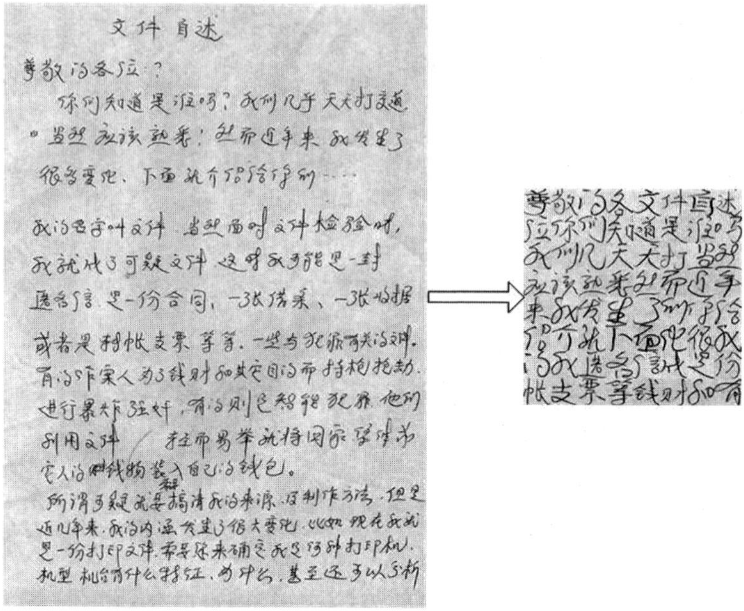

(a) (b)

Fig. 5. An example of preprocessing. (a) One original handwriting image, (b) PHI obtained by our software from this original handwriting image

where, '$*$' stands for the two-dimensional convolution operator, and $\psi(x, y)$ is a two-dimensional wavelet function, which satisfies the "admissibility" condition,

$$c_\psi := \int_{-\infty}^{+\infty} \int_{-\infty}^{+\infty} (\hat{\psi}(\omega_x, \omega_y))^2/(\omega_x^2 + \omega_y^2) d\omega_x d\omega_y < \infty. \quad (7)$$

In order to extract features from an image in different resolutions, the multi-scale wavelet function can be written explicitly as

$$\psi_s(x, y) = (1/s^2)\psi(x/s, y/s), \quad (8)$$

where, s is the scale, and the wavelet transform of $f(x, y)$ at scale s is,

$$W_{s,f}(x, y) = f(x, y) * \psi_s(x, y). \quad (9)$$

Furthermore, some constraints are forced into the "mother" wavelet function so as to guarantee the transform to be non-redundant, complete and form a multi-resolution representation for the original image. A well-known example is Daubechies wavelets (db wavelets), which are the orthonormal bases of compactly supported wavelets [17], from which the pyramid algorithm of wavelet decomposition can be drawn out.

In a certain scale, a one-dimensional filter is first used to convolute the rows of the input image, and one of every two columns is reserved. Then,

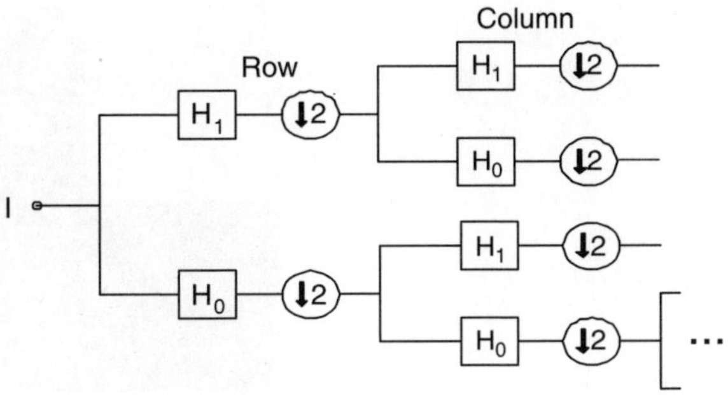

Fig. 6. The pyramidal algorithm of two-dimensional wavelet decomposition. The filters H_0, H_1 are one-dimensional mirror filters. And H_0 is the low-pass filter, H_1 is the high-pass filter. I is the input image

another one-dimensional filter is applied to column convolution on the image, and one of every two rows is reserved. Fig. 6 shows this process.

After we decompose the image into a series of frequency squares (also called wavelet subbands), what we should do next is to find out the feature sets hidden in these wavelet subbands, based on which we can discriminate one image from others.

The Property of Wavelet Coefficients of PHI

The most simple and direct feature of the wavelet coefficients is the energy. Generally, L^1-norm and L^2-norm are selected as measurements of energy. Suppose $\{WI_j(x)\}_{x \in R}$ is collection of wavelet coefficients in the jth wavelet subband, then L^1 and L^2 of this subband are given as:

$$L_j^1 = \frac{1}{R} \sum_{x \in R} |WI_j(x)|, \tag{10}$$

$$L_j^2 = (\frac{1}{R} \sum_{x \in R} WI_j(x)^2)^{1/2}, \tag{11}$$

where, R refers to domain of the jth wavelet subband. In addition, mean and standard derivation are also used as energy features.

The advantage of energy-based models is that only a few parameters are needed to describe an image. Unfortunately, the energy-based models are not sufficient to capture all image properties. It has already been shown that there may be perceptually very different images that have very similar energy features [18]. So we need to find out more effective features to replace the energy features.

By a lot of experimental works, we empirically find out that the marginal statistics of wavelet coefficients within each high-frequency wavelet subband of the PHI are highly non-Gaussian. That is, the margins tend to be much

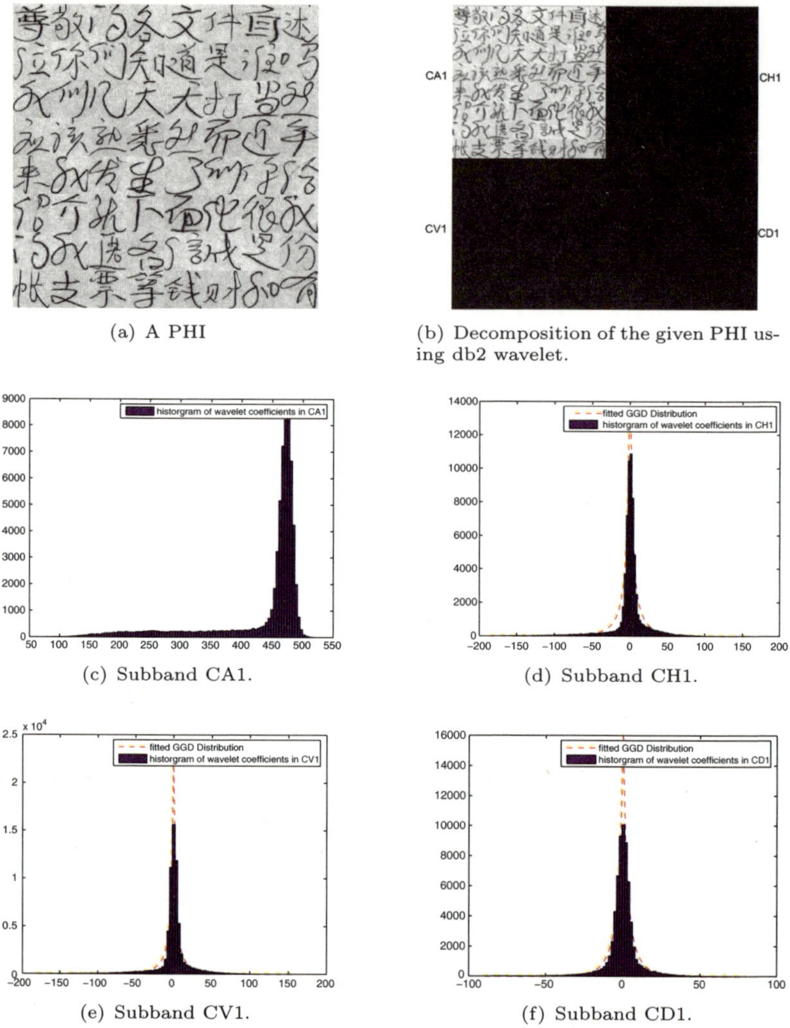

(a) A PHI

(b) Decomposition of the given PHI using db2 wavelet.

(c) Subband CA1.

(d) Subband CH1.

(e) Subband CV1.

(f) Subband CD1.

Fig. 7. The histograms of wavelet coefficients within all wavelet subbands (except for the subband CA1 at the lowest frequency) of the PHI satisfy the GGD distribution

more sharply peaked at zero, with more extensive tails, when compared with a Gaussian of the same variance. This non-Gaussian marginal density can be well-modelled by a generalized gaussian density (GGD) model. An example of our experimental discovery is given in Fig. 7.

The GGD model is given as

$$P(x|\ \{\alpha,\beta\}) = \frac{\beta}{2\alpha\Gamma(1/\beta)} e^{-1(|x|/\alpha)^{\beta}} \tag{12}$$

where, $\Gamma(\cdot)$ is the Gamma function, i.e., $\Gamma(\cdot) = \int_0^\infty e^{-t} t^{Z-1} dt, Z > 0$. The normalization constant is $Z(\{\alpha, \beta\}) = 2\frac{\alpha}{\beta}\Gamma(1/\beta)$. An exponent of $\beta = 2$ corresponds to a Gaussian density, and $\beta = 1$ corresponds to the Laplacian density. The parameter $\alpha > 0$, called scale parameter, describes the standard deviation. α varies monotonically with the scale of the basis functions, with correspondingly higher variance for coarser scale components. The parameter $\beta > 0$, called shape parameter, is inversely proportional to the decreasing rate of the peak. In general, smaller values of β lead to a density that is both more concentrated at zero and has more expansive tails.

GGD model is completely determined by the marginal statistics of the wavelet coefficients with an assumption that the wavelet coefficients within a subband are independent and identically distributed (i.i.d). We must note that the low-frequency wavelet subband cannot be fitted by the GGD model, as shown in the Fig. 7(c).

Estimate the Parameters of GGD Model

The basic idea of our wavelet-based GGD method is to establish corresponding wavelet-based GGD model for a handwriting image, and then the parameters of this model $\{\alpha, \beta\}$ can be regarded as the features of the handwriting. The most important work is to estimate the model parameters $\{\alpha, \beta\}$ according to the input preprocessed handwriting image (PHI).

For a given wavelet subband Y, according to Bayes rule which is optimal in terms of identification error probability, the estimated parameters $\{\hat{\alpha},\hat{\beta}\}$ must be chosen such that they maximize $P(\{\alpha, \beta\}|X)$. Similarly, Bayes theorem dictates that it is equivalent to setting $\{\hat{\alpha},\hat{\beta}\} = argmax_{\{\alpha,\beta\}} P(\{\alpha, \beta\}|X)$. This is maximum likelihood rule.

Define $X = (x_1, ..., x_N)$ is an independent and identically distributed (i.i.d) sequence, which consists of wavelet coefficients in Y. Then the likelihood function of the GGD model in one wavelet subband can be defined as

$$L(X|\ \{\alpha, \beta\}) = \log \prod_{i=1}^{N} P(x_i|\ \{\alpha, \beta\}). \tag{13}$$

According to the Lagrange optimization, we get the following likelihood equations:

$$\frac{\partial L(x|\ \{\alpha, \beta\})}{\partial \alpha} = -\frac{N}{\alpha} + \sum_{1}^{N} \frac{\beta |x_i|^\beta a^{-\beta}}{\alpha}, \tag{14}$$

$$\frac{\partial L(x|\ \{\alpha, \beta\})}{\partial \beta} = \frac{N}{\beta} + \frac{N\Psi(1/\beta)}{\beta^2} - \sum_{i=1}^{N} (\frac{|x_i|}{\alpha}) \log(\frac{|x_i|}{\alpha}), \tag{15}$$

where $\Psi(z) = \Gamma'(z)/\Gamma(z)$.

Since $\beta > 0$ (In (12), it is obvious to know that $1/\beta > 0$ based on requirement of function Γ), the above equations have an unique root in probability. $\hat{\alpha}$, the solution of

$$\frac{\partial L(x|\ \{\alpha, \beta\})}{\partial \alpha} = 0,$$

can be obtained by

$$\hat{\alpha} = (\frac{\beta}{N}\sum_{i=1}^{N}|x_i|^{\beta})^{1/\beta}. \tag{16}$$

Substituting the above equation into (15), we find that the estimation of β is the solution of the following equation,

$$1 + \frac{\Psi(1/\hat{\beta})}{\hat{\beta}} - \frac{\sum_{i=1}^{N}|x_i|^{\hat{\beta}}\log|x_i|}{\sum_{i=1}^{N}|x_i|^{\hat{\beta}}} + \frac{\log(\frac{\hat{\beta}}{N}\sum_{i=1}^{N}|x_i|^{\hat{\beta}})}{\hat{\beta}} = 0. \tag{17}$$

(17) can be numerically solved with a fast algorithm based on the Newton-Raphson iterative procedure and the initial value is given by the moment method [19]. After obtaining β, it is easy to get the estimate value of α from (16).

3.3 Similarity Measurement

To some extent, writer identification is also a multiple hypothesis problem to find out M handwriting images maximizing $P(I_q|\ \theta_j)$, $1 \leq j \leq M$, I_q is the query handwriting image, θ_j is the hypothesis parameter set of training handwriting image I_j. This problem is equivalent to minimizing the Kullback-Leibler Distance (KLD) between the two probability density functions (PDFs) $P(X|\ \theta_q)$ and $P(X|\ \theta_j)$, as is proved in [19] [20]. The definition of the Kullback-Leibler Distance between two PDFs is given as

$$D(P(X|\ \theta_q)||P(X|\ \theta_j)) = \int P(x|\ \theta_q)\log\frac{P(x|\ \theta_q)}{P(x|\ \theta_i)}dx. \tag{18}$$

In GGD model, the hypothesis parameter set $\theta = \{\alpha, \beta\}$. Substituting (12) into (18), after some simple calculations, we find that the KLD between two GGD models is explicitly given by

$$\begin{aligned} D(P(X|\ \{\alpha_1, \beta_1\})\|P(X|\ \{\alpha_2, \beta_2\})) &= \log(\frac{\beta_1 \alpha \Gamma(1/\beta_2)}{\beta_2 \alpha_1 \Gamma(1/\beta_1)}) \\ &+ (\frac{\alpha_1}{\alpha_2})^{\beta_2}\frac{\Gamma((\beta_2+1)/\beta_1)}{\Gamma(1/\beta_1)} - \frac{1}{\beta_1}. \end{aligned} \tag{19}$$

And the KLD between two handwriting images I_1, I_2 is the sum of all the KLDs across all selected wavelet subbands,

$$D(I_1, I_2) = \sum_{i=1}^{K} D(P(X|\{\alpha_1^{(i)}, \beta 1^{(i)}\})||P(X|\{\alpha_2^{(i)}, \beta_2^{(i)}\})), \qquad (20)$$

where K is the number of selected wavelet subbands.

Next, we can generate the identification result according to the similarity distances. The smaller KLD value is, the more similar it is. We only consider the k-nearest neighbor classifier since it is a robust and efficient scheme. That is, identification result is the list of top M handwriting images which are most similar to the query handwriting image. We can know the corresponding writers from the indexes of these top handwritings.

4 Experiments

It must be noted that performance comparison between the existing systems and approaches are very difficult to be established because there does not exist an authoritative handwriting database which can act as a benchmark for performance evaluation and comparison. Most researchers built their own databases. We also have to create an Chinese handwriting database for our experiments. 1000 Chinese handwritings written by 500 persons are collected in our database, with one training handwriting and one query handwriting for each person. All handwritings are scanned into computer with a resolution of 300 dpi. We produce one PHI image from each original handwriting, and totally 1000 PHI images are obtained.

A criterion to evaluation the identification performance of a method is how many handwriting texts are required for identification. Most existing methods for off-line, text-independent writer identification required a full page of text, which generally consists of hundreds of characters or words. In our research, we design the size of PHI and the number of characters contained by PHI to achieve a good balance between the computation cost and the identification accuracy. In our system, each PHI consists of 64 Chinese characters with size 64×64 pixels, arranged in an 8×8 array. Our experiments show such a PHI image not only contains enough writing information to ensure a high identification rate, but also let the identification processing be finished within an acceptable time.

4.1 Identification performance evaluation 1

In our experiments, we make a comparison between the wavelet-based GGD method and the 2-D Gabor model, not only on identification accuracy, but also on computational efficiency. Several combinations of different Gabor frequencies are tested, ranging from 16 to 128. For each spatial frequency, we select 0, 45, 90 and 135 degree as orientations.

For the wavelet-based GGD method, our experiments agree with [21] that the size of the smallest subimages should not be less than 16 pixels $\times$ 16 pixels so that estimated energy values or model parameters would be robust. Therefore, for a PHI of size 512 × 512, the number of wavelet decomposition levels should not be beyond five. According to our experiment records, three level decomposition is sufficient. All identification results shown in this paper are in the case of three level decomposition. We decompose the handwriting image via traditional discrete wavelet transform (DWT), and the used wavelets are Daubechies orthogonal wavelets. Of course, different wavelet filters may lead to different results. While testing all possible wavelet filters and finding which one is the best is out of the scope of this paper.

From a training PHI in the database, two GGD parameters $\{\alpha, \beta\}$ are estimated from each detailed wavelet subband using the MLE described in the previous section. Here we consider that these GGD parameters $\{\alpha, \beta\}$ are the writing features of the PHI, and can be used for the similarity measurement. Thus, in the case of three decomposition, totally 18 parameters are obtained, including 9 $\{\alpha\}$ and 9 $\{\beta\}$.

The evaluation criteria of identification is defined as follows: for each query handwriting, if the training handwriting belonging to the same writer is ranked at the top N matches, we say this is a correct identification, otherwise the identification fails. The identification rate is the percentage of the correct identification. The identification rate certainly changes accordingly when the number of top matches varies. The identification results are recorded in the Table 1 and Fig. 8.

The computational efficiency is measured by the elapsed time of each method. Our programs are implemented in the Matlab environment in PC computer. The software environment of our computer is: Window XP, Matlab 7.0; and the hardware environment is: Intel Pentium IV 2.4GHZ CPU, 512MB RAM. The record of average elapsed time is given in Table 2.

From Table 1, it is clear that in the Gabor method, the more frequencies are combined, the higher identification rate is achieved. Unfortunately, at the

Table 1. Writer identification rate 1 (%)

Number of top matches	Our method	Gabor, f = 16	Gabor, f = 16, 32	Gabor, f = 16, 32, 64, 128
1	39.2	13.4	18.2	32.8
2	45.8	24.6	31.8	39.0
3	54.6	33.8	43.2	49.4
5	62.4	41.4	51.6	56.2
7	69.6	47.2	58.4	64.8
10	77.2	55.0	64.2	71.4
15	84.8	64.6	71.6	79.8
20	92.6	70.8	79.4	85.2
25	97.8	76.2	84.2	91.2
30	100	80.6	87.8	95.6
40	100	86.6	92.8	100

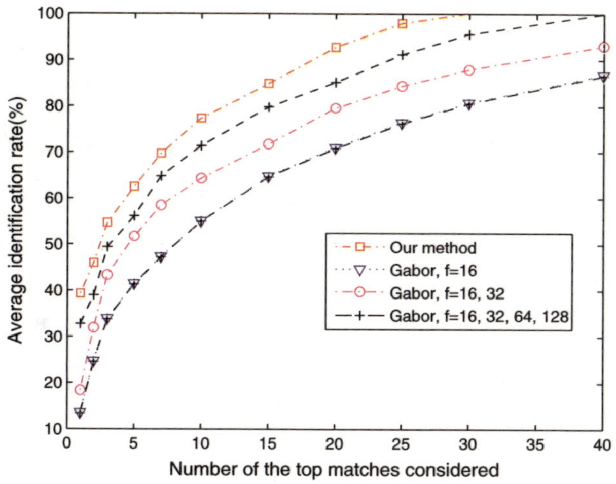

Fig. 8. Identification rate according to the number of top matches considered

Table 2. Average elapsed time 1 for writer identification (second)

Method	Our method	Gabor, f = 16	Gabor, f = 16, 32	Gabor, f = 16, 32, 64, 128
Elapsed time	8.72	53.17	107.03	213.87

same time, the elapsed time also increases greatly. The identification rate of the Gabor model combing four frequencies $f = 16, 32, 64, 128$ is closest to that of wavelet-based GGD method, while its cost time is 24 times of that used in wavelet-based GGD method. The elapsed time of the Gabor model with $f = 16$ is the shortest in the different combinations of Gabor method, however its identification rate is much lower than that of the wavelet-based GGD method. Comprehensively, the wavelet-based GGD method outperforms the Gabor model on both identification performance and the computational efficiency.

4.2 Identification performance evaluation 2

we divide each PHI of 512×512 pixels into 4 non-overlapped sub-PHIs of 256×256 pixels to increase the writing samples for one writer. In this way, we can obtain 8 writing samples for one writer. For each query sub-PHI, only the top $S \geq 7$ matches are considered since there are seven sub-PHIs of the same writer for each query sub-PHI. The identification percentage is the ratio of the number of correct matches within the top S matches to 7. For example, in the case of $S = 10$, the identification rate is $6/7 \times 100\% = 85.71\%$ if 6 correct matches are at the top 10 matches. In this experiment, we do not classify the sub-PHIs into training group and query group. All sub-PHIs are used as

Table 3. Writer identification rate 2 (%)

Number of top matches	Our method	Gabor, f = 16	Gabor, f = 16, 32	Gabor, f = 16, 32, 64, 128
7	29.92	10.17	15.32	22.37
10	34.84	16.65	22.48	27.56
20	44.92	25.82	31.63	37.03
30	53.47	35.73	40.07	44.89
50	64.65	44.25	49.11	54.47
70	73.83	52.19	60.38	66.72
100	81.95	60.39	68.65	74.58
150	87.82	68.48	74.61	80.25
200	95.67	73.54	79.17	88.69
300	99.04	80.26	87.53	93.71

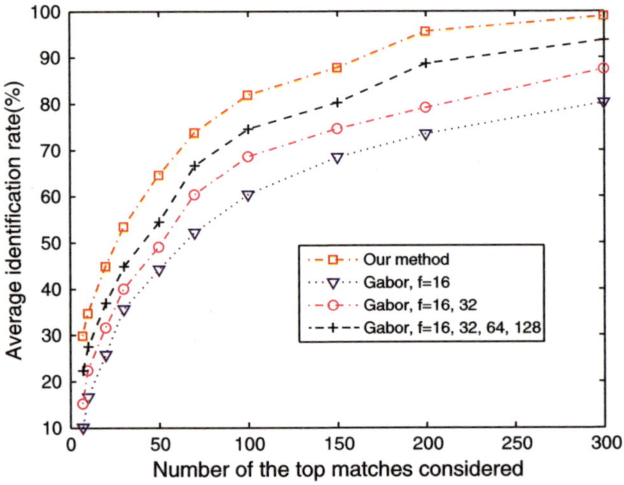

Fig. 9. Identification rate according to the number of top matches considered

a query handwriting, and simultaneously other sub-PHIs except the query one play the role as the training handwritings. The identification rates of the wavelet-based GGD method and the Gabor model with different frequency combinations are offered in Table 3 and Fig. 9. Though the identification rate of the wavelet-based GGD method is not very high in this case, it is still satisfied in view of only 16 Chinese characters are used.

The average elapsed time of our method and 2-D Gabor model in this experiments is offered in Table 4. Combining the results in Table 3 and Table 4, our method still outperforms the 2-D Gabor model in this experiment.

Table 4. Average elapsed time 2 for writer identification (second)

Method	Our method	Gabor, f = 16	Gabor, f = 16, 32	Gabor, f = 16, 32, 64, 128
Elapsed time	0.53	5.01	9.10	17.42

5 Conclusions

A novel approach for off-line, text-independent writer identification based on wavelet transform has been presented in this paper. In this approach, a handwriting document image is firstly preprocessed to generate a PHI image, and then the PHI image is decomposed into several sub-images by DWT. Thereafter, the parameters $\{\alpha, \beta\}$ of generalized Gaussian distribution are extracted from the detailed wavelet sub-images. After obtaining the GGD parameters, Kullback-Leibler Distance is adopted to measure the similarity distance between the feature vectors of query PHI and training PHIs. Unlike most existing methods, this approach is based on the global features of the handwriting images. Experiments on our database consisting of thousands of handwriting images show our approach is highly better than the 2-D Gabor model, which is also based on the global features of handwriting images and widely acknowledged as an efficient method for off-line, text-independent writer identification. It must be noted, our approach is text-independent and hence applicable to other language documents, such as English, Korean, Japanese and Latin Language, etc., since text-independent methods do not care about the content of handwriting documents.

References

1. A.K. Jain. Recent development on biometric authentication. In *Proceeedings of Advanced Study Institute (ASI)*. Hong Kong Baptist University, Hong Kong, 2004
2. M. Benecke. DNA typing in forensic medicine and in criminal investigations: A current survey. *Natur Wissenschaften*, 84(5):181–188, 1997
3. B. Devlin, N. Risch, and K. Roeder. Forensic inference from DNA fingerprints. *Journal of American Statistical Association*, 87(418):337–350, 1992
4. J. Daugman. The importance of being random: Statistical principles of iris recognition. *Pattern Recognition*, 36(2):279–291, 2003
5. A. Jain, L. Hong, and R. Bolle. On-line fingerprint verification. *IEEE Transactions on Pattern Analysis and Machine Intelligence*, 19(4):302–314, 1997
6. S. Srihari, S. Cha, H. Arora, and S. Lee. Individuality of handwriting. *Journal of Forensic Sciences*, 47(4):1–17, 2002
7. H.E.S. Said, T. Tan, and K. Baker. Writer identification based on handwriting. *Pattern Recognition*, 33(1):133–148, 2000
8. Y. Zhu, T. Tan, and Y. Wang. Biometric personal identification based on handwriting. In *Proceedings of the 15th International Conference on Pattern Recognition*, pages 801–804, 2000

9. R. Plamondon and G. Lorrtte. Automatic signature vertification and writer idenfication idenfication – the state of the art. *Pattern Recognition*, 37(12):107–131, 1989
10. E.N. Zois. Morphological wavelform coding for writer identication. *Pattern Recognition*, 33:385–398, 2000
11. C. Hertel and H. Bunke. A set of novel features for writer identification. In *AVBPA*, pages 679–687, 2003
12. A. Schlapbach and H. Bunke. Off-line handwriting identification using HMM based recognizers. In *Proceedings of 17th International Conference on Pattern Recognition*, volume 2, pages 654–658, 2004
13. M. Bulacu, L. Schomarker, and L. Vuurpijl. Writer identification using edge-based directional features. In *Proceedings of the 7th International Conference on Document Analysis and Recognition*, pages 937–941, 2003
14. T.A. Nosary and L. Heutte. Definiting writer's invariants to adapt the recognition task. In *Proceedings of the 5th International Conference on Document Analysis and Recognition*, volume 22, no. 1, pages 765–768, 1999
15. A. Bensefia, A. Nosary, T. Paquet, and L. Heutte. Writer Idenfication by writer's invariants. In *Proceedings of the 8th International Workshop on Frontiers in Handwriting Recognition*, pages 274–279, 2002
16. Z. He. Writer identification using wavelet, contourlet and statistical models. Ph.D Thesis. Hong Kong Baptist University, 2006
17. I. Daubechies. Ten Lectures on wavelets. *SIAM*, 1992
18. E.P. Simoncelli. *Handbook of Video and Image Processing*, 2nd edn. Academic, USA, 2005
19. M.N. Do and M. Vetterli. Wavelet-based texture retrieval using generalized gaussian density and Kullback–Leibler distance. *IEEE Transactions on Image Processing*, 11:146–158, 2002
20. O. Commowick, C. Lenglet, and C. Louchet. Wavelet-based Texture Classification and Retrieval. Technical Report, http://www.tsi.enst.fr/fsi/enseignement/ressources/mti/ReportFinal.html, 2003
21. T. Chang and C.C.J. Kuo. Texture analysis and classfication with tree-structure wavelet transform. *IEEE Transactions on Image Processing*, 2(4):429–441, 1985

Texture Analysis by Accurate Identification of a Generic Markov–Gibbs Model

Georgy Gimel'farb and Dongxiao Zhou

Department of Computer Science, Tamaki Campus,
The University of Auckland, Private Bag 92019, Auckland 1, New Zealand
(g.gimelfarb@auckland.ac.nz; zhoudxnz@gmail.com)

Abstract

A number of applied problems are effectively solved with simple Markov-Gibbs random field (MGRF) models of spatially homogeneous or piecewise-homogeneous images provided that their identification (parameter estimation) is able to focus such a prior on a particular class of images. We propose more accurate analytical potential estimates for a generic MGRF with multiple pairwise pixel interaction and use them for structural analysis and synthesis of stochastic and periodic image textures.

1 Introduction

Probability models, in particular, Markov-Gibbs random fields (MGRF) have gained wide acceptance for solving applied image recognition, analysis, and synthesis problems. Originated in the late nineteen seventies – early eighties (see e.g. [2, 4, 29]), the MGRF models describe images in terms of an explicit spatial geometry and quantitative strength (Gibbs potential) of inter-pixel statistical dependency, or interaction. At present, more and more complex models (e.g. [23, 25] to cite a few) come to the forefront in attempts to better describe various natural images. Nonetheless, far simpler MGRFs with pairwise pixel interaction are still of interest because they solve effectively a reasonably large variety of practical problems.

This chapter overviews potentialities of accurate analytical identification (estimation of parameters) of a generic MGRF with multiple pairwise interaction in its focusing on a class of spatially homogeneous images. Homogeneity is restricted to translation invariance of a set of second-order signal statistics. Unknown model parameters to be estimated from a given training image consist of both a characteristic pixel neighbourhood defining the interaction geometry and the corresponding Gibbs potential. The maximum likelihood estimate (MLE) of the potential has simple but accurate analytical approximations that considerably simplify the identification process.

G. Gimel'farb and D. Zhou: *Texture Analysis by Accurate Identification of a Generic Markov–Gibbs Model*, Studies in Computational Intelligence (SCI) **91**, 221–245 (2008)
www.springerlink.com

As a case in point, we consider structural analysis of stochastic and regular (nearly periodic) image textures, in particular, their description and synthesis by estimating arbitrary shaped elements (textons [14] or texels [12, 13]) and rules of their spatial placement. Accurate model identification helps us to evaluate repetitiveness (regularity) of a texture and select characteristic texels and placement rules. This description leads to a fast technique for realistic texture synthesis-via-analysis called *bunch sampling* [10].

At the analysis stage, a geometric shape and rules of relative spatial arrangement of signal bunches are derived from the analytically identified MGRF model, in particular, from a model-based interaction map (MBIM) of spatially distributed interaction energies. At the synthesis stage, a new texture is generated by placing bunches randomly sampled from the training image into the target image, in accord with the derived rules.

Section 2 describes analytical identification of the MGRF based on more accurate than in [8, 9] potential estimates[1]. The identified model is used in Section 3 for describing a training image in terms of texels. Section 4 describes the bunch sampling algorithm and presents comparative experimental results for various stochastic (aperiodic) and regular (periodic) textures.

2 Identification of a generic Markov-Gibbs model

2.1 Basic notation

Let $\mathcal{Q} = \{0, 1, \ldots, Q-1\}$ be a finite set of scalar image signals q (grey values or colour indices). Let $\mathcal{R} = \{(x, y) : x = 0, 1, \ldots, M-1; y = 0, 1, \ldots, N-1\}$ where (x, y) are integer Cartesian coordinates of pixels be a finite arithmetic lattice of size MN supporting digital images $g : \mathcal{R} \rightarrow \mathcal{Q}$.

Let $\mathcal{N} = \{(\xi_k, \eta_k) : k = 1, \ldots, n\}$ denote a collection of relative coordinate offsets of "neighbours" $(x+\xi, y+\eta)$ interacting, in a probabilistic sense, with a pixel $(x, y) \in \mathcal{R}$. Translation invariant pairwise interaction over the lattice is specified with a fixed, except the lattice borders, neighbourhood $\mathcal{E}_{x,y}$ of each pixel (x, y):

$$\mathcal{E}_{x,y} = \{(x+\xi, y+\eta), (x-\xi, y-\eta) : (\xi, \eta) \in \mathcal{N}\} \wedge \mathcal{R}$$

where each pixel pair $((x, y), (x+\xi, y+\eta))$ is the second-order clique of the neighbourhood graph $\Gamma = \left[\mathcal{R}, \mathcal{E} = \bigcup_{(x,y)\in\mathcal{R}} \mathcal{E}_{x,y}\right]$.

Let $\mathcal{C}_{\xi,\eta} = \{((x, y); (x+\xi, y+\eta)) : (x, y) \in \mathcal{R}; (x+\xi, y+\eta) \in \mathcal{R}\}$ and $\mathcal{C}_{\mathcal{N}} = \{\mathcal{C}_{\xi,\eta} : (\xi, \eta) \in \mathcal{N}\}$ denote a family of all second-order cliques in Γ with the inter-pixel offset (ξ, η) and a subset of the clique families for the neighbourhood $\mathcal{N}$, respectively.

The quantitative interaction in each clique family is given with a Gibbs potential function $V_{\xi,\eta} : \mathcal{Q}^2 \rightarrow \Re = (-\infty, \infty)$ of signal co-occurrences

[1] In a concise form, this identification was also presented in [11].

$(g_{x,y} = q, g_{x+\xi,y+\eta} = s)$; $(q,s) \in \mathcal{Q}^2$, in the cliques. All the families $\mathcal{C}_{\mathcal{N}}$ are represented with the potential vector $\mathbf{V} = \left[\mathbf{V}^{\mathsf{T}}_{\xi,\eta} : (\xi,\eta) \in \mathcal{N}\right]^{\mathsf{T}}$ of length nQ^2 where $\mathbf{V}_{\xi,\eta} = \left[V_{\xi,\eta}(q,s) : (q,s) \in \mathcal{Q}^2\right]^{\mathsf{T}}$ and T denotes transposition.

Let $\mathbf{H}(g) = \left[\mathbf{H}^{\mathsf{T}}_{\xi,\eta}(g) : (\xi,\eta) \in \mathcal{N}\right]^{\mathsf{T}}$ denote an nQ^2-vector of signal co-occurrence histograms $\mathbf{H}_{\xi,\eta}(g) = \left[h_{\xi,\eta}(q,s|g) : (q,s) \in \mathcal{Q}^2\right]^{\mathsf{T}}$ collected over $n = |\mathcal{N}|$ clique families in an image g:

$$h_{\xi,\eta}(q,s|g) = \sum_{\mathcal{C}_{\xi,\eta}} \delta(g_{x,y} - q)\delta(g_{x+\xi,y+\eta} - s); \quad \sum_{(q,s)\in\mathcal{Q}^2} h_{\xi,\eta}(q,s|g) = |\mathcal{C}_{\xi,\eta}|$$

where $\delta(z)$ is the Kronecker function, $\delta(0) = 1$ and $\delta(z) = 0$ if $z \neq 0$.

Let $\mathbf{F}(g)$ be the like nQ^2-vector of scaled empirical probabilities of signal co-occurrences:

$$\mathbf{F}(g) = \frac{1}{|\mathcal{R}|}\mathbf{H}(g) = \left[\rho_{\xi,\eta}\mathbf{F}^{\mathsf{T}}_{\xi,\eta}(g) : (\xi,\eta) \in \mathcal{N}\right]^{\mathsf{T}}$$

where $\mathbf{F}^{\mathsf{T}}_{\xi,\eta}(g) = \left[f_{\xi,\eta}(q,s|g) = \frac{1}{|\mathcal{C}_{\xi,\eta}|} h_{\xi,\eta}(q,s|g) : (q,s) \in \mathcal{Q}^2\right]$ and $\rho_{\xi,\eta} = \frac{|\mathcal{C}_{\xi,\eta}|}{|\mathcal{R}|}$ is the relative size of the family; $\sum_{(q,s)\in\mathcal{Q}^2} f_{\xi,\eta}(q,s|g) = 1$ for all $(\xi,\eta) \in \mathcal{N}$.

2.2 Generic MGRF with pairwise interaction

Given a neighbourhood set $\mathcal{N}$ and potential $\mathbf{V}$, a generic MGRF with multiple pairwise interaction is specified with a Gibbs probability distribution (GPD) factored over the clique families [2, 29]:

$$\begin{aligned} P_{\mathcal{N},\mathbf{V}}(g^\circ) &= \frac{1}{Z_{\mathbf{V}}} \exp\left(\sum_{(\xi,\eta)\in\mathcal{N}} \sum_{\mathcal{C}_{\xi,\eta}} V_{\xi,\eta}(g^\circ_{x,y}, g^\circ_{x+\xi,y+\eta})\right) \\ &= \frac{1}{Z_{\mathbf{V}}} \exp\left(\mathbf{V}^{\mathsf{T}}\mathbf{H}(g^\circ)\right) \equiv \frac{1}{Z_{\mathbf{V}}} \exp\left(|\mathcal{R}|\mathbf{V}^{\mathsf{T}}\mathbf{F}(g^\circ)\right) \end{aligned} \tag{1}$$

where $Z_{\mathcal{N},\mathbf{V}} = \sum_{g\in\mathcal{G}} \exp\left(\mathbf{V}^{\mathsf{T}}\mathbf{H}(g)\right)$ is the partition function for the parent population $\mathcal{G} = \mathcal{Q}^{|\mathcal{R}|}$ of all images.

It is well-known [1] that the MGRF belongs to exponential families of probability distributions so that its likelihood function is unimodal in the space of potentials $\mathbf{V}$ under loose conditions (affine independence of potential and histogram vectors), and its conditional entropy is maximal if marginal probability distributions of signal co-occurrences are equal to empirical probability distributions of these co-occurrences for each clique family $\mathcal{C}_{\xi,\eta}$; $(\xi,\eta) \in \mathcal{N}$.

To identify the entire probability model of Eq. (1), both the characteristic neighbourhood $\mathcal{N}$ and the potential $\mathbf{V}$ are to be estimated from a given training image g°. By this means, the identification can focus on a desired class of images much better than more conventional approaches that are based on a heuristic choice of characteristic pixel neighbourhoods and special parametric functional forms of Gibbs potentials.

2.3 Accurate first approximation of potentials

Given an image g° and a fixed neighbourhood $\mathcal{N}$, the specific log-likelihood of the potential is:

$$\ell_{\mathbf{V}|g^\circ} = \frac{1}{|\mathcal{R}|}\log P_{\mathcal{N},\mathbf{V}}(g^\circ) = \mathbf{V}^\mathsf{T}\mathbf{F}(g^\circ) - \frac{1}{|\mathcal{R}|}\ln\sum_{g\in\mathcal{G}}\exp\left(|\mathcal{R}|\mathbf{V}^\mathsf{T}\mathbf{F}(g)\right) \quad (2)$$

Let $\mathbf{V}_0$ be a potential vector such that gradient ($\nabla\ell$) and the Hessian matrix ($\nabla^2\ell$) of the log-likelihood in the vicinity of $\mathbf{V}_0$ are known:

$$\begin{aligned}\nabla\ell_{\mathbf{V}|g^\circ} &= \mathbf{F}(g^\circ) - \mathbb{E}\{\mathbf{F}(g)|\mathbf{V}\}\\ \nabla^2\ell_{\mathbf{V}|g^\circ} &= -\left(\mathbb{E}\left\{\mathbf{F}(g)\mathbf{F}^\mathsf{T}(g)|\mathbf{V}\right\} - \mathbb{E}\{\mathbf{F}(g)|\mathbf{V}\}\mathbb{E}\{\mathbf{F}^\mathsf{T}(g)|\mathbf{V}\}\right)\\ &= -\mathbf{C}_{\mathbf{F}(g)|\mathbf{V}}\end{aligned} \quad (3)$$

where $\mathbb{E}\{\ldots|\mathbf{V}\}$ and $\mathbf{C}_{\mathbf{F}(g)|\mathbf{V}}$ are the mathematical expectation and the covariance matrix of the scaled empirical probability vectors, respectively, under the GPD of Eq. (1).

Typically, the vector $\mathbb{E}\{\mathbf{F}(g)|\mathbf{V}\}$ of the scaled marginal co-occurrence probabilities and the covariance matrix $\mathbf{C}_{\mathbf{F}(g)|\mathbf{V}}$ are known only when the MGRF of Eq. (1) is reduced to an independent random field (IRF). Starting from such a potential $\mathbf{V}_0$, an approximate maximum likelihood estimate (MLE) of the potential is obtained using the analytical approach introduced in [8, 9]:

Proposition 1. *If the log-likelihood gradient and Hessian of Eq. (3) are known for an image g° and potential $\mathbf{V}_0$, the first approximation of the MLE:*

$$\mathbf{V}^* = \mathbf{V}_0 + \lambda^*\nabla\ell_{\mathbf{V}_0|g^\circ};\ \lambda^* = -\frac{\nabla\ell^\mathsf{T}_{\mathbf{V}_0|g^\circ}\nabla\ell_{\mathbf{V}_0|g^\circ}}{\nabla\ell^\mathsf{T}_{\mathbf{V}_0|g^\circ}\nabla^2\ell_{\mathbf{V}_0|g^\circ}\nabla\ell_{\mathbf{V}_0|g^\circ}} \quad (4)$$

maximises the second-order Taylor series expansion of the log-likelihood

$$\ell_{\mathbf{V}|g^\circ} \approx \ell_{\mathbf{V}_0|g^\circ} + \nabla\ell^\mathsf{T}_{\mathbf{V}_0|g^\circ}(\mathbf{V}-\mathbf{V}_0) + \frac{1}{2}(\mathbf{V}-\mathbf{V}_0)^\mathsf{T}\nabla^2\ell_{\mathbf{V}_0|g^\circ}(\mathbf{V}-\mathbf{V}_0)$$

along the gradient from $\mathbf{V}_0$.

The proof is straightforward. The approximate MLEs of Eq. (4) are centred: $\sum_{(q,s)\in\mathcal{Q}^2} V^*_{\xi,\eta}(q,s) = 0$ for $(\xi,\eta)\in\mathcal{N}$.

The solution in [8, 9] assumes the simplest IRF (denoted IRF_0 below) with zero potential $\mathbf{V}_0 = \mathbf{0}$, i.e. the equal marginal probabilities, $p(q) = \frac{1}{Q}$ and co-occurrence probabilities, $p_{\xi,\eta}(q,s) = \frac{1}{Q^2}$ of independent signals; $(q,s)\in\mathcal{Q}^2$, and equiprobable images in Eq. (1), $P_\mathbf{0}(g^\circ) = \frac{1}{Q^{|\mathcal{R}|}}$. In this case $\ell_{\mathbf{0}|g^\circ} = -\ln Q$.

Let $\mathbf{P}_0$ be the vector of the scaled marginal co-occurrence probabilities for the IRF_0: $\mathbf{P}_0 = \frac{1}{Q^2}\left[\rho_{\xi,\eta}\mathbf{u} : (\xi,\eta)\in\mathcal{N}\right]^\mathsf{T}$ where $\mathbf{u} = [1,1,\ldots,1]^\mathsf{T}$ is the vector of Q^2 unit components. Let $\mathbf{F}_{\mathrm{cn}}(g^\circ) = \left[\rho_{\xi,\eta}\mathbf{F}_{\mathrm{cn};\xi,\eta} : (\xi,\eta)\in\mathcal{N}\right]^\mathsf{T}$ be a vector of the scaled centred empirical co-occurrence probabilities $f_{\mathrm{cn};\xi,\eta}(q,s) = f_{\xi,\eta}(q,s|g^\circ) - \frac{1}{Q^2}$ for the image g°; here,

$$\mathbf{F}_{\mathrm{cn};\xi,\eta}(g^\circ) = \left[f_{\mathrm{cn};\xi,\eta}(q,s|g^\circ) : (q,s) \in \mathcal{Q}^2; \sum_{(q,s)\in\mathcal{Q}^2} f_{\mathrm{cn};\xi,\eta}(q,s|g^\circ) = 0 \right]$$

The log-likelihood gradient is $\nabla \ell_{\mathbf{0}|g^\circ} = \mathbf{F}(g^\circ) - \mathbf{P}_0 = \mathbf{F}_{\mathrm{cn}}(g^\circ)$, and the covariance matrix $\mathbf{C}_{\mathbf{F}(g)|\mathbf{0}}$ is close to the diagonal scaled covariance matrix $\mathbf{C}_{\mathrm{ind}}$ for the independent co-occurrence distributions: $\mathbf{C}_{\mathbf{F}}(g)|\mathbf{0}\} \approx \mathbf{C}_{\mathrm{ind}}$ where $\mathbf{C}_{\mathrm{ind}} = \frac{1}{Q^2}\left(1 - \frac{1}{Q^2}\right) \mathsf{Diag}\left[\rho_{\xi,\eta}\mathbf{u} : (\xi,\eta) \in \mathcal{N}\right]$.

Corollary 1. *The approximate MLE of Proposition 1 in the vicinity of zero potential* $\mathbf{V}_0 = \mathbf{0}$ *is*

$$\mathbf{V}^* = \frac{\mathbf{F}_{\mathrm{cn}}^{\mathsf{T}}(g^\circ)\mathbf{F}_{\mathrm{cn}}(g^\circ)}{\mathbf{F}_{\mathrm{cn}}^{\mathsf{T}}(g^\circ)\mathbf{C}_{\mathrm{ind}}\mathbf{F}_{\mathrm{cn}}(g^\circ)} \mathbf{F}_{\mathrm{cn}}(g^\circ) \approx Q^2 \mathbf{F}_{\mathrm{cn}}(g^\circ) \tag{5}$$

Proof. In line with Eq. (4), the maximising factor λ^* is equal to

$$\frac{\mathbf{F}_{\mathrm{cn}}^{\mathsf{T}}(g^\circ)\mathbf{F}_{\mathrm{cn}}(g^\circ)}{\mathbf{F}_{\mathrm{cn}}^{\mathsf{T}}(g^\circ)\mathbf{C}_{\mathrm{ind}}\mathbf{F}_{\mathrm{cn}}(g^\circ)} = \frac{\sum\limits_{(\xi,\eta)\in\mathcal{N}} \rho_{\xi,\eta}^2 \sum\limits_{(q,s)\in\mathcal{Q}^2} \left(f_{\xi,\eta}(q,s|g^\circ) - \frac{1}{Q^2}\right)^2}{\frac{1}{Q^2}\left(1 - \frac{1}{Q^2}\right) \sum\limits_{(\xi,\eta)\in\mathcal{N}} \rho_{\xi,\eta}^3 \sum\limits_{(q,s)\in\mathcal{Q}^2} \left(f_{\xi,\eta}(q,s|g^\circ) - \frac{1}{Q^2}\right)^2}$$

If $Q \gg 1$ and the lattice $\mathcal{R}$ is sufficiently large to make $\rho_{\xi,\eta} \approx 1$ for all clique families, the factor is reduced to $\lambda^* \approx Q^2$. □

The approximate MLEs of Eq. (5) can be directly compared to the actual Gibbs potentials for a general-case IRF which, in contrast to IRF_0, has an arbitrary marginal probability distribution of signals $\mathbf{P}_{\mathrm{irf}} = (p_{\mathrm{pix}}(q) : q \in \mathcal{Q})$. The general-case IRF is represented like Eq. (1) with the centred pixel-wise potential vector $\mathbf{V}_{\mathrm{pix}} = \left[V_{\mathrm{pix}}(q) : q \in \mathcal{Q}; \sum_{q\in\mathcal{Q}} V_{\mathrm{pix}}(q) = 0\right]^{\mathsf{T}}$ of length Q:

$$P_{\mathrm{irf}}(g) = \frac{1}{Z_{\mathrm{irf}}} \exp\left(\sum_{(x,y)\in\mathcal{R}} V_{\mathrm{pix}}(g_{x,y}) \right) \equiv \frac{1}{Z_{\mathrm{irf}}} \exp\left(|\mathcal{R}|\mathbf{V}_{\mathrm{pix}}^{\mathsf{T}}\mathbf{F}_{\mathrm{pix}}(g)\right) \tag{6}$$

where $\mathbf{F}_{\mathrm{pix}}(g) = [f_{\mathrm{pix}}(q|g) : q \in \mathcal{Q}]^{\mathsf{T}}$ is the vector of empirical marginal probabilities of signals over an image g, and Z_{irf} is the partition function which has in this case an obvious analytical form: $Z_{\mathrm{irf}} = \left(\sum_{q\in\mathcal{Q}} \exp V_{\mathrm{pix}}(q)\right)^{|\mathcal{R}|}$.

The actual MLE $\mathbf{V}_{\mathrm{pix}}$ of the potential for the IRF in Eq. (6) such that $P_{\mathrm{irf}}(g^\circ) = \mathbf{F}_{\mathrm{pix}}(g^\circ)$ for the training image g° and its first approximation $\mathbf{V}_{\mathrm{pix}}^*$ obtained much as in Corollary 1 are, respectively,

$$\begin{aligned} V_{\mathrm{pix}}(q) &= \ln f_{\mathrm{pix}}(q|g^\circ) - \tfrac{1}{Q} \sum_{\kappa\in\mathcal{Q}} \ln f_{\mathrm{pix}}(\kappa|g^\circ) \text{ and} \\ V_{\mathrm{pix}}^*(q) &= \tfrac{Q^2}{Q-1}\left(f_{\mathrm{pix}}(q|g^\circ) - \tfrac{1}{Q}\right);\ q \in \mathcal{Q} \end{aligned} \tag{7}$$

Table 1 presents both the estimates in a particular case when one intensity, q', has the empirical probability $f_{\mathrm{pix}}(q'|g^\circ) = f$ and all remaining intensities are equiprobable, $f_{\mathrm{pix}}(q|g^\circ) = \frac{1-f}{Q-1}$; $q \in \mathcal{Q}\backslash q'$. The estimates are given in function of Q and the relative probability $\beta = \frac{f(Q-1)}{1-f}$. For small Q, both the estimates are close to each other except for $f \approx 1$. But for larger Q, the approximate MLE of Corollary 1 considerably exceeds the actual one. Thus, the approximation may become intolerably inaccurate for the MGRFs.

More accurate first approximation of the MLE is obtained if the starting potential $\mathbf{V}_0$ in Proposition 1 produces a general-case IRF_{g° rather than the IRF_0, the general-case IRF_{g° being identified by using the empirical marginal probability distribution $\mathbf{F}_{\mathrm{pix}}(g^\circ)$ of signals in the training image g°:

Proposition 2. *Assume* $\sum_{s\in\mathcal{Q}} f_{\xi,\eta}(q,s|g) = \sum_{s\in\mathcal{Q}} f_{\xi,\eta}(s,q|g) = f_{\mathrm{pix}}(q|g)$ *for all* $(\xi,\eta) \in \mathcal{N}$. *Then for the large lattices* $\mathcal{R}$, *the potential*

Table 1. Approximate ("e"), $V_{\mathrm{pix}}(q')$, and actual ("a"), $V^*_{\mathrm{pix}}(q')$, MLE of the centred potentials specifying the general-case IRF for the relative probability $\beta = \frac{f(Q-1)}{1-f}$ if $f_{\mathrm{pix}}(q'|g^\circ) = f$ and $f_{\mathrm{pix}}(q|g^\circ) = \frac{1-f}{Q-1}$ for $q \in \mathcal{Q}\backslash q'$.

		Relative probabilities β												
Q		1.0	2.0	5.0	10	20	50	100	200	500	10^3	10^4	10^5	∞
2	e	0.0	0.67	1.33	1.64	1.81	1.92	1.96	1.98	1.99	2.0	2.0	2.0	2.0
	a	0.0	0.35	0.80	1.15	1.50	1.96	2.30	2.65	3.11	3.45	4.61	5.76	∞
	f	0.50	0.67	0.83	0.91	0.95	0.98	0.99	1.0	1.0	1.0	1.0	1.0	1.0
2^2	e	0.0	0.80	2.00	2.77	3.3	3.7	3.84	3.92	3.97	3.98	4.0	4.0	4.0
	a	0.0	0.52	1.21	1.73	2.25	2.93	3.45	3.97	4.66	5.18	6.91	8.63	∞
	f	0.25	0.4	0.63	0.77	0.87	0.94	0.97	0.98	0.99	1.0	1.0	1.0	1.0
2^3	e	0.0	0.89	2.67	4.24	5.63	6.88	7.40	7.69	7.87	7.94	7.99	8.0	8.0
	a	0.0	0.61	1.41	2.01	2.62	3.42	4.03	4.64	5.44	6.04	8.06	10.1	∞
	f	0.13	0.22	0.42	0.59	0.74	0.88	0.93	0.97	0.99	0.99	1.0	1.0	1.0
2^4	e	0.0	0.94	3.20	5.76	8.69	12.1	13.8	14.8	15.5	15.8	16.0	16.0	16.0
	a	0.0	0.65	1.51	2.16	2.81	3.67	4.32	4.97	5.83	6.48	8.63	10.8	∞
	f	0.06	0.12	0.25	0.40	0.57	0.77	0.87	0.93	0.97	0.99	1.0	1.0	1.0
2^5	e	0.0	0.97	3.56	7.02	11.9	19.4	24.2	27.6	30.1	31.0	31.9	32.0	32.0
	a	0.0	0.67	1.56	2.23	2.90	3.79	4.46	5.13	6.02	6.69	8.92	11.2	∞
	f	0.03	0.06	0.14	0.24	0.39	0.62	0.76	0.87	0.94	0.97	1.0	1.0	1.0
2^6	e	0.0	0.98	3.76	7.89	14.7	27.8	38.9	48.4	56.7	60.2	63.6	64.0	64.0
	a	0.0	0.68	1.58	2.27	2.95	3.85	4.53	5.22	6.12	6.80	9.07	11.3	∞
	f	0.02	0.03	0.07	0.14	0.24	0.44	0.61	0.76	0.89	0.94	0.99	1.0	1.0
2^7	e	0.0	0.99	3.88	8.41	16.5	35.4	55.8	77.9	102.0	113.0	126.0	128.0	128.0
	a	0.0	0.69	1.60	2.28	2.97	3.88	4.57	5.26	6.17	6.85	9.14	11.4	∞
	f	0.01	0.02	0.04	0.07	0.14	0.28	0.44	0.61	0.80	0.89	0.99	1.0	1.0
2^8	e	0.0	1.00	3.94	8.69	17.7	41.1	71.4	112.0	169.0	204.0	250.0	255.0	256.0
	a	0.0	0.69	1.60	2.29	2.98	3.90	4.59	5.28	6.19	6.88	9.17	11.5	∞
	f	0.0	0.01	0.02	0.04	0.07	0.16	0.28	0.44	0.66	0.80	0.98	1.0	1.0

$$\mathbf{V}_0 = \left[V_{\mathrm{irf}:\xi,\eta}(q,s) = \frac{1}{2\rho_{\xi,\eta}|\mathcal{N}|}\left(V_{\mathrm{pix}}(q) + V_{\mathrm{pix}}(s)\right) : (q,s) \in \mathcal{Q}^2; (\xi,\eta) \in \mathcal{N}\right]^{\mathsf{T}}$$

combining the scaled potential values $V_{\mathrm{pix}}(q)$ *of Eq. (7) reduces the MGRF of Eq. (1) asymptotically to the general-case* IRF_{g° *with the marginal signal probability distribution* $\mathbf{P}_{\mathrm{irf}} = \mathbf{F}_{\mathrm{pix}}(g^\circ)$.

Proof. The assumption holds precisely for the marginal signal co-occurrence and marginal signal distributions, and therefore it is asymptotically valid for the empirical distributions, too, if the lattice $\mathcal{R}$ is sufficiently large to ignore deviations due to border effects. In this case the normalised MGRF exponent is as follows: $\sum\limits_{(\xi,\eta)\in\mathcal{N}} \rho_{\xi,\eta} \sum\limits_{(q,s)\in\mathcal{Q}^2} V_{\mathrm{irf}:\xi,\eta}(q,s) f_{\xi,\eta}(q,s|g) = \sum\limits_{q\in\mathcal{Q}} V_{\mathrm{pix}}(q) f_{\mathrm{pix}}(q|g)$.

□

If the potential $\mathbf{V}_0$ from Proposition 2 is used in Proposition 1, the resulting general-case IRF_{g° has the specific log-likelihood

$$\ell_{\mathbf{V}_0|g^\circ} = \sum_{q\in\mathcal{Q}} f_{\mathrm{pix}}(q|g^\circ) \ln f_{\mathrm{pix}}(q|g^\circ)$$

and the co-occurrence probabilities $p_{\xi,\eta}(q,s) = f_{\mathrm{pix}}(q|g^\circ) f_{\mathrm{pix}}(s|g^\circ)$ for all $(q,s) \in \mathcal{Q}^2$ and $(\xi,\eta) \in \mathcal{N}$. Therefore, the vector of length nQ^2 of the expected scaled probabilities is $\mathbb{E}\{\mathbf{F}(g)|\mathbf{V}_0\} = \mathbf{P}(g^\circ) \equiv [\rho_{\xi,\eta}\boldsymbol{\phi}(g^\circ) : (\xi,\eta) \in \mathcal{N}]^{\mathsf{T}}$ where $\boldsymbol{\phi}(g^\circ) = \left[f_{\mathrm{pix}}(q|g^\circ) f_{\mathrm{pix}}(s|g^\circ) : (q,s) \in \mathcal{Q}^2\right]$.

Let $\Delta_{\xi,\eta;q,s}$ and $\mathrm{var}_{q,s}$ denote the difference between the empirical signal co-occurrence probability for g° and the co-occurrence probability for the IRF_{g° and the variance of the latter probability, respectively:

$$\begin{aligned} \Delta_{\xi,\eta;q,s} &= f_{\xi,\eta}(q,s|g^\circ) - f_{\mathrm{pix}}(q|g^\circ) f_{\mathrm{pix}}(s|g^\circ); \\ \mathrm{var}_{q,s} &= f_{\mathrm{pix}}(q|g^\circ) f_{\mathrm{pix}}(s|g^\circ)\left(1 - f_{\mathrm{pix}}(q|g^\circ) f_{\mathrm{pix}}(s|g^\circ)\right) \end{aligned}$$

The gradient $\nabla \ell_{\mathbf{V}_0|g^\circ} = \mathbf{F}(g^\circ) - \mathbf{P}(g^\circ) \equiv \boldsymbol{\Delta}(g^\circ)$ of the log-likelihood is the nQ^2-vector of the scaled differences:

$$\boldsymbol{\Delta}^{\mathsf{T}}(g^\circ) = \left[\left[\rho_{\xi,\eta}\Delta_{\xi,\eta;q,s} : (q,s) \in \mathcal{Q}^2\right] : (\xi,\eta) \in \mathcal{N}\right]$$

and the covariance matrix $\mathbf{C}_{\mathbf{F}(g)|\mathbf{V}_0}$ is closely approximated by the scaled diagonal matrix $\mathbf{C}_{\mathrm{irf}} = \mathsf{Diag}\left[\rho_{\xi,\eta}\boldsymbol{\psi}(g^\circ) : (\xi,\eta) \in \mathcal{N}\right]$ where $\boldsymbol{\psi}(g^\circ)$ is the Q^2-vector of the variances: $\boldsymbol{\psi}(g^\circ) = \left[\mathrm{var}_{q,s} : (q,s) \in \mathcal{Q}^2\right]^{\mathsf{T}}$. It is easy to prove

Proposition 3. *The first approximation of the potential MLE in the vicinity of the point* $\mathbf{V}_0$ *from Proposition 2 in the potential space is* $\mathbf{V}^* = \mathbf{V}_0 + \lambda^*\boldsymbol{\Delta}(g^\circ)$ *with the maximising factor*

$$\lambda^* = \frac{\boldsymbol{\Delta}^{\mathsf{T}}(g^\circ)\boldsymbol{\Delta}(g^\circ)}{\boldsymbol{\Delta}^{\mathsf{T}}(g^\circ)\mathbf{C}_{\mathrm{irf}}\boldsymbol{\Delta}(g^\circ)} = \frac{\sum\limits_{(\xi,\eta)\in\mathcal{N}} \rho^2_{\xi,\eta} \sum\limits_{(q,s)\in\mathcal{Q}^2} \Delta^2_{\xi,\eta;q,s}}{\sum\limits_{(\xi,\eta)\in\mathcal{N}} \rho^3_{\xi,\eta} \sum\limits_{(q,s)\in\mathcal{Q}^2} \mathrm{var}_{q,s}\Delta^2_{\xi,\eta;q,s}} \tag{8}$$

Now for all the signal cardinalities Q the actual MLE for the IRF and its approximation in Proposition 3 completely agree so that the approximation is much closer to the actual MLE than the previous one in Corollary 1.

The identified MGRF involves both the first- and second-order potentials and signal statistics. It is easily seen that the following relationships hold:

Corollary 2. *The MGRF model with the approximate potential estimate of Proposition 3 is represented by the GPD*

$$P_{\mathcal{N},\mathbf{V}^*}(g) = \frac{1}{Z_{\mathcal{N},\mathbf{V}^*}} \exp\left(\sum_{(x,y)\in\mathcal{R}} V_{\text{pix}}(g_{x,y}) + \sum_{(\xi,\eta)\in\mathcal{N}} \sum_{\mathcal{C}_{\xi,\eta}} \widetilde{V}_{\xi,\eta}(g_{x,y}, g_{x+\xi,y+\eta}) \right)$$

$$= \frac{1}{Z_{\mathcal{N},\mathbf{V}^*}} \exp\left(|\mathcal{R}| \left(\mathbf{V}_{\text{pix}}^{\mathsf{T}} \mathbf{F}_{\text{pix}}(g) + \lambda^* \mathbf{\Delta}^{\mathsf{T}}(g^\circ)\mathbf{F}(g) \right) \right)$$

where $\mathbf{V}^* = \left[\mathbf{V}_{\text{pix}}^{\mathsf{T}}; \widetilde{\mathbf{V}}^{\mathsf{T}} = \lambda^* \mathbf{\Delta}^{\mathsf{T}}(g^\circ) \right]$ *with the second-order potential values* $\widetilde{V}_{\xi,\eta}(q,s) = \lambda^* \rho_{\xi,\eta} \Delta_{\xi,\eta;q,s}$ *for all* $(\xi,\eta) \in \mathcal{N}$ *and* $(q,s) \in \mathcal{Q}^2$.

2.4 Model-based interaction maps (MBIM)

In a generic MGRF model of Eq. (1), the pixel neighbourhood $\mathcal{N}$ to be estimated can be arbitrary except that the longest interactions to be recovered depend on the size of a training image. Let $\mathcal{W} = \{(\xi,\eta) : |\xi| \le \xi_{\max}; |\eta| \le \eta_{\max}\}$ be a set of inter-pixel coordinate offsets such that the longest anticipated interaction is smaller than $(\xi_{\max}, \eta_{\max})$, i.e. $\mathcal{N} \subset \mathcal{W}$. To capture nearly periodic geometry of interactions, $\mathcal{W}$ should cover at least a few repetitions.

Given a training image g°, the interaction structure is observed and extracted using a MBIM [9] containing specific energies of the clique families: $\text{MBIM}(g^\circ) = \{e_{\xi,\eta}(g^\circ) : (\xi,\eta) \in \mathcal{W}\}$ where each energy

$$e_{\xi,\eta}(g^\circ) = \frac{1}{|\mathcal{R}|} \sum_{c_{\xi,\eta}^{x,y} \in \mathcal{C}_{\xi,\eta}} V_{\xi,\eta}(g^\circ_{x,y}, g^\circ_{x+\xi,y+\eta}) = \rho_{\xi,\eta} \mathbf{V}_{\xi,\eta}^{\mathsf{T}} \mathbf{F}_{\xi,\eta}(g^\circ)$$

is computed with the approximate second-order potential $\mathbf{V}_{\xi,\eta}$ of Corollary 1 or $\widetilde{\mathbf{V}}_{\xi,\eta}$ of Corollary 2.

Empirically, $\xi_{\max} \le M/3$ and $\eta_{\max} \le N/3$ in order to have clique families of the comparable relative sizes $\rho_{\xi,\eta}$ for estimating the potential. To meet the computational restrictions and have statistically meaningful estimates with relatively small training images (e.g., $128 \times 128 - 256 \times 256$), the signal range in our experiments is reduced to $Q = 16$ grey levels.

Figure 1 presents grey-coded MBIMs for a few training samples of stochastic and regular textures from [3, 20] (the darker the pixel, the higher the energy). Obvious correspondence between periodicity of the texture and the spatial structure of the MBIM gives grounds to estimate first from the MBIM the neighbourhood $\mathcal{N}$ and then use it to derive texels and their placement

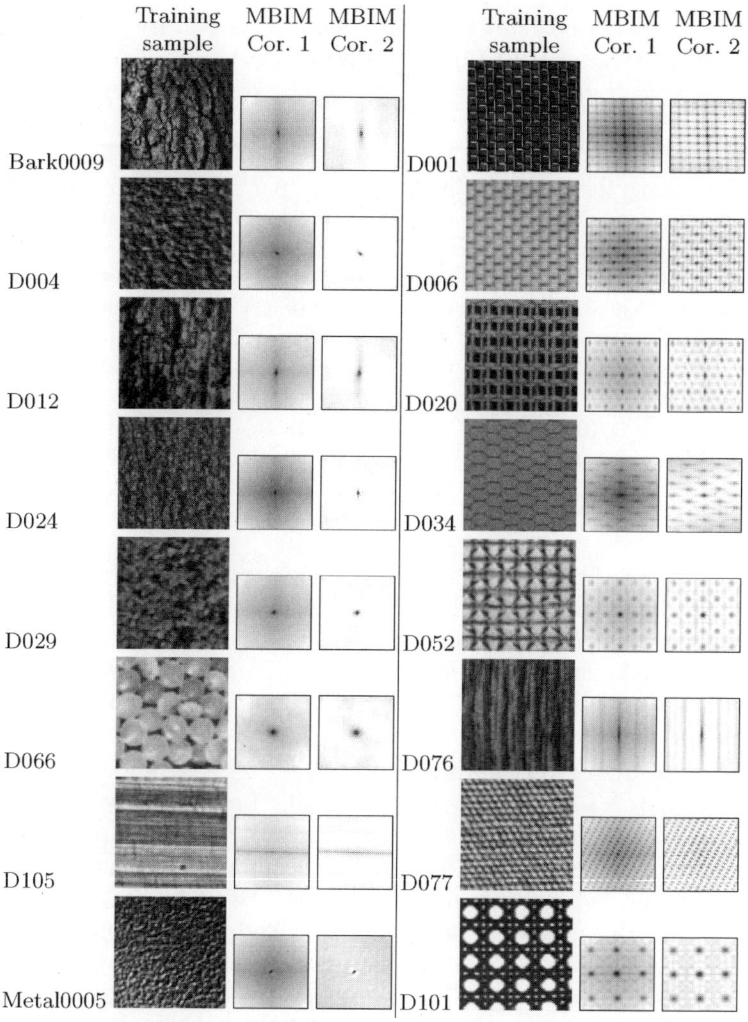

Fig. 1. Stochastic (left) and regular (right) Brodatz [3] and MIT VisTex [20] textures: training samples 128 × 128 and their grey-coded MBIMs 85 × 85 ($\xi_{\max} = \eta_{\max} = 42$) for the potentials of Corollary 1 and Proposition 3

rules [10]. Important structural properties of each texture are revealed because the vast majority of the clique families in the MBIM have low energies indicating almost independent pixels and only a small characteristic group with higher energies actually impacts upon the spatial pattern of the texture.

Energy histograms in Table 2 show the non-characteristic majority of the clique families form a dominant peak whereas a relatively small characteristic families form a high-energy "tail". The latter has to be separated from the dominant part in order to estimate $\mathcal{N}$.

Table 2. Empirical probability distributions of the partial energies for the second-order potentials of Corollary 1 (a) and Corollary 2 (b).

Relative energy bins in the range $[\min\{e_{\xi,\eta}(g^\circ)\}\ldots\max\{e_{\xi,\eta}(g^\circ)\}]$									
0.05	0.15	0.25	0.35	0.45	0.55	0.65	0.75	0.85	0.95
Bark0009: range (b) $[-0.03\ldots1.6]$; (a) $1.3\ldots8.3$									
b: 0.98	0.014	0.002	0.001	0.001	0.0003	0.0	0.0	0.0	0.0003
a: 0.14	0.73	0.12	0.01	0.001	0.0008	0.0003	0.0	0.0	0.0003
D029: range (b) $[-0.03\ldots1.6]$; (a) $[3.3\ldots18.9]$									
b: 0.99	0.004	0.001	0.001	0.0003	0.0003	0.0	0.0003	0.0000	0.0003
a: 0.09	0.63	0.27	0.01	0.002	0.0003	0.001	0.0	0.0003	0.0003
D034: range (b) $[-0.02\ldots0.7]$; (a) $[24.6\ldots84.2]$									
b: 0.56	0.39	0.04	0.01	0.001	0.0000	0.0003	0.0	0.0	0.0003
a: 0.04	0.26	0.39	0.23	0.08	0.01	0.001	0.001	0.0	0.0003
D101: range (b) $[-0.03\ldots1.61]$; (a) $[5.4\ldots33.5]$									
b: 0.74	0.20	0.03	0.02	0.01	0.003	0.002	0.0	0.0003	0.0003
a: 0.13	0.62	0.21	0.02	0.01	0.004	0.002	0.0006	0.0003	0.0003

According to Fig. 1 and Table 2, the more accurate potential of Proposition 3 results in more distinct separation between the characteristic and non-characteristic energies. Due to this separation, our experiments below use a simple heuristic threshold from [8, 9]: $\theta = e_{\text{ave}} + k\sigma$ where e_{ave} is the average energy over the MBIM, σ is the standard deviation of the energies, and the factor k is chosen empirically. The characteristic neighbourhoods $\mathcal{N} = \{(\xi,\eta) : (\xi,\eta) \in \mathcal{W}; e_{\xi,\eta}(g^\circ) > \theta\}$ for $k = 4$ together with the second-order potentials $\widetilde{\mathbf{V}}_{\xi,\eta} = (\widetilde{V}_{\xi,\eta}(q,s) : (q,s) \in \mathcal{Q}^2)$ of Corollary 2 for the clique family $\mathcal{C}_{\xi,\eta}$ with the maximum energy in the MBIM are shown in Fig. 2. Here, the coordinate offsets $(\xi_k,\eta_k) \in \mathcal{N}$ of the neighbours are indicated by black points in the (ξ,η)-plane representing the set $\mathcal{W} = \{-40 \le \xi \le 40; -40 \le \eta \le 40\}$, and the potential values are grey-coded by mapping their range $[v_{\min} = \min_{(q,s)\in\mathcal{Q}^2} V_{\xi,\eta}(q,s);\ v_{\max} = \max_{(q,s)\in\mathcal{Q}^2} V_{\xi,\eta}(q,s)]$ onto the grey range [0(black); 255(white)]. All the potentials shown in Fig. 2 correspond to the families of the nearest neighbours: $\mathcal{C}_{0,1}$ for D001, D004, D006, D029, D034, D052, D066, D077, D101, D105, and 'Metal0005' and $\mathcal{C}_{1,0}$ for 'Bark0009', D12, D20, D24, and D76. But as Fig. 3 suggests, other clique families in the neighbourhood $\mathcal{N}$ chosen for each particular training texture have closely similar patterns of the potential values, though the patterns become progressively more and more "smeared-out" and their range of values gradually decreases.

Because most of the empirical energy distributions for the MBIMs are almost unimodal, i.e. positively skewed with a single peak at the lower energy end as in Table 2, unimodal thresholding [22] with no heuristic parameters results in a very similar separation of the MBIM's energies.

Characteristic neighbours form local clusters in the (ξ,η)-plane of the MBIM that reveal both the exact periodicity and statistical deviations from it in the training texture. Stochastic and regular (periodic) textures can be

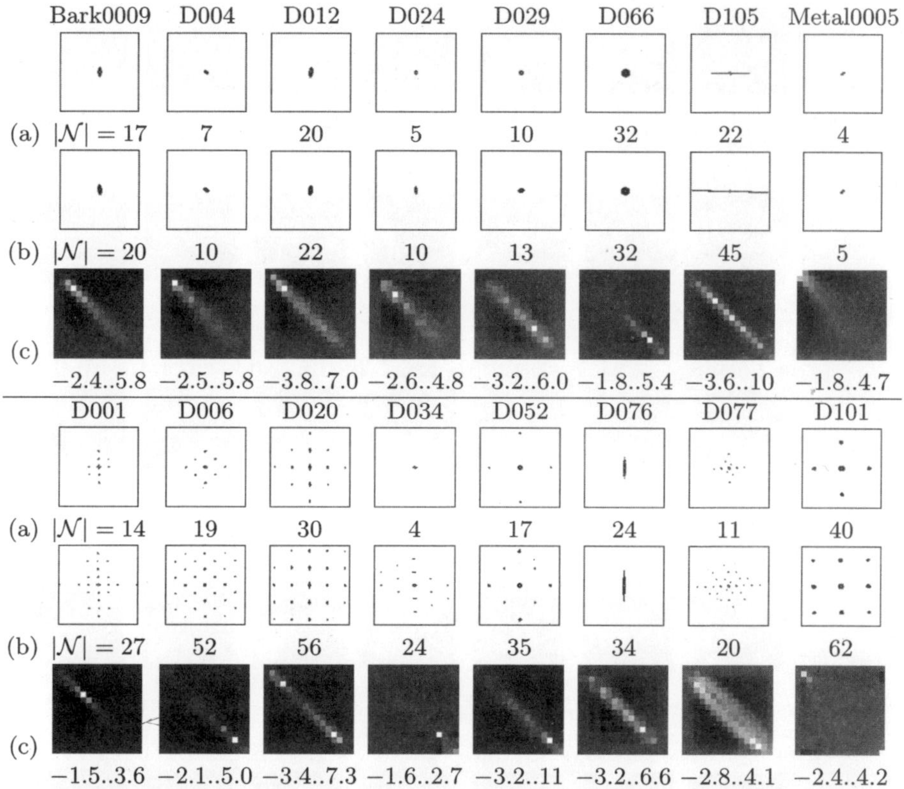

Fig. 2. Neighbourhoods $\mathcal{N}$ in the (ξ, η)-plane with the left-to-right ξ-axis, $-40 \leq \xi \leq 40$, and the top-down η-axis, $-40 \leq \eta \leq 40$, estimated for stochastic and regular textures in Fig. 1 with the simple thresholding of their MBIMs for the second-order potentials of Corollary 1 (a) and 2 (b) and the grey-coded potential (c) of Corollary 2 for the most energetic clique family (the left-to-right q-axis, $q = 0, 1, \ldots, 15$, and the top-down s-axis, $s = 0, 1, \ldots, 15$; black-to-white mapping of the shown potential range $v_{\min}..v_{\max}$ with the 8×8 square per value $V_{\xi,\eta}(q, s)$ in the potential picture). Notice that the estimate (a) fails for D034

classified in line with their different cluster patterns. The former like D004, D012, D024, or D066 in Fig. 1 have a single prominent central cluster indicating that only short-range pixel interactions dominate in these textures. In contrast, the regular textures such as D001, D006, D020, or D076 in Fig. 1, have also a number of prominent peripheral clusters spatially distributed in a structured manner reflecting the translational invariance of pixel interaction.

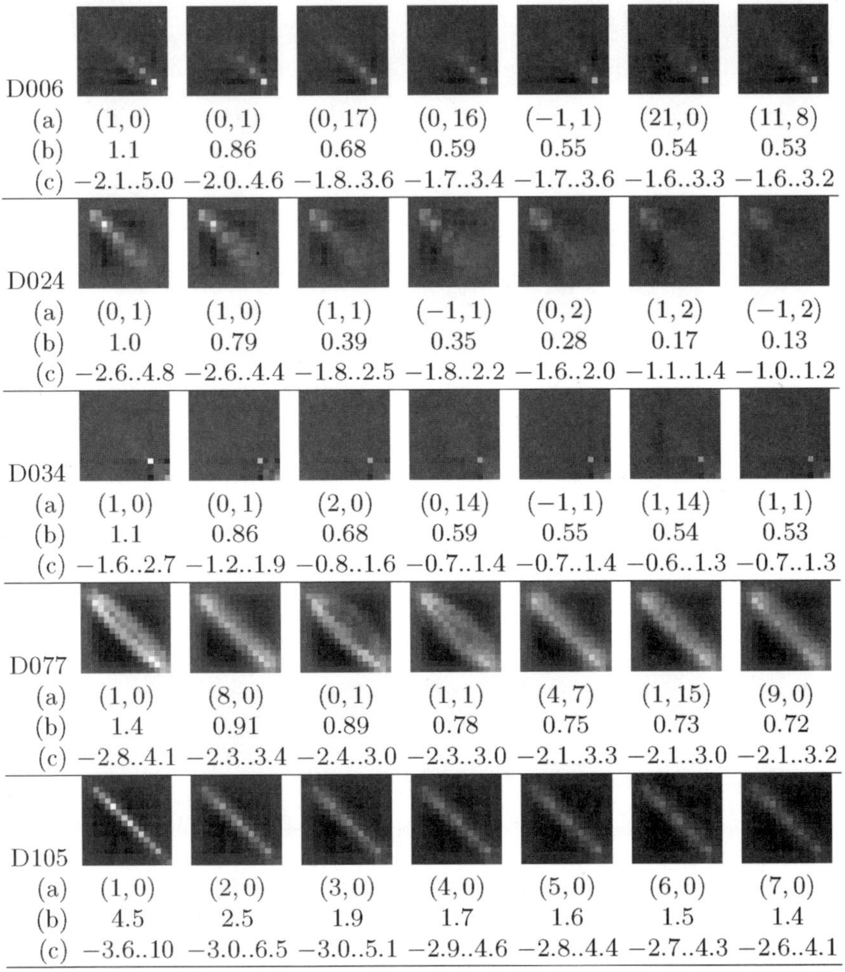

Fig. 3. Grey-coded potentials $\widetilde{\mathbf{V}}_{\xi,\eta}$ for the seven top-rank clique families: (a) (ξ,η); (b) $e_{\xi,\eta}(g^\circ)$, and (c) the range $v_{\min}..v_{\max}$ of the potential values

3 Characteristic Structure and Texels

Structural approach to texture analysis [12, 13, 31] suggests that each texture is built from a subset of elements (texels) repetitively placed to an image lattice in accord with certain rules of spatial interplay. Texels and their mutual arrangement determine local and global properties of a texture, respectively, and obviously differ for stochastic and regular textures.

For simplicity, we assume a uniform geometric structure for all the texels in a texture, i.e. the texels for a translation invariant texture have the same arbitrary and not necessarily continuous geometric structure varying for different textures, and each individual texel is distinguished by the combination of signals over its area. Such texel-based description can be obtained in line with the structural identification of the MGRF model of the training texture in Eq. (1) as follows:

1. Form the MBIM using the approximate potentials of Proposition 3.
2. Select a geometric structure of the texels by taking account of the higher-energy clique families in the MBIM.
3. Derive the placement rules for these texels.

The identified structure combines the most probable translation invariant pairwise interactions. The simplest approach to estimate such a structure is based on the largely heuristic thresholds for the partial energies in [7, 8] (see also Section 2). But this thresholding takes no account of statistical interplay between the clique families. An alternative sequential approach in [30] selects each next characteristic clique family by comparing the training GLCHs to those for an image sampled from the MGRF with the currently estimated neighbourhood. But the image sampling (generation) and re-collection of the GLCHs at each step make the process too computationally complex, namely, $0.5|\mathcal{N}|^2$ repetitions of the image generation – model identification cycle of complexity $O(\max\{|\mathcal{R}||\mathcal{N}|T, |\mathcal{R}||\mathcal{W}|\})$ to find the neighbourhood of the size $|\mathcal{N}|$. Here, T is an expected number of steps[2] to generate a single MGRF sample using the MCMC process of pixel-wise stochastic relaxation.

Geometric structure of texels for a regular texture

Structural identification is simplified in this case by the observation that most energetic clique families selected by the simplest thresholding form isolated clusters (or blobs) in the MBIM shown in Fig. 2. The blobs are easily segmented using simple classical connected component labelling [21].

Some of the clusters relate to "secondary" interactions that appear due to statistical interplay of the "primary" ones closest to the MBIM's origin. Only these primary nearest neighbours relate to a single repetition in the nearly periodic pattern of the texture and specify the desired characteristic structure. From the statistical viewpoint, the clique family with the maximum partial energy in each primary cluster represents the locally most characteristic interaction whereas other clique families in that cluster that are also likely but less probable reflect local variations of the interaction over the texture. Thus, the local maximum of each primary cluster indicates a representative neighbour to include into the characteristic structure. Their compounding parallelogram could act as a cell of an underlying repetitive guiding grid reflecting the periodicity of the texture.

The resulting texel has a rather simple structure with only a few pixels, e.g. six pixels for the texture D034. Such simple texel is not particularly meaningful by itself and even is not noticeable in a texture. However, multiple

[2] At least, theoretically T grows exponentially with $|\mathcal{R}|$ although in practice it is typically limited to a few hundred steps.

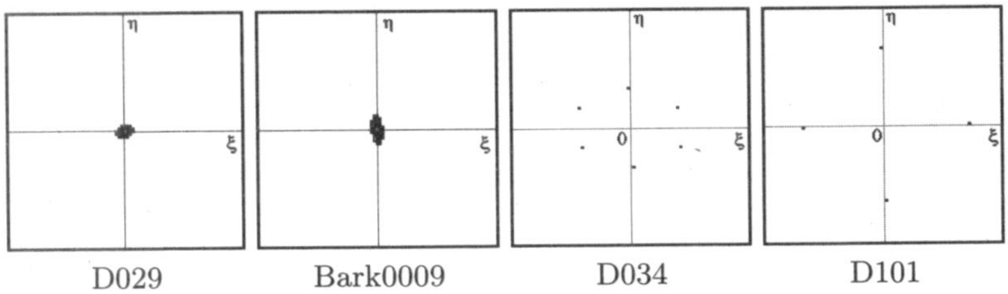

Fig. 4. Texels for stochastic textures Brodatz D029 [3] and MIT VisTex Bark0009 [20] and regular textures Brodatz D034 and D101 [3]

occurrences of these simple structures in accord with the guiding grid produce various textures similar to the training image.

Geometric structure of texels for a stochastic texture

In this case, the shape of the single central cluster defines the desired structure. Each such texel behaves like a seed, and a texture is generated by randomly scattering the seeds over the image lattice. Figure 4 exemplifies geometric shapes of the texels estimated by thresholding the MBIM. Each texture has its own geometric structure of the texels.

Placement rules for bunch sampling

The geometric structure of the texels acts as a sampling mask. Texels are retrieved from a training texture by superposing the mask at certain locations in the texture and extracting the selected groups of signals. By changing the locations, the texels with different signal combinations are retrieved. Since our masks are of arbitrary shape, these texels look like *bunches* of signals and this process is called bunch sampling in [10]. Because the training texels repeat many times over a synthetic texture, a special placement rule has to model spatial relations between the individual texels, i.e. how the locations of multiple occurrences of one texel or different texels are spatially related.

In a stochastic texture, the texels appear with no explicit placement rules and have mostly weak spatial interrelations. Therefore, a synthetic texture of this class can be formed by sampling, repeating, and randomly placing the distinct training texels.

Contrastingly, a regular texture involves strict placement rules reflecting the underlying strong periodicity. Because of modelling translation invariant pixel interaction, only translation symmetry in textures of this class is taken into account for deriving the placement rules for texels. For a large number of natural regular textures, there exists a underlying placement grid guiding the repetition of the texels. Each grid cell is a compact bounding parallelogram around the texel mask with parameters $\psi = (\theta_x, \theta_y, m, n)$ where θ_x and θ_y are guiding angles that orient the cell sides with respect to the image coordinate

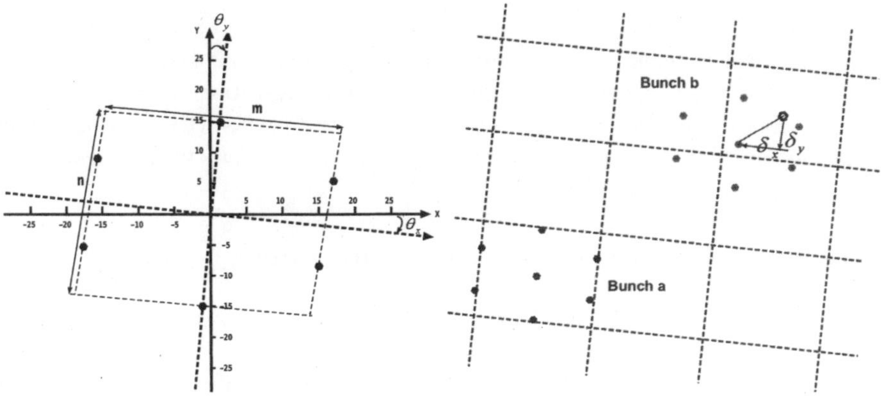

Fig. 5. Parameters $\psi = (\theta_x, \theta_y, m, n)$ of the placement grid for and the tessellation of the regular texture (D034) with this grid indicating the relative shifts of the texels (bunches), e.g., $(0, 0)$ for Bunch a and (δ_x, δ_y) for Bunch b

axes, and m and n are the side lengths, i.e. the mask spans along the orientation directions. The parallelogram is computed using an invariant fitting algorithm proposed in [27]. Generally, the grids with the parallelogram cells represent any of the five possible types of translational repetitiveness known in the theory of wallpaper groups [24].

Figure 5 shows the six-pixel texel mask and the bounding parallelogram for the Brodatz D034 texture as well as the resulting placement grid tessellating the training image to guide the bunch sampling. Each six-pixel texel is associated with a relative shift, (δ_x, δ_y), of its centre, (x, y), with respect to the origin made coincident with the centre of the closest grid cell:

$$\begin{aligned} \delta_x &= \ \ (x \cdot \cos\theta_x + y \cdot \sin\theta_y) \bmod m \\ \delta_y &= (-x \cdot \sin\theta_x + y \cdot \cos\theta_y) \bmod n \end{aligned} \tag{9}$$

The placement rule is to repeat each training texel at arbitrary locations having the same relative shift with respect to the placement grid. Due to an infinite number of absolute image locations with the same relative shift, the rule reflects translational symmetry of regular textures. It also suggests that an arbitrarily large texture can be synthesised by expanding the image lattice in line with the infinite placement grid.

Texel Selection

Since most of regular textures are not precisely periodic, signal variations in the training texels with the same relative shift may hinder the overall repetitiveness. To ensure the latter, all the training texels with the same relative shift are replaced with the single Bayesian maximum marginal posterior (MMP) estimate using empirical probability distributions of image signals in each pixel of the superposed texels. The MMP estimate provides the most expected texel at each relative shift to be used for texture synthesis.

A slightly more detailed analysis of the signal distributions makes it possible to exclude outliers while form a group of the texels with random signal variations such that do not impact the repetitiveness of a synthetic texture.

4 Texture Synthesis by Bunch Sampling

The bunch sampling forms a synthetic texture texel by texel. At each step, a texel is first sampled from a training texture using the estimated mask. Then the texel is copied into a synthetic image with due account of the placement rule, i.e. the locations of the texel in the training and synthetic images have the same relative shift with respect to the estimated placement grids randomly translated with respect to the image lattices. The sampling is repeated until the entire lattice of the synthetic texture is fully covered by the texels. Since each step is independent of any previous step in retrieving the texels and selecting the locations, the synthesis is non-causal.

Because the placement rule is absent for a stochastic texture, the synthesis may result in signal collisions when a next texel has to be placed into a region occupied in part by the previous texels. A simple heuristic rule of preserving the already placed signals resolves most of the collisions in a visually satisfactory manner [10].

At present most of the texture analysis methods are limited only to grey level images. The MGRF model of Eq. (1) is also restricted to $Q = 16$ grey levels in our experiments because otherwise the model identification becomes computationally too expensive.

A possible way to build the MGRF model of a colour texture is to consider each colour channel as a separate grayscale image and account for both intra- and inter-channel pairwise interaction [30]. But independent processing of separate colour channels for different pixels will necessarily result in false colours of a synthetic texture. Thus, the bunch sampling is limited to those texture features that are invariant to colour palettes of the images. A colour texture is converted to the grayscale intensity image for analysis, i.e. estimation of the texel structure and placement rules. At the synthesis stage, the original colour texture is used as a source of the colour texels.

For most of the colour textures in our experiments, this simple extension of the bunch sampling produces generally good results. But since hue and saturation are neglected, it may fail when the repetitive patterns are formed by interplay between the colours.

Bunch sampling is generally fast comparing to other major synthesis techniques. Time complexity of the analysis stage is quadratic in the training lattice size, $O(|\mathcal{R}|^2)$; the MBIM construction is quadratic, $O(|\mathcal{R}||\mathcal{W}|) = O(|\mathcal{R}|^2)$, if $|\mathcal{W}|$ is a fixed fraction of $|\mathcal{R}|$ as assumed in Section 2.4, and the spatial analysis of the MBIM is linear, $O(|\mathbf{W}|) = O(|\mathcal{R}|$. The synthesis complexity is linear in the synthetic image size, $O(|\mathcal{R}_{\text{syn}}|)$ if the time for forming each pixel is constant due to pre-building a hash table with the constant query time to store signals for each texel with a particular relative shift. The hash table is built in linear time, $O(|\mathcal{R}|)$, by scanning the training lattice pixel by pixel, and usually $|\mathcal{R}| < |\mathcal{R}_{\text{syn}}|$.

Figures 6–8 show synthetic images obtained for some spatially homogeneous greyscale and colour regular and stochastic training textures with the bunch sampling. Visual fidelity is preserved in most of the synthetic textures so that the identified texels and placement grids describe adequately the corresponding training images. In these examples the most characteristic GLCHs selected as sufficient statistics for texture modelling are efficient in representing global texture patterns and reproducing their periodicity.

However, bunch sampling cannot directly mimic local geometric deformations and signal deviation typical for weakly homogeneous textures such that the geometric structure of the texels varies across the image. Since the variations are usually random and individual due to imperfect cameras or defects of materials, global signal statistics are ineffective means for modelling them. Figure 9 presents results of the bunch sampling for four typical weakly-homogeneous textures. In these examples only the rectified, or homogenised synthetic textures D003 (Crocodile skin) and 'Fabrics0010' are obtained, whereas distinctive individual local deformations of the original images are not preserved. The repetitive global structure of the texture 'Parade' is also reproduced although details of each single soldier are obviously missed, as well as long strokes in the training texture 'Paint0411' are cut short to the average in the synthetic texture.

All the examples show the bunch sampling is only able to rectify a weakly-homogeneous texture. This problem is caused by the initial assumption that all the texels in a training texture have the same geometric structure. The single structure averages variations of all the texels in both the structure and signal configurations. As a result, the bunch sampling produces the idealised (precisely periodic) regular textures but fails dramatically on a class of textures with aperiodic irregular shapes and/or arbitrary placement of local elements, e.g. pebbles and brick tiles shown in Fig. 10. In these cases, only the second-order signal statistics are simply unable to adequately model the training images.

5 Comparisons and conclusions

The obtained results show that the accurately identified generic MGRF model leads to a structural texture description characterising a training texture by geometric structure and placement rules of its distinct texels. The description suggests a fast texture synthesis technique, the bunch sampling. Below we overview in brief several other texel-based texture analysis methods and non-parametric sampling methods of texture synthesis in order to compare them to our synthesis-by-analysis technique.

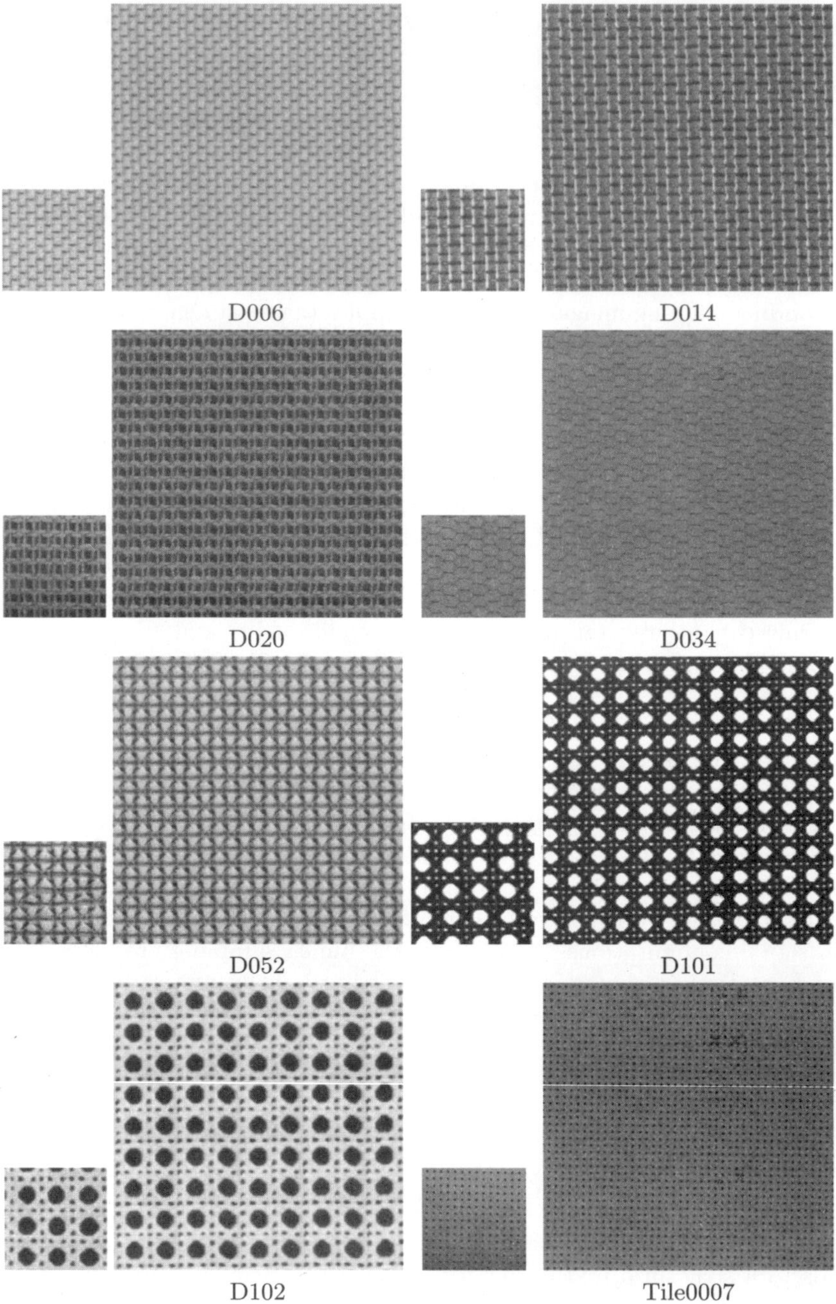

Fig. 6. Synthesis of regular textures from [3, 20] with the bunch sampling (the training 128×128 and synthetic 360×360 images)

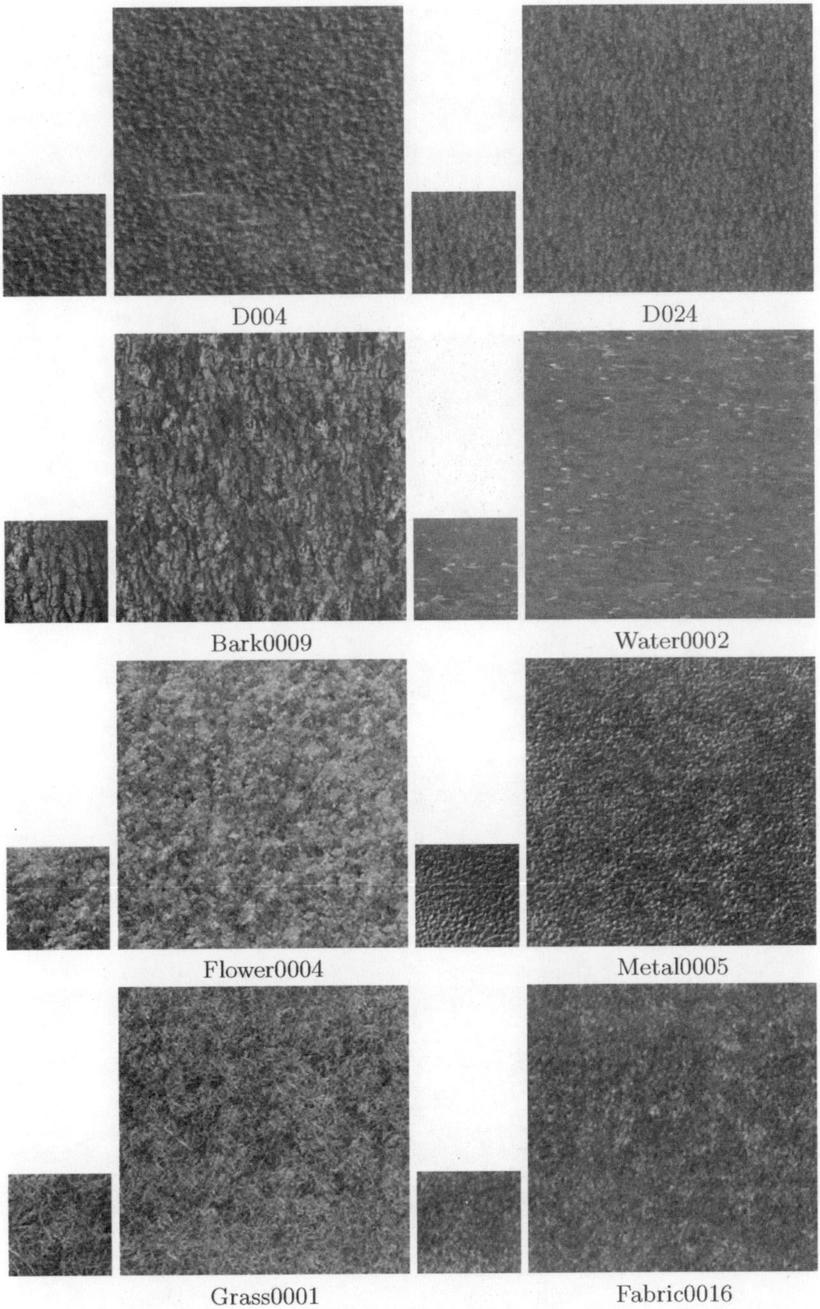

Fig. 7. Synthesis of stochastic textures from [3, 20] with the bunch sampling (the training 128×128 and synthetic 360×360 images)

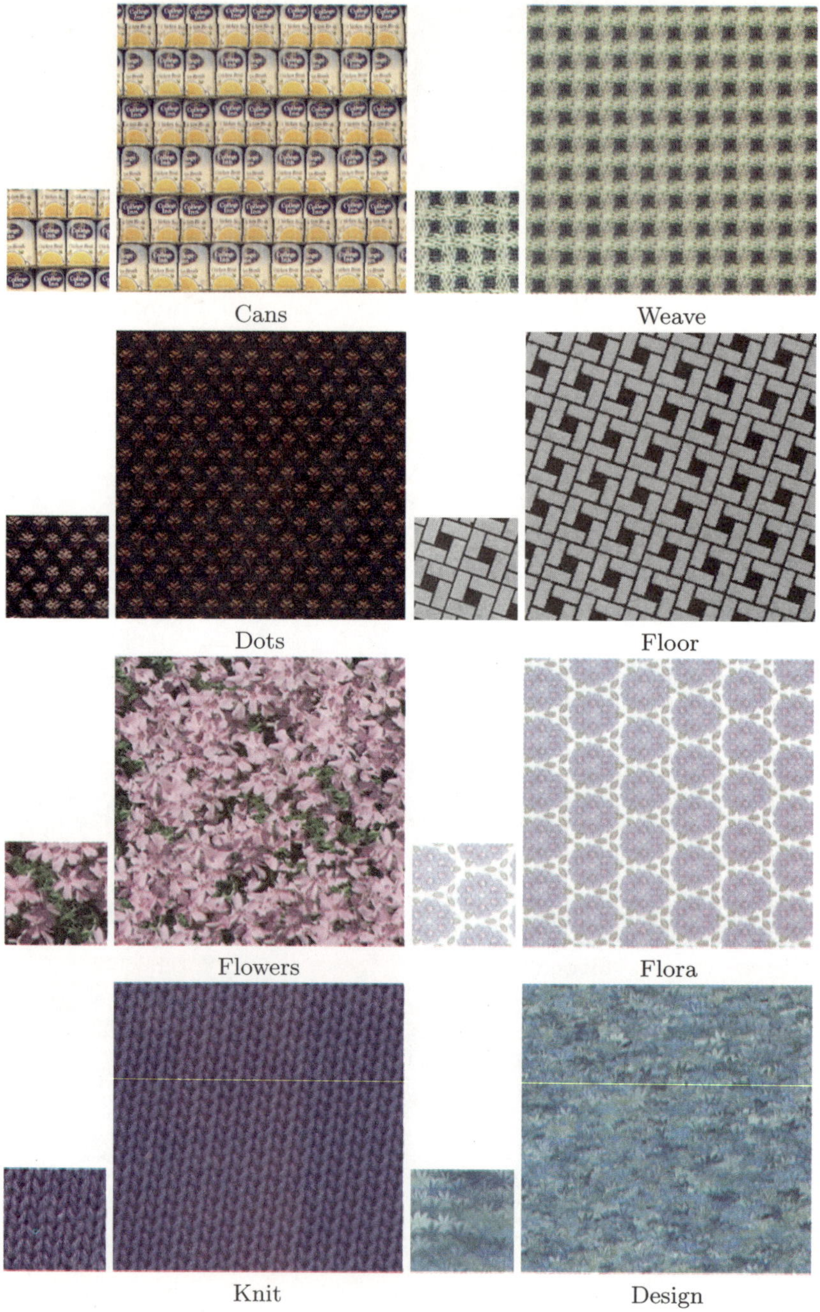

Fig. 8. Synthesis of colour textures from [17] with the bunch sampling (the training 128×128 and synthetic 360×360 images)

D003 [3] Fabrics0010 [20]

Parade (from the Web) Paint0411 [20]

Fig. 9. Bunch sampling of weakly homogeneous textures

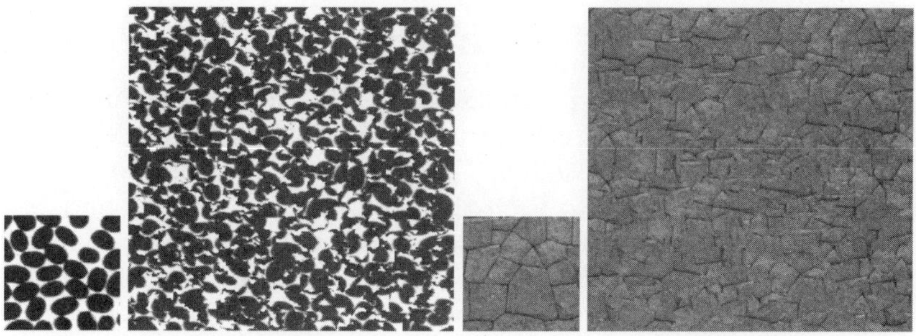

D075 [3] Brick0004 [20]

Fig. 10. Textures that cannot be synthesised with the bunch sampling

Texel-based texture representation

Instead of the term 'texel', the majority of the known works use the alternative term 'texton' coined by Julesz [14] to refer to small specific objects or areas such that they comprise together a texture and only a difference in textons or in their density can be detected pre-attentively by human early visual system. Motivated by the Julesz's texton theory, a few recent works [16, 26] employ a filter-based spectral analysis to relate textons to centres of clusters of filter responses over a stack of the training images. Conceptually, each texton represents a particular spectral feature describing repetitive patterns of a texture and acts as a feature descriptor in the spectral domain. All the textons form a global texton dictionary, or a feature space to characterise a texture by an empirical probability distribution of the textons, i.e., the frequency with which each texton in the dictionary occurs in the texture. A nearest-neighbour classifier with similarity metrics based on the chi-square distance between the texton distributions can classify textures into different categories. But since only the occurrences of the textons are taken into account, spatial information about relationships between the textons is completely lost in the description.

A texton-based generative model of image in [31] contains local constructs at three levels: pixels, image bases, and textons. An image base is a group of pixels forming a micro geometric element like a circle or a line. A texton is defined as a mini-template consisting of a number of image bases of some geometric and photometric configurations. Typically, the textons are meaningful objects such as stars or birds that could be observed in an image. The probability model with parameters $\Theta = \{\Psi, \Pi, \kappa\}$ is specified as follows:

$$\Pr(g^\circ; \Theta) = \int \Pr(g^\circ|\mathbf{B}; \Psi) \Pr(\mathbf{B}|\mathbf{T}; \Pi) \Pr(\mathbf{T}; \kappa) d\mathbf{B} d\mathbf{T} \tag{10}$$

where Ψ and Π denote global base and texton maps containing all the image bases or textons, repectively, in the entire configuration space of images, $\mathbf{B}$ and $\mathbf{T}$ are the base and textons maps, respectively, specific to a particular image g°, and the probability distribution $\Pr(\mathbf{T}; \kappa)$ accounts for the textons and their spatial relationships in the image g°.

The MLE of the model parameters Θ or the estimates minimising the Kullback-Leibler divergence are learned using a data-driven MCMC algorithm. Due to the complex likelihood function, the experiments were limited in several aspects in order to keep the problem tractable: (*i*) only a small number of image bases in the global base map, e.g. only a few Laplacian-of-Gaussian and Gabor filters as the base functions; (*ii*) the independent textons for simplicity, and (*iii*) only very simple textures with *a priori* obvious image bases and textons. In both this and a majority of other known texton-based approaches, spatial relationship among the textons is either neglected or otherwise is too difficult to represent. In contrast to them, our synthesis-by-analysis approach provides much simpler but yet more complete texture description.

Related texture synthesis methods

Today's mainstream of texture synthesis methods [5, 6, 15, 17, 28] exploit non-parametric techniques. Although the bunch sampling resembles them in that image signals retrieved from the training image are used to build synthetic textures, these methods adopt a very different approach to circumvent time-consuming stages of model creation and identification. Actually, the entire stage of building a texture model from image statistics is skipped and instead the synthesis relies on local neighbourhood matching to constrain the selection of the training signals and replicate the texture features. But without a model, these methods are unaware of the global structure of textures and encounter problems in determining the proper size and shape of the local neighbourhood for capturing features at various scales. In this case, user intervention is necessary to provide an adequate neighbourhood that could vary from one texture to another. If a texture is built sequentially, pixel by pixel, the accumulated errors eventually may destroy the desired pattern.

Because textures are generated using only local constraints, the non-parametric methods may outperform the bunch sampling in catching local variations of stochastic textures but are weak in reproducing regular textures. The alternative synthesis of regular textures in [19] in its basic idea is similar to the bunch sampling. But the periodic structure of a training texture is recovered from translational symmetries of an autocorrelation function. Since the latter describes statistics of pairwise signal products over the clique families, the interaction structure is presented in a less definite way than with more general statistics of pairwise signal co-occurrences. Another difference from the bunch sampling is that the method in [19] uses large image tiles as construction units for texture synthesis. Each tile is cut out from the training image and then placed into the synthetic texture in line with the estimated placement grid. As a result, the overlapping regions have to be blended at seams to avoid visual disruption.

Figure 11 compares results of the synthesis of the colour texture 'Mesh' using the bunch sampling, the non-parametric sampling in [5, 6, 15, 28], and the method proposed in [18]. Although this texture is periodic, the periodicity

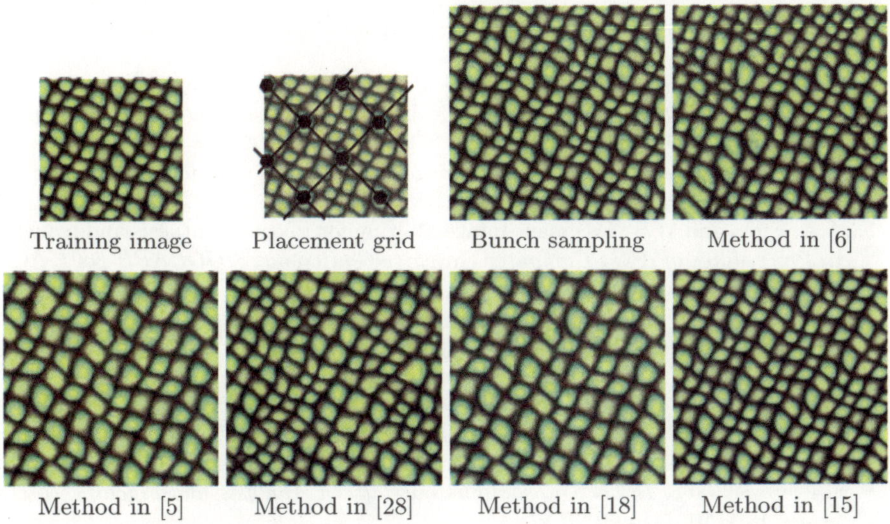

Fig. 11. Bunch sampling comparing to other texture synthesis methods (part of the images are taken from [18])

is accurately discovered only by the bunch sampling (the placement grid estimated from the MBIM is also shown in Fig. 11). Therefore, only the accurate identification of the MGRF allows us to describe this training image as a perfect regular texture and generate its periodic replicas using the bunch sampling. All other methods that exploit the similarity of the local neighbourhoods perceive the same training image as a weakly-homogeneous texture and mostly lose its actual periodicity in the synthetic versions.

Since each natural texture possesses both global and local features, neither the model-based bunch sampling nor the non-parametric approaches can solve alone the texture modelling problem. One might expect that texture analysis and synthesis methods would become more efficient after combining accurate global and local texture representations. The described simple analytical identification of the MGRF model provides global descriptions of the training textures. Simultaneously, it is of immediate interest to other areas of signal modelling, analysis, and synthesis where the MGRF priors with pairwise interaction are used.

Acknowledgments

This work was partially supported in 2002 - 2005 by the Royal Society of New Zealand under the Marsden Fund Grant UOA122 of G.G. and in 2003 - 2006 by the Tertiary Education Commission of New Zealand under the Top Achiever Doctoral Scholarship of D.Z.

References

1. O. Barndorff-Nielsen. *Information and Exponential Families in Statistical Theory.* Wiley, New York, 1978
2. J.E. Besag. *Journal of Royal Statistical Society*, B48:192–236, 1974
3. P. Brodatz. *Textures: A Photographic Album for Artists and Designers.* Dover, New York, 1966
4. G.R. Cross and A.K. Jain. *IEEE Transactions on Pattern Analysis and Machine Intelligence*, 5:25–39, 1983
5. A.A. Efros and W.T. Freeman. Image quilting for texture synthesis and transfer. In *Proceedings of ACM Computer Graphics Conference SIGGRAPH 2001.* ACM Press, New York, 2001
6. A.A. Efros and T.K. Leung. Texture synthesis by non-parametric sampling. In *Proceedings of 7th International Conference on Computer Vision (ICCV 1999)*, volume 2. IEEE CS Press, Los Alamitos, 1999
7. G.L. Gimel'farb. Non-Markov Gibbs texture model with multiple pairwise pixel interactions. In *Proceedings of the 13th International Conference on Pattern Recognition (ICPR 1996)*, volume B. IEEE CS Press, Los Alamitos, 1996
8. G.L. Gimel'farb. *IEEE Transactions on Pattern Analysis and Machine Intelligence*, 18:1110–1114, 1996
9. G.L. Gimel'farb. *Image Textures and Gibbs Random Fields.* Kluwer Academic, Dordrecht, 1999

10. G. Gimel'farb and D. Zhou. Fast synthesis of large-size textures using bunch sampling. In *Proceedings of the Image and Vision Computing New Zealand (IVCNZ 2002)*. Wickliffe, Dunedin 2002
11. G. Gimel'farb and D. Zhou, Accurate identification of a Markov–Gibbs model for texture synthesis by bunch sampling. In *Proceedings of the International Conference on Computer Analysis of Images and Patterns (CAIP 2007)*, volume 4673, LNCS. Springer, Berlin Heidelberg New York, 2007
12. R.M. Haralick. *Proceedings of IEEE* 67:786–804, 1979
13. R.M. Haralick and L.G. Shapiro. *Computer and Robot Vision*, volume 2. Addison-Wesley, Reading, 1993
14. B. Julesz. *Nature*, 290:91–97, 1981
15. V. Kwatra, A. Schidl, I.A. Essa, et al. Graphcut textures: Image and video synthesis using graph cuts. In *Proceedings of the ACM Computer Graphics Conference SIGGRAPH 2003*. ACM Press, New York, 2003
16. T.K. Leung and J. Malik. *International Journal of Computer Vision* 43:29–44, 2001
17. L. Liang, C. Liu, and H.Y. Shum. Real-time texture synthesis by patch-based sampling. Technical Report MSR-TR-2001-40, Microsoft Research, 2001
18. Y. Liu and W.-C. Lin. Deformable texture: the irregular-regular-irregular cycle. In *Proceedings of the 3rd International Workshop Texture Analysis and Synthesis (Texture 2003)*, Heriot-Watt University, Edinburgh, 2003
19. Y. Liu, Y. Tsin, and W. Lin. *International Journal of Computer Vision* 62:145–159, 2005
20. R. Picard, C. Graszyk and S. Mann et al. *VisTex Database*. MIT Media Lab, Cambridge, USA, 1995
21. A. Rosenfeld and J.L. Pfaltz. *Journal of ACM* 13:471–494, 1966
22. P. Rosin. *Pattern Recognition* 34:2083–2096, 2001
23. S. Roth and M.J. Black. Fields of experts: A framework for learning image priors. In *Proceedings of IEEE CS Conference on Computer Vision Pattern Recognition (CVPR 2005)*, volume 2. IEEE CS Press, Los Alamitos, 2005
24. D. Schattschneider. *American Mathematical Monthly* 85:439–450, 1978
25. A. Srivastava, X. Liu and U. Grenander. *IEEE Transactions on Pattern Analysis Machine Intelligence* 24:1200–1214, 2002
26. M. Varma and A. Zisserman. Classifying images of materials: Achieving viewpoint and illumination independence. In *Proceedings of the European Conference on Computer Vision (ECCV 2002)*, pt. III, volume 2352, LNCS. Springer, Berlin Heidelberg New York, 2002
27. K. Voss and H. Suesse. *IEEE Transactions on Pattern Analalysis and Machine Intelligence* 19:80–84, 1997
28. L. Wei and M. Levoy. Fast texture synthesis using tree-structured vector quantization. In *Proceedings of the ACM Computer Graphics Conference, SIGGRAPH 2000*. ACM Press/Addison Wesley, Longman, New York, 2000
29. G. Winkler. *Image Analysis, Random Fields and Dynamic Monte Carlo Methods*. Springer, Berlin Heidelberg New York, 1995
30. A. Zalesny and L.J. Van Gool. A compact model for viewpoint dependent texture synthesis. In *Proceedings of the Second European Workshop on 3D Structure from Multiple Images of Large-Scale Environments (SMILE 2000: Revised Papers)*. Springer, Berlin Heidelberg New York, 2001
31. S.C. Zhu, C.E. Guo, Y. Wang, and Z. Xu. *International Journal of Computer Vision* 62:121–143, 2005